Industrial Processing of Fruits and Vegetables

The Authors

Dr. U.D. Chavan obtained his B. Sc. and M. Sc. (Agri. in Biochemistry) degrees from Mahatma Phule Krishi Vidyapeeth Rahuri in 1985 and 1987 respectively. He received his Ph.D. degree in Food Science from Memorial University of Newfoundland St. John's Canada in 1999. He has done International training on "Global Nutrition 2002" at Uppsala University Uppsala, Sweden in 2002. He also attended follow-up International workshop on "Global Nutrition 2002" at Hanoi, Vietnam in 2002. Dr. Chavan visited Denmark, Finland, Ireland, France, Switzerland, Poland, Spain, Vietnam, Thailand, England, and U.S.A. under "Global Nutrition 2002" Programme sponsored by Swedish International Development Agency (SIDA). During this programme he worked on human nutritional diet and disorders as well as on genetically modified organisms (GMO).

Dr. Chavan worked as Senior Research Assistant in the Department of Biochemistry and Food Science and Technology at Mahatma Phule Krishi Vidyapeeth Rahuri from 1988 to 2000. During his Ph.D., he worked as Technician/Research Associate at Atlantic Cool Climate Crop Research Center and Agriculture and Agri-Food Canada. He has guided 17 students for M.Sc. (Agri.) in Biochemistry and Food Science and Technology. From 2000 to 2004 he worked as Assistant Professor of Biochemistry at Mahatma Phule Krishi Vidyapeeth Rahuri. Now he is working as a Senior Cereal Food Technologist in the Department of Food Science and Technology at Mahatma Phule Krishi Vidyapeeth, Rahuri, Dist, Ahmednagar, Maharashtra.

Dr. Chavan received Korgavkar Trust Fellowship, under-graduate and post-graduate merit scholarship as well as Senior Research Fellowship from ICAR, New Delhi and United States Department of Agriculture U.S.A. He received University Graduate Fellowship from Memorial University of Newfoundland St. John's Canada for his Ph.D. programme. Dr. Chavan also received international Scholar Award and Excellentiam Award for his Ph.D. research work from Memorial University of Newfoundland St. John's Canada. He has received a "Certificate of Appreciation" from the U.S. Department of Agriculture in 1992 for his work on processing of groundnut under the guidance of Dr. S. S. Kadam (Principle Investigator). He has written 101 research papers and 95 popular articles. He has authored three books in Marathi, two books in English and two book chapters in English. He has been awarded "Literary Award" for his best book on "Growth Regular" on agriculture during 1997. Dr. Chavan was selected as a best group leader and best presentation for "Global Nutrition 2002" by SIDA. He has been awarded "Life Time Achievement Award 2004" for his outstanding contribution in post-harvest technology of fruits and vegetables and allied fields by United Writers' Association of India. He has been elected as a Fellow of United Writers' Association of India in 2004. Dr. Chavan received pride of nation and Maharashtra Gunigan Ratna Award for 2006. He also received Jewels of India Award for 2006 for his contribution in the field of Food Science and Technology. He has been awarded "Literary Award" for his best book on "Sorghum Grain Processing" on agriculture during 2009-2010. He is recipient of "Krishi Goorav Award-2011" from Bhartiya Krishak Samaj, Maharashtra State, Nasik. He also contributed in the development of crop varieties in wheat NIAW-917 and in Sorghum Phule Anuradha (RSV-423), Phule Chitra (RSV 1546), Phule Vasudha (RSV 423), Phule Revati (RSV 1006), SPV-1626 and Phule Panchami (RSPOV-3) for Sorghum pops.

Dr. J.V. Patil obtained his B.Sc., M.Sc. and Ph.D. degrees (Agri. in Botany, Cytogenetics and plant breeding) from Mahatma Phule Krishi Vidyapeeth Rahuri in 1979, 1981 and 1992 respectively.

Dr. Patil worked on several positions since 1981 as Senior Research Assistant, Jr. Genetist-2, Associate Professor, and Plant breeder, Professor of Botany and now from 2007 he is working as a Head Department of Agricultural Botany and holding additional charge of Sorghum Breeder at Mahatma Phule Krishi Vidyapeeth Rahuri. During his service currier he worked on several administrative posts in the University. He has guided 12 students for M.Sc. (Agri.) and 7 students for Ph.D. in the discipline of Agricultural Botany.

Dr. Patil received Merit Scholarships during his under and postgraduate study programme. He also received *Baliraja*: Late Annasaheb Shinde Smruti Krishi Sanshodhak Gaurao Puraskar (2004) and "*Vasantrao Naik Krishi Puraskar" (2005)* from Government of Maharashtra. He has written 73 research papers, 48 technical publications and 58 popular articles. He also organized 13 farmers' rallies, and given 27 radio talks and 10 TV programmes. He has developed Girna (safflower), Virat, Vaibhav, Varun, Vipula, Digvijay, Rajas (chickpea), Phule Amruta, Uttara, CSV 19 SS, Chitra, CMS 1409 A/B, RSV 423, (sorghum) crop varieties during his service currier. He also gave 5 recommendations for farming various crops on farmers' field. He also developed several innovative agro techniques such as planting methods, and sprinkler method in irrigated chickpea, technique for in-situ rainwater conservation in *Rabi* sorghum etc. Now he is working as a Director of National Research Center for Sorghum, Rajendranagar, Hyderabad from August, 2010.

Industrial Processing of Fruits and Vegetables

U. D. Chavan
Senior Cereal Food Technologist
Department of Food Science and Technology
Mahatma Phule Krishi Vidyapeeth, Rahuri – 413 722
Dist: Ahmednagar, Maharashtra

J.V. Patil
Director
Directorate of Sorghum Research
Rajendranagar – 500 030
Hyderabad, A.P.

2013
Daya Publishing House®
A Division of
Astral International Pvt. Ltd.
New Delhi – 110 002

ISBN 9789351240761

Published by : **Daya Publishing House®**
A Division of
Astral International Pvt. Ltd.
– ISO 9001:2008 Certified Company –
4760-61/23, Ansari Road, Darya Ganj,
New Delhi-110 002
Ph. 011-43549197, 23278134
E-mail: info@astralint.com
Website: www.astralint.com

Laser Typesetting : **Classic Computer Services**
Delhi - 110 035

Printed at : **Salasar Imaging Systems**
Delhi - 110 035

PRINTED IN INDIA

Acknowledgement

Authors are thankful to the Hon. Dr. T.A. More, Vice-Chancellor, Mahatma Phule Krishi Vidyapeeth, Rahuri for his critical and fruitful suggestions while the preparation of "*Industrial Processing of Fruits and Vegetables*" book. We are thankful to those scientists who worked on development of various processing technology, value-added products from fruits and vegetables and their contribution is acknowledged in this book as well their sources are cited in the reference section. Authors are also thankful to all scientists and friends those helped directly or indirectly while preparation of this manuscript. We would like to express over profound thanks to our family members for their constant support, patience and devotion, which inspired us to complete this manuscript.

U.D. Chavan

J.V. Patil

Foreword

Latest advances are fast transforming postharvest management from a means of subsistence to an organized industry. This has stimulated a great deal of interest among widely diversified group of scientists. In this era of high yielding crop varieties and multiple cropping techniques, global competition in the quality of food products has put every country on the alert to improve upon postharvest techniques and value addition sector. In India, the processing of only 2 per cent of the total produce of fruits and vegetables is at present undertaken as compared to 70-80 per cent in the developed countries. However, the Government of India has fixed a target of 10 per cent of such produce to be processed at the earliest. Also, the actual producer is not getting his due share from the prevailing market prices.

The Indian Council of Agricultural Research and the State Agricultural Universities have made concentrated efforts in increasing productivity of fruits and vegetables. In recent years, the Government of Maharashtra has launched an ambitious Horticulture Development Programme to increase area under fruit crops by providing grant-in-aid to the growers. It is envisaged that the area covered under fruit crops in Maharashtra would be 10 lakh hectares in the next few years. However, very little attention has been paid to reduce the post harvest losses that occur due to poor handling, storage and transportation. According to reports of Swaminathan Committee, post harvest losses of fruits and vegetables in India vary from 20-30 per cent of production with estimated value of Rs. 1500-2000 crores annually. If these losses are controlled by using improved post harvest technology, the supplies of fresh fruits and vegetables will be increased to the extent of their existing losses. This will help to increase the per capita availability of fruits and vegetables.

Recent developments in nutrition have created greater demands for fresh fruits and vegetables. The processing technology that increases the functionality without changing their fresh like properties has significant role in modern fruit and vegetable industry. This book addresses factors that are involved in maintaining the quality of fresh fruits and vegetables after harvest. It is vital that consumers in health-conscious society would be provided with food products which retain their quality characteristics leading to more healthy individuals.

Fruits and vegetables, being perishable, are processed into several products such as jams, jellies, marmalade, syrups, squashes and canned, dehydrated and fermented products by employing different methods of processing. This book incorporates information on principles involved in preparation of various products as well as methodology employed in home scale as well as industrial processing of fruits and vegetables. I hope, this information will be useful for students in the field of horticulture, food science and nutrition and professionals in food industries.

The ever-increasing scientific research on the subject has resulted in a large body of knowledge which can be found scattered in various scientific journals and technical reports and is not easily accessible. The available material needs to be assembled and briefed in order that the students/technologists/agriculturists/horticulturists/engineers and others actively engaged in quantitative studies can have it as a ready reference. This book provides a comprehensive but concise treatise on the subject matter.

J.K. Chavan

Ex. Head, Food Science and Technology

MPKV, Rahuri

Preface

Processing of agricultural produce and their postharvest management is an emerging area in the present scenario encompassing a widely diversified group of scientists. India has now attained the stage where apart from enhanced production of crops, emphasis is being laid on processing and value addition of fruits and vegetables crops. However, the utilization of fruits and vegetables for processing in the organized and un-organized sectors is around 2 per cent of the total production. Given the huge potential for preparation of processed food products and limitation of the domestic market in consuming these, it is imperative that concerted efforts are made to export processed food for which the private, public and co-operative sectors, all have to play their rightful role. Value addition is not only a science, it is an art too.

Fruits and vegetables are important constituents of our diet which provide dietary fibers, vitamins, and minerals in addition to proteins, carbohydrates and a small amount of lipids. These contribute more than 50 per cent requirement of vitamin A. The dietary fibers aid in digestion, bowel movement and utilization of more concentrated foods in human diet. Fruits and vegetables are highly perishable commodities. Losses to the extent of 20-30 per cent occur during postharvest handling of fruits and vegetables. If these losses are controlled by using improved postharvest technology, the supplies of fresh fruits and vegetables can be increased to the extent of their existing losses. This will certainly help to increase per capita availability of fruits and vegetables.

The postharvest technology of fruits and vegetables includes harvesting, sorting, grading, packaging, transportation, storage and processing. The consumption of fruits and vegetables in fresh, unprocessed form is economically cheaper and nutritionally more desirable. The minimal processing involves treatment to these commodities without Changing wholesomeness and nutritional properties of fresh produce. It includes sorting, grading and packaging of fresh produce so that consumer can get processed product having physical, nutritional and aesthetic properties similar to that of fresh produce. In recent years, there are significant developments in canning, freezing and dehydration of fruits and vegetables. Novel product in tetra pack like Frooti has captured the Indian market as ready-to-serve beverage. Similarly, canned corn and canned mixed beans have become popular in Indian market. Ketchup and sauces are also produced and marketed under various trade names. The processed products like jams and jellies have become integral part of breakfast served in middle class families of our society. Considering the importance of fresh as well as processed products, an attempt has been made to present developments in processing technology of fruits and vegetables in this book. I hope, this will be useful book to under-graduate as well as post-graduate students in agriculture, food technology, horticulture, home science and mostly industrial sectors.

U.D. Chavan

J.V.Patil

Contents

1

Fruits and Vegetables in Human Nutrition

A. Introduction

The world's hunger is widespread and growing. Billions of people suffer from severe under nutrition in Asian, African, and Latin American countries, and from malnutrition in North American and Western European countries. There is nearly a 20-year gap in life expectancy between the rich and poor nations. Rapid population growth throughout the world necessitates a matching increase in for the food supplies to provide nutritionally adequate diet to everyone. Eradication of hunger and malnutrition depends not only on rapid increase in the agricultural productivity, but also on efficient use of farm products, especially in developing countries. Development of appropriate processing, transportation, and marketing technologies are important to make efficient use of farm products.

In the present day society, more and more people live in urban areas far away from the sources of fresh or natural foods. This has resulted in increased dependence on processed food products. This has forced the governments in these countries to explore the possibilities of processing of fruits and vegetables into more durable products, to maintain their supply during when, and in places where, the fresh fruits and vegetables would normally be unavailable. Modern food processing technologies have also expanded the farm produce market for growers in countries where the production greatly exceeds demand.

Fruits and vegetables are highly perishable food products. Water loss and postharvest decay account for most of the losses. These have been estimated to be more than 20 - 30 per cent in the tropics and subtropics. The problems of postharvest technology of fruits are critical in the battle to minimize wastage and to extract the maximum potential from the harvested crops. Wastage of fruits is so high in some instances that between the field and the consumer, bountiful amounts of highly nutritious crops are reduced to heaps of refuse. Reduction of high postharvest losses of fruits and vegetables includes proper harvesting, handling, low temperature storage and environmental control (controlled/modified and hypobaric storage), irradiation and use of chemicals and fungicides as well as packaging techniques, and processing of fruits and vegetables into suitable products.

B. Nutritional Significance

Fruits and vegetables play an important role in human nutrition particularly in supplying certain constituents in which other food materials are deficient. Fruits and vegetables contribute vitamins and minerals and the bulk of the dietary fiber to the diets of human beings. They neutralize the acid substances produced in the course of digestion of meat, cheese, and other high-energy foods. The value of fruits and vegetables as dietary fiber has increasingly been realized in recent years. The possible beneficial effects of fiber, vitamin C, β-carotene, and the vitamin B-complex derived from fruits on human health are being reexamined with the objective of minimizing certain important human diseases related to lack of fiber and antioxidant properties of some vitamins and minerals in diets. Ascorbic acid (vitamin C) is the principal vitamin supplied by fruits and vegetables in the diet. About 90 per cent of a person's dietary vitamin C requirement is obtained from fruits and vegetables. In addition, fruits and vegetables are rich source of β-carotene, thiamine, niacin, folic acid and Vitamin A.

The β-carotene is essential structural component of the eye, and a prolonged deficiency of this vitamin may lead to blindness. These are also rich in calcium, magnesium, phosphorus, and iron, which are not present in many other food materials in quantities sufficient for the needs of the body.

Fruits and vegetables are particularly rich in pectin and gums. Pectin has been shown to delay gastric emptying, which may lead to favourable changes in glycaemic responses to foods. Pectin may also induce a satiety effect. It acts as a general intestinal regulator and detoxifying agent. It is used as a standard additive in commercial baby food formulations. Pectin has been shown to have antibacterial, antiviral, wound-healing, and metal-binding properties. It is shown to be effective in reducing cholesterol levels.

A substantial proportion of carbohydrates in fruits and vegetables are present as dietary fiber in the form of cellulose, hemicelluloses, pectin substances, and lignin. It is not digested and passed through the human intestinal system, as people are not capable of secreting the digestive enzymes necessary to break down these polymer complexes into simple units in the forms absorbed by the intestinal tract. Human body lacks enzymes such as cellulase, hemicellulase, and pectinase. Fibre was once considered to be an unnecessary component in the human diet although it was thought to relieve constipation. The epidemiological evidence obtained so far shows that dietary fiber can be the panacea to cure several human diseases, especially in the affluent societies.

Fruits and vegetables contain some flavour compounds. These include sugars, amino acids, organic acids, volatiles such as aromatic hydrocarbons, aldehydes, acetals, ketones, alcohols, ester and sulphur compounds when used along with other food items. These flavouring compounds make food more palatable Onions and garlic are used as flavouring agents in variety of dishes, soups, sausage and curries. The pungency of garlic is due to volatile sulphur compound, diallyl disulphide which is produced by the action of allinase enzyme on the amino acid allicin present in garlic. Several volatile sulphur compounds are responsible for characteristic flavour of cole crops. Dimethyl trisulphide has been indicated as a major aroma component in cooked cole vegetables.

C. Therapeutic Properties

Many therapeutic drugs in use in modern medicine originated as plant extracts. Certain fruit and vegetable components exert pharmacological or therapeutic effects. Limonin and nomilin (and other limonoids) are present in citrus fruits such as orange, lemon, β-carotene may also play a role in the prevention of some forms of cancer.

Fruits and vegetables are very well known sources of different types of flavonoids which include flavonones, flavones, flavonols, anthocyanins, catechins and biflavans. Apart from fat soluble tocopherols, flavonoids are commonest and most active antioxidant compounds in human food, being active in hydrophillic as well as in lipophillic systems. This has considerable practical impact on nutrition. It protects food stuff from oxidative deterioration, prolongs their shelf life and keeping quality and improves taste, acceptability and wholesomeness of mixed dishes by inhibiting the oxidation of accompanying lipids. Flavonoids increase or stabilize biological activity of ascorbic acid. All fruits and vegetables rich in flavonoids are marked by the unusual stability of their vitamin C. The ascorbic acid stabilizing effect of flavonoids results in increased production of ascorbic acid-2-sulfate which improves membrane stability by synthesis of sulfated mucopolysaccharides and enhances cholesterol excretion. Ellagic acid is a naturally occurring phenolic constituent present in fruits, especially strawberries and other berries and has been shown to be effective as an anti mutagen and anticarcinogen and a potential inhibitor of common inducers of cancer.

Onion and garlic are known to reduce low-density lipoprotein and induce the formation of high-density-lipoprotein. The increase in high-density lipoprotein is more evident from consumption of white and yellow onion and is reduced with cooking. The allicin which is produced in garlic when tissue is injured by cutting or crushing has been reported to have hypocholesterolaemic action.

Cole vegetables contain a group of compounds known as indoles which have been recently linked with the prevention of cancers of colon, rectum and breast. Some of the species of *Cucumis, Luffa, Coccinia,* and *Momordica* exhibit varying degrees of bitterness in fruits, leaves and twigs. This is caused by terpenes called momordicins. Owing to its bitter principle, the bitter gourd is used in Ayurvedic system of medicine. A hypoglycemic ingredient cheratin isolated from bitter gourd lowers blood sugars. Sapogenin compound called diosgenin present in some species of yam is used for production of cortisome and contraceptive drugs.

D. Processed Products

Fruits and vegetables are subjected to minimal processing such as sorting, grading, packaging and premarketing treatments such as precooling, and waxing to provide a convenient product with few or no preservatives. The processed products exhibit as near as the fresh like properties of original produce. Innovative techniques, being investigated include the use of high temperature short time treatment or low dose of irradiation to produce a product with fresh like quality and greater shelf-stability than fresh item.

Fruits and vegetables are processed into several products such as juices, jams, jellies, wines, dried fruits, canned products by employing different methods of food preservation. There are some losses of nutrients during processing. Once processed, the nutrient content of the product remains relatively stable. Hence, processing conditions should be such that processed products will suffer from minimum loss of nutrients compared to raw material. Considering the nutritional and pharmacological Significance of fruits and vegetables and their products, they have a major role in nutrition and health of human population.

2

Principles of Preservation of Fruits and Vegetables

A. Introduction

The basic preservation techniques such as canning, freezing, dehydration, salting, pickling, and freeze-drying are designed to change perishable foods into more usable form and to prevent undesirable changes caused by microorganisms, or by chemical, physical and biochemical reactions in the food itself. Equally important is the retention of texture, color, nutritive value and palatability of foods. Hence, an ideal processing technology would involve prevention of growth of microorganisms and at the same time, retention of flavour, texture, nutritive value and palatability of the product.

B. Nature of Spoilage

1. Microbial Spoilage

One of the major objectives of any processing technology is to prevent microbial spoilage of foods by destroying microorganisms that are present and then eliminating recontamination by outside microorganisms. Heat is still most commonly used to destroy microorganisms. Recontamination by outside microorganisms in processed food is prevented or delayed by altering the pH of the medium and removal of oxygen etc. However, foods cannot be stored indefinitely because of other changes induced by the processing.

2. Physical, Chemical and Biochemical Changes

The most common deteriorative reactions occurring in food is oxidation of stored fats, vitamins, and flavouring compounds. Another enzymatic change that reduces the shelf-life of stored product is breakdown of pectins and other cell wall materials

Figure 2.1: Enzymatic browning in fruits and vegetables

Fruits and Vegetables

↓

Cutting and Slicing

↓

Enzymes, Oxygen, Metals, Substrates, Polyphenols, Tannins

↓

Browning

(Figure 2.1). As these changes are catalyzed by one or more enzymes, (existing and de novo synthesis) they can be retarded by heat destruction. Removal of molecular oxygen from containers can also reduce the rate of oxidation.

Certain enzymatic and nonenzymatic changes occurring in food also shorten their shelf-life and cause browning, discoloration, and may impart off-flavour to the product (Figure 2.1). Recent developments in preservation techniques have made it possible to retain most of the desirable chemical and physical properties of the product even after extended storage.

3. Nutritive Losses

In addition to qualitative changes, reduction in nutritive value occurs during processing and storage of processed fruits and vegetables. The best example is the reduction in ascorbic acid content in processed fruits and vegetables. An ideal processing technology, therefore, must prevent such nutritive changes in the product.

C. Principles of Preservation

The food is preserved for a short or long duration. The short term preservation can be achieved by taking care in handling the produce to low temperature storage and removing air from the atmosphere surrounding the produce. It is possible that in long term preservation, a product may keep well for a pretty long period and even for an indefinite period as far as microbial spoilage is concerned but slow chemical changes however, may render the food inedible if kept for an unduly long period. Long term preservation can be achieved by canning, drying and dehydration, fermentation and by using chemical preservatives. The basic six principles, on which various methods of preservation are based, are as follows:

1. Moisture Removal

- Drying
- Dehydration
- Concentration
- Intermediate moisture processing
- Freeze-drying

The metabolic activities of microorganisms require plenty of free water. Removal of the biologically active water through drying or dehydration stops the growth of microorganisms. It also reduces the rate of enzymatic activity and biochemical reactions. The effect of water removal on nutrients is relatively small if dehydration temperature is moderate and food is subsequently packaged. Freeze dehydration offers decisive advantages in nutrient preservation over dehydration at elevated temperature.

2. Heat Treatment

- ✰ Blanching
- ✰ Pasteurization
- ✰ Sterilization

Heat treatment results in denaturation of proteins *i.e.* inactivation of the microbial and native food enzymes. Pasteurization frees the food from human pathogens and most vegetative microorganisms whereas sterilization means the destruction of all viable microorganisms. Heat sterilization severely affects labile vitamins and reduces the nutritional quality of proteins mainly through Maillard reactions.

3. Low-temperature Treatment

- ✰ Cold storage
- ✰ Refrigeration
- ✰ Freezing
- ✰ Dehydro-freezing

Low temperature treatment particularly freezing is a relatively harmless method of food preservation. Low temperature inhibits microbial growth and slows down the rate of chemical and enzymatic reactions. Many vegetables are blanched before freezing in order to avoid undesirable quality changes due to enzymatic activity at freezer temperature. Vitamin losses are minimal compared to other methods of food preservation. Most of the fruits and vegetables are preserved at low temperature.

4. Acidity Control

- ✰ Fermentation
- ✰ Acidic additive

Spoilage of low acid foods is relatively rapid. The growth of food spoilage organism is significantly inhibited in an acidic environment. Acidic fermentation lowers the pH of carbohydrate containing foods by producing lactic acid. The acidity of some foods may be increased by acidic preservatives such as vinegar or citric acids, producing some inhibitory effect on spoilage. Juices, pickles, sauerkraut are some of the examples of preservation of food by acidity control.

5. Use of Preservatives

- ✰ SO_2
- ✰ Benzoic acid

- Sorbic acid
- Sugars and salts

Chemical additives can substantially contribute to the preservation of foods by providing inhibitory environment for microbial growth and for enzymatic and chemical reactions. Several preservatives such as sodium benzoate, potassium metabisulfite, sorbic acid etc. are commonly used for preservation of foods. The use of sugars and salt creates unfavourable environment for growth of microorganisms by reducing water activity of the foods. Sugars are utilized mainly in preparation of jams, jellies, marmalades and preserves. The salts are commonly used for pickling purpose. Cucumber, cabbage and bamboo shoots or fruits like unripe mangoes, and lime etc. are commonly preserved by salting process.

6. Irradiation

Irradiation is the recent method of food preservation. The free radical mechanism of irradiation destroys microorganisms but is also detrimental to nutrients particularly vitamins. Gamma rays are commonly used for preservation of vegetables. Cobalt-60 is the radiation source generally preferred for irradiation of foods including vegetables. Potatoes, onions are commonly preserved by radiation treatment.

3

Raw Material for Processing

A. Introduction

The processor is primarily interested in color, aroma, flavour and texture whereas the grower is more concerned with yield, disease resistance and ease of cultivation and harvesting. A processed product of excellent quality cannot be made from poor quality raw material. Quality characteristics of product may be divided into three categories; sensory, hidden and quantitative. The sensory characteristics of quality include color, gloss, size, shape, defects, odour, and taste which the consumer can evaluate with his senses. Those characters which consumers cannot evaluate with his senses are nutritive value, presence of harmful adulterants and presence of toxic substances. The quality of finished product depends upon the quality of raw material used for processing.

B. Color

Color increases the attractiveness of the product. Color is also associated with flavour, texture, nutritive value, maturity and wholesomeness. Colored fruits when harvested at firm ripe stage should be fully and uniformly colored. Three major classes of pigments occur in fruits and vegetables; the chlorophyll, carotenoids and xanthophyll and anthocyanins. The characteristic yellow color of mango, citrus and pineapple is mainly due to carotene and xanthophyll esters. There are great demands for juice with deep orange color especially in concentrates and bases. Hence, varieties yielding deep orange color are preferred for processing. Red color in fruits and vegetables is mainly due to anthocyanins. This color gradually passes out into syrup or brine used in Canning. Hence, color retention in canned fruits or vegetables is important in processing. Guava, lychee, banana and broad beans may turn brown or pink during processing due to the presence of leucoanthocyanin, and this appears to differ with variety. The varieties which retain their original color during processing are preferred.

C. Size and Shape

Grading of the raw material into various size and shape categories is usually one of the first steps in food processing operations. Size grading is done mainly to facilitate succeeding operations such as cutting, peeling, to obtain uniformity in the product, and to provide consumers with the, preferred size. It may also be an indirect means of grading for other quality characteristics; for example small peas, okra, and cucumbers for pickling are usually less mature and consequently more tender and desirable. Size is of major interest to the grower as it is directly proportional in certain crops, *e.g.*, pineapple, to the yield per hectare. Certain sizes are in greater demand than others for specific purposes; for example, the desirable long shreds/cuts can be made only from large onions. Evenly-sized materials are easier to handle in large quantities, and promote less wastage, rapid production and high quality.

Shape of the raw produce sometimes determines the suitability for processing. To reduce losses during mechanical trimming and handling, the shape of a fruit or a vegetable should readily lend itself to such processes. Selection and breeding of raw produce primarily for a suitable shape is yet to be attempted in most crops.

D. Texture

Texture and consistency, the structural features of fruits and vegetables, are attributes of prime importance. Texture characteristics involve touch sensations, which determine the firmness, softness (yielding quality), juiciness, grittiness, fibrousness, and mealiness of a fruit or a vegetable. In general, these factors can be readily measured with such instruments as tenderometers, texture meters, puncture meters, succulometers, fibrometers and pressure testers. In addition, certain physico-chemical tests are employed successfully for measuring esthetic properties. Certain vegetable crops, such as corn, tend to accumulate solids as they approach maturity (become less desirable); moisture determination may be used to measure this quality characteristic. Determination of alcohol-insoluble solids is particularly suitable for vegetables such as peas, sweet corn, and limabeans. Fruit specific gravity, which is related to the alcohol-insoluble solids content of fruits, can be used to identify this stage.

Fruits and vegetables used for processing should be firm enough to withstand necessary heat treatment. The firmer canned product given by "pear" tomatoes as compared with "round" tomatoes has been attributed to differences in pectin substances. Consistency or viscosity is of importance in such products as juices, jams, jellies, and preserves. Some juices (*e.g.*, orange juice) are consumed in a naturally cloudy state, and juices of apple, grape, and other berries have traditionally been consumed in a clear condition and the problem has been to clarify them effectively and to maintain them in a brilliant condition throughout their storage life. However, in recent years, unclarified or opalescent juices have also come into production. The juices of some fruits lack much of the flavour and color of the fruit itself, as these are associated with insoluble cellular material. Fruits rich in carotenoids, such as mango, papaya and tomato, are usually converted into pulpy products by disintegrating the whole fruit to give purees or nectars.

E. Flavour

Flavour distinguishes one food from another. It is difficult to evaluate instrumentally, and is still measured largely by subjective methods such as taste panels or the profile method. Though many deciduous tropical and subtropical fruits have attractive flavours, the aroma components tend to be heat-sensitive. In passion fruits, the distinctive aroma and flavour are very elusive, and are lost during heat treatment. Various methods are adapted during processing of citrus to enhance the flavour. Incorporation of fine pulp and small quantities of peel oil in juices and addition of fresh juice to the concentrate before packing are used.

F. Defects

Most defects are still largely evaluated by the consumer's eye, though in some cases instruments may be used. The presence of defects frequently lowers the grade of products which are otherwise of very high quality. Defects may be caused by heredity or unfavourable environmental conditions. The green coloration in ripe tomatoes, the tendency of some types of sweet potatoes to disintegrate and peas to split on canning is some of the examples of defects caused by genetic factors. Insects and microorganisms are major sources of defects in fruits and vegetables. Microorganisms not only reduce the crop yield but also deform the produce and cause internal defects such as off-coloring, corky tissue, mould and rot. The wholesomeness of the product decreases with such infections. Mechanical defects may occur during handling. Bruises upset normal biochemical reactions, causing discoloration, off-flavour and more rapid deterioration.

G. Nutritive Value

Succulent fruits and tubers high in total solids generally contain more total stored food values. They contain more nutrients as they approach maturity, although the proportion of each changes. On the other hand, many vegetables become fibrous and woody as their total dry matter increases, and there is a decline in the digestible carbohydrates, proteins, minerals and vitamins.

H. Toxicity

Various chemical compounds are used extensively in fruit and vegetable production. Commercial canning and juice extraction operations may remove the residues in tomatoes, but sometimes these persist. Persistent insecticides belonging to the chlorinated hydrocarbon group such as aldrin, dieldrin, and DDT, cause problems due to their preferential absorption by crops such as carrots, radish and potato. Edible tissues may accumulate amounts even beyond permissible limits. These residues may lead to bitter or musty flavour in the processed products, and present a health hazard.

I. Yield of Finished Product

The yield of finished product mostly depends upon the raw material used for processing. The good quality raw material gives maximum and acceptable final finished product. Good quality raw material eliminates much wastage.

Postharvest Physiology of Fresh Fruits and Vegetables

Harvested fruits and vegetables continue to maintain physiological systems and sustain metabolic processes that were present before harvest. Losses of water and substrates used in respiration can no longer be replaced and deterioration of the product begins.

Maturation, ripening and senescence induce many changes in fruits and vegetables. Although a strict physiological distinction between fruit ripening and senescence is unclear, ripening hastens the onset of senescence and the probability of cell injury and death. Fruit ripening involves many complex changes, including seed maturation, color changes, abscission from the parent plant, tissue softening, volatile production, wax development on skin, and changes in respiration rate, ethylene production, tissue permeability, carbohydrate composition, organic acids and proteins. Product respiration, transpiration and ethylene production are major factors contributing to the deterioration of fresh fruits and vegetables.

Gas Exchange

1. Respiration

The energy is generated by respiration, which is the oxidative catabolism of carbohydrates. Respiration can be considered a series of enzymatic reactions, involving three pathways: (*i*) glycolysis (glucose to pyruvate), (*ii*) the tricarboxylic acid cycle (pyruvate to CO_2), and (*iii*) Oxidative phosphorylation [reduced nicotinamide adenine dinucleotide (NADH) + reduced flavin adenine dinucleotide (FADH) to adenosine triphosphate (ATP).

$$C_6H_{12}O_6 + 6O_2 + 36\ ADP \rightarrow 6CO_2 + 6H_2O + 36\ ATP$$

Measuring Gas Exchange

1. *Static method*: A specific gas composition is generated around an object and the gas flow is closed for a specific period of time. Gas composition is measured at the beginning and end of the period.
2. *Flow through method*: A specific gas composition is generated around an object and the gas composition of the inward and outward is measured.
3. *Modified atmosphere method (MA method)*: Using a package with film of known O_2 and CO_2 permeability, the equilibrium concentrations that develop are measured. Gas exchange rates can then be calculated.

Respiration Quotient

The ratio of moles of CO_2 produced pre mole of O_2 consumed is called the respiratory quotient (RQ).

Fermentation

The main source of CO_2 production by plant tissue is respiration. At low O_2 or high CO_2 atmospheres, however, fermentation becomes increasingly important. The main fermentative metabolites found in plant tissues are ethanol, acetaldehyde and lactic acid. Ethanol fermentation is the catabolism of pyruvate to ethanol; the combination of glycolysis and fermentation can be expressed as:

$$C_6H_{12}O_6 + 2ADP \rightarrow 2CO_2 + 2C_2H_5OH + 2ATP$$

Transpiration

Transpiration is a mass-transfer process in which water vapour moves from the surface of fruit or vegetable to the surrounding air. This process of moisture loss induces wilting, shrinkage, and loss of firmness and crispness of fruits and vegetables and thus adversely affects the appearance, texture, flavour and mass of produce. Most fruits and vegetables loss their freshness after 3 to 10 per cent mass loss. Transpiration is considered to be the primary cause of postharvest loss and poor quality in leafy vegetables, such as lettuce, chard, spinach, cabbage and green onion and is considered the major cause of commercial and physiological deterioration in citrus fruits.

Factors affecting transpiration:

1. Water vapour pressure deficit
2. Temperature
3. Storage environment
4. Product respiration
5. Size, shape and surface of a commodity
6. Maturity and ripening

Controlling Transpiration

Transpiration of fresh fruits and vegetables can be reduced by minimizing the water vapour pressure difference between the produce and the air and/or by increasing the resistance of the product by storage conditions, packaging (modified humidity packages, MHP) and surface coatings (commercial carnauba waxes).

Ethylene [C_2H_4]

Ethylene can induce plant responses at very low concentrations (0.1μl/L or less) and has a wide variety of effects on virtually all stages of a plant life cycle (*e.g.*, germination, vegetative growth, flowering, fruiting, abscission, ripening, senescence and dormancy). Ethylene is sometimes called the "ripening gas" because its most important postharvest effect is the acceleration of ripening and senescence. However, there are other postharvest effects of ethylene, including induction or suppression of potato sprout growth, loss of chlorophyll (degreening) and induction or suppression of disease resistance. Some of these effects are desirable and some are undesirable, depending on the intended use of the stored plant product.

Production of Ethylene

Biological ethylene production: In flowering plants, the formation of ethylene from methionine major source of ethylene, via the ACC pathway is now recognized as the major source of ethylene, whereas non-flowering plants do not possess ACC oxidase and produce ethylene from other unknown pathway when they are stressed or damaged. The rate of ethylene synthesis is controlled by the activity of ACC synthase, but it can sometimes be limited by ACC oxidase activity.

Non-biological ethylene production:

1. Incomplete combustion of organic fuels
2. Smoke production
3. Kerosene and illuminating gases

Ethylene, Respiration and Fruit Ripening

Respiration rate highest in an immature vegetables and fruits. One group has a pattern similar to vegetables and is called non-climacteric fruits. The other group called climacteric fruits exhibits a temporary respiratory increase, which occurs when the fruit reaches full maturation and size and is entering the ripening stage. In climacteric fruits, the respiration climacteric and subsequent ripening and senescence are associated with a coincident increase in endogenous ethylene production. Exogenous ethylene application; no increase in respiration in climacteric fruits; but increase in respiration in non-climacteric fruits.

Ethylene Addition

Ethylene gas is explosive at concentration between 2.7–36 Kpa. Ethylene (2-chloroethane phosphoric acid) commercially available under various trade names, ethyrel, florel, cerone, prep and CEPA in liquid form. If pH increase above 5 ethylene gas produce from above solutions.

Ethylene Removal

Potassium permanganate ($KMnO_4$) oxidizes ethylene to CO_2 and H_2O. Ultraviolet (UV) lamps or Ozone (O_3) generators oxidize ethylene. Ethylene adsorbed by activated or brominated charcoal. Cold storage reduces ethylene production. Soil bacteria that use ethylene as a biochemical substrate. Hollow fiber technology and pressure swing adsorption as well as a modified pressure swing adsorption called vacuum swing adsorption which keeps ethylene level below 1µl/L.

4

Aonla

Phyllanthus emblica (syn. *Emblica officinalis*), the Indian gooseberry, or *aamla*, is a deciduous tree of the Phyllanthaceae family. It is known for its edible fruit of the same name.

The tree is small to medium in size, reaching 8 to 18 m in height, with a crooked trunk and spreading branches. The branchlets are glabrous or finely pubescent, 10–20 cm long, usually deciduous; the leaves are simple, subsessile and closely set along branchlets, light green, resembling pinnate leaves. The flowers are greenish-yellow. The fruit are nearly spherical, light greenish yellow, quite smooth and hard on appearance, with six vertical stripes or furrows.

Ripening in autumn, the berries are harvested by hand after climbing to upper branches bearing the fruits. The taste of Indian gooseberry is sour, bitter and astringent, and it is quite fibrous. In India, it is common to eat gooseberries steeped in salt water and turmeric to make the sour fruits palatable. It is also used to straighten hair.

Aonla is the fruit of a deciduous plant, of the same name, grown in many states of India. Its English name is 'Indian Gooseberry' and it has been associated with a number of health benefits. Aonla is very good source of Vitamin C and has been found to have great antioxidant properties. Apart from helping in curing a number of ailments, it also boosts the immune system of an individual. In fact, aonla is regarded as a widely-used herb, in case of ayurvedic medicines, owing to its high nutritional value. In the following lines, we have provided information on the nutritional benefits of eating aonla/Indian gooseberry.

Indian gooseberry has undergone preliminary research, demonstrating *in vitro* antiviral and antimicrobial properties. There is preliminary evidence *in vitro* that its extracts induce apoptosis and modify gene expression in osteoclasts involved in rheumatoid arthritis and osteoporosis. It may prove to have potential activity against some cancers. One recent animal study found treatment with *E. officinalis* reduced severity of acute pancreatitis (induced by L-arginine in rats). It also promoted the spontaneous repair and regeneration process of the pancreas occurring after an acute attack.

Experimental preparations of leaves bark or fruit have shown potential efficacy against laboratory models of disease, such as for inflammation, cancer, age-related renal disease, and diabetes.

A human pilot study demonstrated a reduction of blood cholesterol levels in both normal and hypercholesterolemic men with treatment. Another recent study with alloxan-induced diabetic rats given an aqueous amla fruit extract has shown significant decrease of the blood glucose, as well as triglyceridemic levels and an improvement of the liver function caused by a normalization of the liver-specific enzyme alanine transaminase activity.

Although fruits are reputed to contain high amounts of ascorbic acid (vitamin C), 445 mg/100g, the specific contents are disputed, and the overall antioxidant strength of amla may derive instead from its high density of tannins. The fruit also contains other polyphenols: flavonoids, kaempferol, ellagic acid and gallic acid.

Medicinal Use

In traditional Indian medicine, dried and fresh fruits of the plant are used. All parts of the plant are used in various Ayurvedic/Unani medicine (*Jawarish amla*) herbal preparations, including the fruit, seed, leaves, root, bark and flowers. According to Ayurveda, aamla fruit is sour (*amla*) and astringent (*kashaya*) in taste (*rasa*), with sweet (*madhura*), bitter (*tikta*) and pungent (*katu*) secondary tastes (*anurasas*). Its qualities (*gunas*) are light (*laghu*) and dry (*ruksha*), the postdigestive effect (*vipaka*) is sweet (*madhura*), and its energy (*virya*) is cooling (*shita*).

According to Ayurveda, aamla balances all three doshas. While aamla is unusual in that it contains five out of the six tastes recognized by Ayurved, it is most important to recognize the effects of the "virya", or potency, and "vipaka", or post-digestive effect. Considered in this light, aamla is particularly helpful in reducing *pitta* due to its cooling energy and balances both Pitta and *vata* by virtue of its sweet taste. The *kapha* is balanced primarily due to its drying action. It may be used as a *rasayana* (rejuvenative) to promote longevity, and traditionally to enhance digestion (*dipanapachana*), treat constipation (*anuloma*), reduce fever (*jvaraghna*), purify the blood (*raktaprasadana*), reduce cough (*kasahara*), alleviate asthma (*svasahara*), strengthen the heart (*hrdaya*), benefit the eyes (*chakshushya*), stimulate hair growth (*romasanjana*), enliven the body (*jivaniya*), and enhance intellect (*medhya*).

In Ayurvedic polyherbal formulations, Indian gooseberry is a common constituent, and most notably is the primary ingredient in an ancient herbal *rasayana* called *Chyawanprash*. This formula, which contains 43 herbal ingredients as well as clarified butter, sesame oil, sugar cane juice, and honey, was first mentioned in the Charaka Samhita as a premier rejuvenative compound. In Chinese traditional therapy, this fruit is called *yuganzi*, which is used to cure throat inflammation. *Emblica officinalis* tea may ameliorate diabetic neuropathy. In rats it significantly reduced blood glucose, food intake, water intake and urine output in diabetic rats compared with the non diabetic control group.

Herbal Medicines

Demand for aonla in the herbal medicine sector comes from the Ayurveda, Sidha and Unani systems of medicines, which are the major wings of Indian System of Medicine and Homeopathy (ISM&H). While these traditional branches of ISM&H have a strong presence in Indian social and cultural fabric, they are making a comeback as health aids. Presently, the share of Ayurveda is Rs. 2,500 crore (US$ 555.6 million) in the Rs. 12,500 crore (US$ 277.8 million) pharmaceutical industry (www.eximbankindia.com). This study has attempted to assess the demand of aonla by the Indian Ayurvedic medicine producers. It is seen that the industry is largely based on 400 plant species and aonla is one of the important ingredients. However, there is a mushrooming of local producers, with and unorganized set-up. In 1999-2000, there were 7,843 licensed pharmacies of ISM&H, along with a number of unlicensed companies. Out of this, about 4,700 pharmacies are based on Ayurveda. The consumption pattern varies within the industry due to differences in size of units. These pharmacies range from large national drug houses such as Dabur, Zandu, Ayurvaidyshala Kottakal, Himalayan Drug Company, etc., State government run pharmacies catering to government hospitals to small size Vaids preparing their own herbal medications. Wizmin (2002) gathered the data of demand of aonla through a simple questionnaire. The major Ayurvedic and herbal drug houses were contacted to provide their annual aonla requirements. The responses were aided with the analysis of the destination data from the Pratapgarh market, and the information provided by the report on Medicinal Plants by Task Force. These processors in different forms, such as fresh fruit, pulp, whole sundry and powder require aonla. To arrive at the demand, their green equivalent has been used. Annual demands of fresh or green aonla for some companies are as follows.

Annual Requirement of Aonla by Major Pharmaceutical Companies

Company	*Tonnes*
Zandu Pharmaceuticals	5, 000
Dabur	5, 000
Ayurvaidyashala, Kottakal	650
Shri Dhaootpapeshwar Ltd., Mumbai	150
Baidynath Ltd., Jhansi	65
Natural Remedies, Bangalore	25
Gufic health care Ltd., Mumbai	20
Gurukul Kangri, Haridwar	10
Sandu Brothers	10
Bharat Pharm, Alwar	10
Ajanta Pharmacy	4
Total	**10, 944**

Zandu and Dabur are the major players in the market of herbal medicines, requiring a total of 10,000 t of aonla every year. Besides them, there are about 35 large size units. The average annual aonla requirement of these units is about 200 t and each of their collective requirements is about 7,000 t per annum. Therefore, the total demand of aonla by these large size companies is about 17,000 t per annum. Ayurvedic pharmacies are mostly clustered in U.P., Maharashtra, Rajasthan, Gujarat and Kerala. The aonla requirement generated by Ayurvedic drug producers is the highest in Maharashtra and Kerala. In Kerala itself, there are about 1,000 old pharmacies. In all, there are about 4,500 smaller size Ayurvedic pharmacies. Taking only 1.5 t per annum requirement, the total demand of aonla by smaller ISM&H pharmacies is estimated to be about 6,700 t or say 7,000 t. Adding all the three segments, the demand of aonla by the herbal pharmacy segment will be about 25,000 t per year.

While this is the present status of demand, experts feel that this segment offers tremendous future potential. Some of the domestic and multinational companies engaged in manufacturing synthetic drugs, have joined this herbal bandwagon. Others have expressed their plan to do the same. A few plant-based drugs have successfully passed the clinical tests and FDA approval. International pharmaceutical giant. Merck, in April 2001 announced its aonla based products, with patent for one of the active ingredients of aonla. Pharmaceutical industry has made massive investments on pharmacological, clinical and chemical researches all over the world and efforts to discover more potent plant drugs are underway. India is viewed to have excellent potential for getting into plant-based medicines, due to raw material availability and traditional knowhow. The genetic and pharmacological make-up of aonla is still an area of active research. All the research and development is driven by research carried out by multinational pharmaceutical industry but there are no reports of active perusal of this route by Indian companies. Since the aonla based phyto-medicines are still in the research and development stage, there is no substantial demand of aonla presented by them. Their present requirement is understood to be of dehydrated aonla and some exporters are believed to be supplying aonla in salts and standardized extracts. One-day group discussion on issues in developing a sustainable industry based on aonla and aonla products was conversed by the Centre for Advancement of Sustainable Agriculture (CASA) in collaboration with Central Institute for Subtropical Horticulture, Lucknow, Biosys Ltd. and Rothamsted International. The discussion revealed that there is quite a large array of opportunities to develop and market new products including food items/additive, cosmetic preparations, ayurvedic/herbal preparations, etc.

Culinary Use

Particularly in South India, the fruit is pickled with salt, oil, and spices. Aamla is eaten raw or cooked into various dishes. In Andhra Pradesh, tender varieties are used to prepare *dal* (a lentil preparation), and *amle ka murabbah,* a sweet dish indigenous to the northern part of India (wherein the berries are soaked in sugar syrup for a long time till they are imparted the sweet flavor); it is traditionally consumed after meals.

Other Uses

Popularly used in inks, shampoos and hair oils, the high tannin content of Indian gooseberry fruit serves as a mordant for fixing dyes in fabrics. Aonla shampoos and hair oil are traditionally believed to nourish the hair and scalp and prevent premature grey hair.

Nutritional Value of Indian Gooseberry/Aonla

Indian Gooseberry comprises of the following nutrients:

- ☆ Dietary Fiber
- ☆ Vitamin C
- ☆ Calcium
- ☆ Phosphorus
- ☆ Iron
- ☆ Carotene
- ☆ Vitamin B Complex
- ☆ Protein
- ☆ Carbohydrates

Indian Gooseberry has also been found to be low in Saturated Fats, Cholesterol and Sodium, making it good for health.

Nutritional Benefits of Eating Aonla

- ☆ Being rich in Vitamin C, aonla has great antioxidant properties. It helps in fighting free radicals that are responsible for ageing and cell degeneration.
- ☆ Aonla is known to be good for the health of hair, skin and even eyes.
- ☆ Aonla ensures better digestion of food and also balances stomach acid.
- ☆ Indian gooseberry is associated with fortification of liver and nourishment of the brain, resulting in better mental functioning.
- ☆ Aonla is believed to support the heart and strengthen the lungs.
- ☆ Aonla regulates elimination and enhances fertility.
- ☆ Indian gooseberry acts as a body coolant and also flushes out toxins.
- ☆ Aonla has been found to increase vitality and improve muscle tone.
- ☆ Aonla power improves immunity and adds to the physical strength of an individual.
- ☆ Indian gooseberry is said to help in the treatment of hemorrhage, menprrhagia, leucorrhoea and discharge of blood from uterus.
- ☆ Aonla increases red blood cell production and in turn, strengthens teeth and nails.
- ☆ Aonla has been found to be pretty useful in controlling blood sugar.

Indian gooseberry is said to help in the treatment of the following ailments also:

- ☆ Anemia
- ☆ Cerebral, Gastro and Cardiovascular Illness
- ☆ Constipation

- ✰ Fever
- ✰ Indigestion
- ✰ Insomnia
- ✰ Liver Disorder
- ✰ Respiratory Problems
- ✰ Sun Burns
- ✰ Sun Strokes
- ✰ Urinary Problems

Nutritional Value of Aonla Fruits

Constituents	*Amount*
Moisture, per cent	84.36
Protein, per cent	0.88
T.S.S, Brix	14.00
Reducing sugars, per cent	2.37
Non-reducing sugars, per cent	0.74
Total Sugars, per cent	3.11
Fat, per cent	0.10
Fiber, per cent	1.9-3.4
Carbohydrates, per cent	14.10-21.89
Acidity, per cent	2.24
pH	2.50
Pectin, per cent	0.88
Tannin, per cent	0.55
Ascorbic acid, mg/100g	571.60
Iron, mg/100g	1.20
Calcium, per cent	0.012-0.050
Phosphorus, per cent	0.020-0.026
Nicotinic acid, mg/100g	0.20
Vitamin C, mg/100g	200-1814
Carotene, mg/100g	0.01
Thiamine, mg/100g	0.03
Riboflavin, mg/100g	0.05
Niacin, mg/100g	0.18
Tryptophan, mg/100g	3.00
Methionine, mg/100g	2.00
Lysine, mg/100g	17.00

Figure 4.1: Flow sheet for preparation of aonla toffee

Aonla fruits fully ripe

↓

Seed removal and shredding

↓

Homogenization of pulp

↓

Addition of sugar, fat, milk powder (750g, 100g, 100g/Kg Pulp)

↓

Heating upto 80 °Brix

↓

Addition of salt and citric acid (5g, 2g)

↓

Heating upto 82 °Brix

↓

Spreading in aluminium trays (1 cm thick; already smeared with fat)

↓

Cooling

↓

Cutting into attractive pieces (1.5 × 2.5 cm)

↓

Rapping into metallic paper

↓

Storage

Figure 4.2: Flow sheet for preparation of aonla pulp

Fully mature fruits
↓
Washing
↓
Heating in boiling water, 10 min
↓
Separate segments and stone/seed
↓
Add water equal to the wt. of segments
↓
Pass through pulping machine
↓
Pulp
↓
Heating pulp to 75 °C for 5 min
↓
Add potassium metabisulphite (2g/kg pulp)
↓
Filling and capping the sterilized bottles
↓
Storage

Figure 4.3: Flow sheet for preparation of jam from aonla pulp

Aonla pulp
↓
Mixing with sugar
↓
Cooking
↓
Addition of citric acid
↓
Judging of end point (68 per cent, TSS)
↓
Filling into clean sterilized bottles
↓
Capping
↓
Storage

Figure 4.4: Flow sheet for preparation of ready to serve from aonla pulp

Aonla Pulp
↓
Mixing sugar, water, citric acid
↓
Homogenization
↓
Bottling
↓
Crown corking
↓
Pasteurization
↓
Cooling
↓
Storage

Figure 4.5: Flow sheet for preparation of squash from aonla pulp

Aonla pulp
↓
Mixing with syrup solution
↓
Addition of 350 ppm SO_2
↓
Filling into clean sterilized bottles
↓
Capping
↓
Storage

Ingredients	*Squash*	*Syrup*
Pulp per cent	40–45	40–50
TSS per cent	50	68
Acidity per cent	1.0	1.2

Figure 4.6: Flow sheet for preparation of aonla candy

Mature aonla fruits

↓

Washing

↓

Pricking

↓

Dipping in 2 per cent salt solution for 24 h

↓

Washing

↓

Dipping in 2 per cent alum solution for 24 h

↓

Washing

↓

Blanching in boiling water for 5 min

↓

Dipping in 40 per cent syrup solution for 24 h

↓

Dipping in 50 per cent syrup solution for 24 h

↓

Dipping in 60 per cent syrup solution for 24 h

↓

Dipping in 70 per cent syrup solution for 3 days

↓

Draining

↓

Drying the fruits under shade upto 15 per cent moisture or 75 per cent TSS

↓

Wrapping with powder sugar

↓

Packing in 400 gauge polythene pouches (1 kg capacity)

↓

Storage

Figure 4.7: Flow sheet for preparation of aonla chutney

Fully mature aonla fruits

↓

Steam for 15 min in cooker

↓

Separate segments, seeds

↓

Grind using mixer

↓

Fine pulp

↓

Mix with ingredients

↓

Cook the mixture to desired consistency

↓

Add glacial acetic acid

↓

Filling glass bottles

↓

Cooling

↓

Storage

Ingredients	*Quantity*
Fruits	1.25 kg
Salt	50 g
Sugar	1 kg
Red chilli	10 g
Ginger	15 g
Hot spices	25 g
Acetic acid	10 ml

Figure 4.8: Flow sheet for preparation of aonla pickle

Fully mature fruits
↓
Washing
↓
Blanching for 10 min
↓
Separation of segments and seeds
↓
Draining of moisture (sun drying 2 h)
↓
Mixing with ingredients (fried)
↓
Packing
↓
Storage

Ingredients	*Quantity*
Aonla segments	1.0 kg
Salt	150 g
Turmeric powder	10 g
Red chilli	10 g
Nigella seeds	10 g
Fenugreek powder	30 g
Oil	300 ml

Figure 4.9: Flow sheet for preparation of aonla shreds

Fully mature aonla fruits

↓

Washing

↓

Grating into shreds

↓

Mixing with 4 per cent salt

↓

Drying upto 15 per cent moisture

↓

Packing

↓

Storage

Figure 4.10: Flow sheet for preparation of aonla juice

Selection of fully mature aonla fruits

↓

Washing

↓

Preparation of fruits (cutting, removing seeds)

↓

Incubation with pectinase enzyme (0.2 per cent, 37 °C, 6 h)

↓

Extraction of juice

↓

Clarification of juice

↓

Pure juice

Figure 4.11: Flow sheet for preservation of aonla juice

Pure aonla juice

↓

Pasteurization (80 °C, 20 min)

↓

Addition preservative KMS 750 ppm

↓

Hot filling in sterilized containers

↓

Capping

↓

Storage

Figure 4.12: Flow sheet for preparation of spicey aonla chutney

Aonla Fruits (Fully mature)

↓

Washing

↓

Boiling for 25-30 min

↓

Removal of seed

↓

Making fine pulp

↓

Addition of one half its weights of water; sugar and salt and cooking gently

↓

Rest of process is similar to that of sweet mango chutney

Sl.No.	*Item*	*Quantity*	*Sl.No.*	*Item*	*Quantity*
1.	Aonla	1.25 Kg	8.	Cinnamon	10 g
2.	Salt	50 g	9.	Cardamom	10 g
3.	Sugar	1 kg	10.	Aniseed (powder)	10 g
4.	Onion (chopped)	50 g	11.	Black pepper (powder)	10 g
5.	Ginger (chopped)	15 g	12.	Cumin	10 g
6.	Garlic (chopped)	15 g	13.	Vinegar	100 ml
7.	Red chilli powder	10 g			

Figure 4.13: Flow sheet for preparation of aonla sauce

Aonla (Fully mature fruits)

↓

Washing

↓

Boiling for 25-30 min

↓

Separation into segments and removal of seed

↓

Addition of double its weight of water

↓

Pulp extraction

↓

Straining of pulp

↓

Rest of process similar to that of tomato sauce

Sl.No.	*Item*	*Quantity*	*Sl.No.*	*Item*	*Quantity*
1.	Aonla pulp	1Kg	6.	Hot spices	10 g
2.	Salt	14 g	7.	Garlic (chopped)	5 g
3.	Sugar	50 g	8.	Red chilli powder	5 g
4.	Onion (chopped)	50 g	9.	Vinegar/acetic acid	40 ml
5.	Ginger (chopped)	10 g	10.	Sodium benzoate	700mg/kg

5

Apple

The apple is the pomaceous fruit of the apple tree, species *Malus domestica* in the rose family (Rosaceae). It is one of the most widely cultivated tree fruits, and the most widely known of the many members of genus *Malus* that are used by humans. Apple grows on small, deciduous trees that blossom in the spring and produce fruit in the fall. The tree originated in Western Asia, where its wild ancestor, *Malus sieversii*, is still found today. Apples have been grown for thousands of years in Asia and Europe, and were brought to North American by European colonists. Apples have been present in the mythology and religions of many cultures, including Norse, Greek and Christian traditions. In 2010, the fruit's genome was decoded, leading to new understandings of disease control and selective breeding in apple production.

There are more than 7,500 known cultivars of apples, resulting in a range of desired characteristics. Different cultivars are bred for various tastes and uses, including in cooking, fresh eating and cider production. Domestic apples are generally propagated by grafting, although wild apples grow readily from seed. Trees are prone to a number of fungal, bacterial and pest problems, which can be controlled by a number of organic and non-organic means.

At least 55 million tonnes of apples were grown worldwide in 2005, with a value of about $10 billion. China produced about 35 per cent of this total. The United States is the second-leading producer, with more than 7.5 per cent of world production. Iran is third, followed by Turkey, Russia, Italy and India. Apples are often eaten raw, but can also be found in many foods (especially desserts) and drinks. Many beneficial health effects have been found from eating apples; however, the seeds are slightly poisonous and two forms of allergies are seen to various proteins found in the fruit.

The apple forms a tree that is small and deciduous, reaching 3 to 12 meters (9.8 to 39 ft) tall, with a broad, often densely twiggy crown. The leaves are alternately arranged simple ovals 5 to 12 cm long and 3–6 centimetres (1.2–2.4 in) broad on a 2 to

5 centimetres (0.79 to 2.0 in) petiole with an acute tip, serrated margin and a slightly downy underside. Blossoms are produced in spring simultaneously with the budding of the leaves. The flowers are white with a pink tinge that gradually fades, five petaled, and 2.5 to 3.5 centimetres (0.98 to 1.4 in) in diameter. The fruit matures in autumn, and is typically 5 to 9 centimetres (2.0 to 3.5 in) in diameter. The center of the fruit contains five carpels arranged in a five-point star, each carpel containing one to three seeds, called pips.

Maturation and Harvest

Cultivars vary in their yield and the ultimate size of the tree, even when grown on the same rootstock. Some cultivars, if left unpruned, will grow very large, which allows them to bear much more fruit, but makes harvesting very difficult. Mature trees typically bear 40–200 kilograms (88–440 lb) of apples each year, though productivity can be close to zero in poor years. Apples are harvested using three-point ladders that are designed to fit amongst the branches. Dwarf trees will bear about 10–80 kilograms (22–180 lb) of fruit per year.

Storage

Commercially, apples can be stored for some months in controlled-atmosphere chambers to delay ethylene-induced onset of ripening. The apples are commonly stored in chambers with higher concentrations of carbon dioxide with high air filtration. This prevents ethylene concentrations from rising to higher amounts and preventing ripening from moving too quickly. Ripening continues when the fruit is removed. For home storage, most varieties of apple can be held for approximately two weeks when kept at the coolest part of the refrigerator (*i.e.* below 5 °C). Some types, including the Granny Smith and Fuji, can be stored up to a year without significant degrade.

Human Consumption

Apples are often eaten raw; except for the seeds, which are slightly poisonous (see below); the whole fruit including the skin is suitable for human consumption. Varieties bred for this purpose are termed dessert or table apples.

Apples can be canned or juiced. They are milled to produce apple cider (non-alcoholic, sweet cider) and filtered for apple juice. The juice can be fermented to make cider (alcoholic, hard cider), ciderkin, and vinegar. Through distillation, various alcoholic beverages can be produced, such as applejack, Calvados, and apfelwein, Pectin and apple seed oil may also be produced.

Apples are an important ingredient in many desserts, such as apple pie, apple crumble, apple crisp and apple cake. They are often eaten baked or stewed, and they can also be dried and eaten or reconstituted (soaked in water, alcohol or some other liquid) for later use. Puréed apples are generally known as apple sauce. Apples are also made into apple butter and apple jelly. They are also used (cooked) in meat dishes.

- ☆ In the UK, a toffee apple is a traditional confection made by coating an apple in hot toffee and allowing it to cool. Similar treats in the US are candy apples (coated in a hard shell of crystallized sugar syrup), and caramel apples, coated with cooled caramel.

- ☆ Apples are eaten with honey at the Jewish New Year of Rosh Hashanah to symbolize a sweet new year.
- ☆ Farms with apple orchards may open them to the public, so consumers may themselves pick the apples they will buy.

Sliced apples turn brown with exposure to air due to the conversion of natural phenolic substances into melanin upon exposure to oxygen. Different cultivars vary in their propensity to brown after slicing. Sliced fruit can be treated with acidulated water to prevent this effect. Organic apples are commonly produced in the United States. Organic production is difficult in Europe, though a few orchards have done so with commercial success, using disease-resistant cultivars and the very best cultural controls. The latest tool in the organic repertoire is a spray of a light coating of kaolin clay, which forms a physical barrier to some pests, and also helps prevent apple sun scald.

Apple Allergy

One form of apple allergy, often found in northern Europe, is called birch-apple syndrome, and is found in people who are also allergic to birch pollen. The allergy is caused by a protein in apples that is similar to birch pollen, and people affected by this protein can also become allergic to other fruits, nuts and vegetables. Reactions, which are called oral allergy syndrome (OAS), generally involve itching and inflammation of the mouth and throat, but in rare cases can also include life-threatening anaphylaxis. This reaction only occurs when raw fruit is consumed - the allergen is neutralized in the cooking process. The variety of apple, maturity and storage conditions can change the amount of allergen present in individual fruits. Long storage times can increase the amount of proteins that cause birch-apple syndrome.

In other areas, such as the Mediterranean, people have adverse reactions to apples because of their similarity to peaches, including a close relationship between the allergens of the two fruits. This form of apple allergy also includes OAS, but often has more severe symptoms, such as vomiting, abdominal pain and urticaria, and can be life-threatening. Individuals with this form of allergy can also develop reactions to other fruits and nuts. Cooking does not break down the protein causing this particular reaction, so affected individuals cannot each either raw or cooked apple. Freshly harvested, over-ripe fruits tend to have the highest levels of the protein that causes this reaction. Breeding efforts have yet to produce a hypoallergenic fruit for either of the two types of apple allergy.

Health Benefits

The proverb *"An apple a day keeps the doctor away"* addressing the health effects of the fruit, dates from 19th century Wales. Research suggests that apples may reduce the risk of colon cancer, prostate cancer and lung cancer. Compared to many other fruits and vegetables, apples contain relatively low amounts of vitamin C, but are a rich source of other antioxidant compounds. The fiber content, while less than in most other fruits, helps regulate bowel movements and may thus reduce the risk of colon cancer. They may also help with heart disease, weight loss, and controlling cholesterol. The fiber contained in apples reduces cholesterol by preventing reabsorption, and (like most fruits and vegetables) they are bulky for their caloric content. However, apple seeds are mildly poisonous, containing a small amount of amygdalin, a cyanogenic glycoside. It usually is not enough to be dangerous to humans, but can deter birds. There is evidence from laboratory experiments that apples possess phenolic compounds which may be cancer-protective and demonstrate antioxidant activity. The predominant phenolic phytochemicals in apples are quercetin, epicatechin, and procyanidin B2.

Apple juice concentrate has been found to increase the production of the neurotransmitter acetylcholine in mice, providing a potential mechanism for the "prevention of the decline in cognitive performance that accompanies dietary and genetic deficiencies and aging." Other studies have shown an "alleviation of oxidative damage and cognitive decline" in mice after the administration of apple juice. Researchers at the Chinese University of Hong Kong discovered that fruit flies that were fed an apple extract lived 10 per cent longer than other flies that were fed a normal diet.

Apples, with Skin (Edible Parts)

Nutritional value per 100 g (3.5 oz)	
Energy	218 kJ (52 kcal)
Carbohydrates	13.81 g
Sugars	10.39 g
Dietary fiber	2.4 g
Fat	0.17 g
Protein	0.26 g
Water	85.56 g
Vitamin A equiv.	3 μg
Thiamine (vit. B_1)	0.017 mg
Riboflavin (vit. B_2)	0.026 mg
Niacin (vit. B_3)	0.091 mg
Pantothenic acid (B_5)	0.061 mg
Vitamin B_6	0.041 mg
Folate (vit. B_9)	3 μg
Vitamin C	4.6 mg
Calcium	6 mg
Iron	0.12 mg
Magnesium	5 mg
Phosphorus	11 mg
Potassium	107 mg
Zinc	0.04 mg

Figure 5.1: Flow sheet for preparation of apple jam

Apple fruits fully mature

↓

Cleaning, Cutting and Homogenization of pulp

↓

Addition of sugar (1:0.75)

↓

Heating (30 min)

↓

Adding pectin 0.5 per cent

↓

Heating (68 °brix)

↓

Jam

↓

Filling in sterilized glass bottles

↓

Cooling

↓

Waxing

↓

Capping and sealing

↓

Storage

Figure 5.2: Flow sheet for preparation of apple toffee

Apple fruits fully ripe

↓

Peeling/Cutting into small pieces

↓

Homogenization of pulp

↓

Addition of sugar, fat, milk powder (750g, 100g, 100g/Kg Pulp)

↓

Heating upto 80 °Brix

↓

Addition of salt and citric acid (5g, 2g)

↓

Heating upto 82 °Brix

↓

Spreading in aluminium trays (1 cm thick; already smeared with fat)

↓

Cooling

↓

Cutting into attractive pieces (1.5 × 2.5 cm)

↓

Rapping into metallic paper

↓

Storage

Figure 5.3: Flow sheet for preparation of apple sauce

Mature apples
↓
Washing
↓
Peeling
↓
Removal of core and seeds
↓
Making fine pulp
↓
Straining of pulp
↓
Putting spice bag in pulp and pressing occasionally
↓
Cooking pulp with one third quantity of sugar
↓
Cooking to one third of original volume of pulp
↓
Removal of spice bag (after squeezing in pulp)
↓
Adding remaining sugar and salt
↓
Cooking to one third its original volume
↓
Addition of vinegar/acetic acid and preservative
↓
Filling hot into sterilized bottles
↓
Crown corking
↓
Pasteurization (at 85-90 °C, 30 min)
↓
Labeling and Storage
↓
Marketing

Sl.No.	*Item*	*Quantity*	*Sl.No.*	*Item*	*Quantity*
1.	Apple pulp	1Kg	8.	Cinnamon	15 g
2.	Salt	10 g	9.	Cardamom	15 g
3.	Sugar	250 g	10.	Aniseed (powder)	15 g
4.	Onion (chopped)	200 g	11.	Clove	5 No.
5.	Ginger (chopped)	100 g	12.	Vinegar/acetic acid	50 ml
6.	Garlic (chopped)	50 g	13.	Sodium benzoate	700mg/kg
7.	Red chilli powder	10 g			

6

Apricot

The apricot, *Prunus armeniaca*, is a species of *Prunus*, classified with the plum in the subgenus *Prunus*. The native range is somewhat uncertain due to its extensive prehistoric cultivation.

Seeds or kernels of the apricot grown in central Asia and around the Mediterranean are so sweet that they may be substituted for almonds. The Italian liqueur amaretto and amaretti biscotti are flavoured with extract of apricot kernels rather than almonds. Oil pressed from these cultivar kernels, and known as 'Oil of Almond', has been used as cooking oil. Kernels contain between 2.05 per cent and 2.40 per cent hydrocyanic acid, but normal consumption is insufficient to produce serious effects.

Medicinal and Non-Food Uses

Cyanogenic glycosides (found in most stone fruit seeds, bark, and leaves) are found in high concentration in apricot seeds. Laetrile, a purported alternative treatment for cancer, is extracted from apricot seeds. Apricot seeds "were used against tumors as early as A.D. 502. In England during the seventeenth century, apricot oil was also used against tumors, swellings, and ulcers". In 2005, scientists in the Republic of Korea found that treating human prostate cancer cells with amygdalin induces programmed cell death. They concluded that "amygdalin may offer a valuable option for the treatment of prostate cancers". A 2006 systematic review by the Cochrane Collaboration concluded: "The claim that aetrile has beneficial effects for cancer patients are not supported by data from controlled clinical trials. This systematic review has clearly identified the need for randomised or controlled clinical trials assessing the effectiveness of aetrile or amygdalin for cancer treatment." Given the lack of evidence, laetrile has not been approved by the U.S. Food and Drug Administration. The U.S. National Institutes of Health evaluated the evidence separately and concluded that clinical trials of amgydalin showed little or no effect against cancer. For example, a 1982 trial of 175 patients found that tumor size had increased in all but one patient. The authors reported that "the hazards of amygdalin therapy were evidenced in several patients by symptoms of cyanide toxicity or by blood cyanide levels approaching the lethal range." The study concluded that "Patients exposed to this agent should be instructed about the danger of cyanide poisoning, and their blood cyanide levels should be carefully monitored. Amygdalin (Laetrile) is a toxic drug that is not effective as a cancer treatment".

In Europe, apricots were long considered an aphrodisiac, and were used in this context in William Shakespeare's *A Midsummer Night's Dream*, and as an inducer of childbirth, as depicted in John Webster's *The Duchess of Malfi*.

Due to their high fiber to volume ratio, dried apricots are sometimes used to relieve constipation or induce diarrhea. Effects can be felt after eating as few as three.

Apricots, raw

Nutritional value per 100 g (3.5 oz)	
Energy	201 kJ (48 kcal)
Carbohydrates	11 g
Sugars	9 g
Dietary fiber	2 g
Fat	0.4 g
Protein	1.4 g
Vitamin A equiv.	96 μg
β-carotene	1094 μg
Vitamin C	10 mg
Iron	0.4 mg

Apricots, dried

Nutritional value per 100 g (3.5 oz)	
Energy	1,009 kJ (241 kcal)
Carbohydrates	63 g
Sugars	53 g
Dietary fiber	7 g
Fat	0.5 g
Protein	3.4 g
Vitamin A equiv.	180 μg
β-carotene	2163 μg
Vitamin C	1 mg
Iron	2.7 mg

Research shows that of any food, apricots possess the highest levels and widest variety of carotenoids. Carotenoids are antioxidants that may help to prevent heart disease, reduce "bad cholesterol" levels and protect against cancer. Although initial studies suggested that antioxidant supplements might promote health, later large clinical trials did not detect any benefit and suggested instead that excess supplementation may be harmful. In traditional Chinese medicine, apricots are considered helpful in regenerating body fluids, detoxifying, and quenching thirst

Figure 6.1: Flow diagram of canning process of apricot fruits

Washing apricot fruits

↓

Sorting/Grading

↓

Pitting/Slicing

↓

Filling/Exhausting/Sealing

↓

Retorting/Cooling

↓

Storage

Figure 6.2: Flow diagram for apricot dehydration (dry-blanch-dry)

Fresh apricots fruits

↓

Washing

↓

Halve, pit, spread on trays or belt

↓

Sulphur (sulphur house, spray or dip)

↓

Dry to 50 per cent of fresh weight

↓

Blanch

↓

Dehydrate to final moisture

↓

Store to equilibrate

↓

Packing

↓

Store/Ship

7

Bael

Bael belongs to the same family as the orange; it has a thin hard woody shell that turns yellow when ripe. The pulp of the fruit is orange-colored, with numerous, hairy seeds embedded in it. The fruit's medicinal value is very high when it just begins to ripen therefore it is best eaten fresh just as it ripens. The bael fruit's claim to fame rests most on its medicinal properties. The bael fruit tree which is commonly grown in India is found in near most temples as the leaves of the tree bark is traditionally used as sacred offering to Lord Shiva. The tree is also regarded as sacred in Indonesia and Malaysia. In Florida only the small, hard-shelled type is known and this has to be sawed open, cracked with a hammer, or flung forcefully against a rock. Fruits of this type are standard for medicinal uses rather than for consuming as normal food.

- ✰ Origin: India
- ✰ Varieties: wild (small), cultivated (larger size)
- ✰ Dimension of Tree: height: up to 40 to 50 ft (12-15 m)
- ✰ Dimension of Fruit: shape: round or oval; diameter: 5 - 12.5 cm; weight 1-2.5 kg.
- ✰ Rind/Pulp: thin, hard, woody rind. When unripe the rind color is grayish green and yellow when ripe.
- ✰ Culinary uses: eaten fresh; also prepared as sun-dried slices, pickled, sherbet, marmalade and syrup. The young leaves and shoots are eaten as a vegetable in Thailand and used to season food in Indonesia.

Common name of Bael fruit in different countries:

- ✰ English name: Stone Apple
- ✰ India: Bael, Bel or Sirphal
- ✰ Thailand: Matum
- ✰ Malaysia and Indonesia: Maja

- ☆ Australia: Bengal quine
- ☆ French: Oranger du Malabar
- ☆ Portuguese: Marnelos

Health Benefits

- ☆ Bael has a high tannin content which makes it an effective cure for dysentery and cholera. There is as much as 9 per cent tannin in the pulp of wild fruits, less in the cultivated types. The rind contains up to 20 per cent . The leaves also contain tannin.
- ☆ Bael is beneficial in the digestive system and it is also a mild laxative.
- ☆ A decoction of the unripe fruit with ginger and fennel is said to be effective for the treatment of hemorrhoids.
- ☆ The juice of its leaves mixed with honey, can relieve catarrh and fever.
- ☆ The pulp is used in the treatment of Vitiligo.
- ☆ The leaf decoction is said to alleviate asthma.
- ☆ A hot poultice of the leaves is considered an effective treatment for various inflammations.
- ☆ The bark decoction is administered in cases of malaria.
- ☆ The fruit, roots and leaves have antibiotic activity.
- ☆ The root, leaves and bark are used in treating snakebite.
- ☆ An infusion from bael leaves is an effective remedy for peptic ulcer.

Nutritive Value per 100 g of Edible Portion:

Nutrient	*Amount*
Water	54.96-61.5 g
Carbohydrates	28.11-31.8 g
Protein	1.8-2.62 g
Fat	0.2-0.39 g
Vitamin A	55 mg
Vitamin C	8-60 mg
Thiamine	0.13 mg
Riboflavin	1.19 mg
Niacin	1.1 mg

Toxicity

- ☆ The leaves are said to cause abortion and sterility in women.
- ☆ Tannin ingested frequently and in quantity over a long period of time, is antinutrient and carcinogenic.

Shelf Life

Harvested bael fruits can stored for 2 weeks outside the refrigerator and 4 in a refrigerator. Thereafter, mold is likely to develop at the stem-end and any crack in the rind.

Other Uses

- ☆ From the bael flowers cologne is obtained by distillation.
- ☆ Artists add fruit pulp to their water color, and it may be applied as a protective coating on paintings.
- ☆ In rural areas, the fruit, which has binding qualities, is mixed with lime plaster for water proofing wells and is added to cement for building walls in villages.
- ☆ Bael fruit is use to eliminate the scum in vinegar making.

Its Usefulness and Therapeutic Value

1. The medicinal value of Bael fruit is enhanced due to presence of Tanin, the evaporating substance in its rind. The rind contains 20 per cent and the pulp has only 9 per cent of Tanin. This substance helps to cure diabetes.
2. *Treatment of Asthma* - Grind 5 gm of Bael leaves. Add 1 spoon of honey. Take orally in morning and evening for relief.
3. *Cure of Anemia* - Extract the pulp of Bael. Dry it and grind it to powder form. Add one spoonful of this powder to boiled cow milk. Also add some sugar candy. Take this dose twice a day in morning and evening for a long period.
4. *Fractures* - Extract the pulp of raw bael. Dry it and grind it to powder form. Mix 10 gm of powder, 50 gm of pure ghee, ½ spoonful of turmeric powder with a glass of Luke warm water, stir well. Take it orally twice a day.
5. *Healing of Wound* - Take rind, root, leaves, fruit pulp of equal quantity. Grind them to extract juice. Add 10 gm of honey. Drink it.
6. *Swollen Joints* - Few bael pulps mixed with hot mustard oil to be applied on the affected area twice a day during morning and evening for relief.
7. *High Blood Pressure* - Bael leaves to be taken every morning. The juice of bael leaves added with honey can also be taken every morning.
8. *Jaundice* - Extract juice of 100 nos of soft bael leaves. Mix it with the powder of 10 Nos of black pepper. Take the mixture every morning and evening. Added to this take at least five glasses of sugarcane juice daily after meals.
9. *Diarrhoea* - Dry raw bael fruit pulp. Take out seeds. The dried pulp if consumed quickly tops loose motion.
10. *Troubles During Pregnancy* - One spoon of raw fruit pulp if taken twice a day stops frequent vomiting nausea during pregnancy. Little sugar candy may be added to the pulp for taste.
11. *Typhoid* - Our body becomes weak due to high fever. Grind 200 bael leaves. Boil in one cup of water till it becomes thick. Take this paste with a little honey twice or thrice a day.

12. *Healthy Mind and Brain* - Ripe bael fruit taken with fresh cream (butter) and sugar candy powder sharpens concentration and intelligence. One can make bael squash and keep it at home thus preserving and utilizing surplus ripe bael fruit for off season. Bael Squash - Collect well ripened Bael fruit. Clean it and collect the pulp. Add lemon juice to the pulp in the proportion of six table spoonful of lemon juice to one kg of pulp. Boil this mixture in water - One liter per kg of pulp. Stir well while boiling. Now strain the mixture though steel strainer. Take one kg of sugar for one kg of pulp and boil it in ½ liter of water for fifteen minutes. Add the strained Bael juice to the sugar pulp. Add preservative potassium metabisulphite 1/4 teasponful to the sugar solution when it is cooled. Keep it in a sterilized bottle. Seal the cap with paraffin wax and keep it for future use. Except Bael seeds everything is useful for consuming. Taking 5-6 leaves regularly keeps the body and mind healthy because of its medicinal qualities it cures a lot of aliments. That is why Bael is also known as *Mahaphala* or Great fruit.

Figure 7.1: Flow sheet for preparation of bael toffee

Bael fruits fully ripe

↓

Peeling removing outer coat

↓

Homogenization of pulp

↓

Addition of sugar, fat, milk powder (750g, 100g, 100g/Kg Pulp)

↓

Heating upto 80 °Brix

↓

Addition of salt and citric acid (5g, 2g)

↓

Heating upto 82 °Brix

↓

Spreading in aluminium trays (1 cm thick; already smeared with fat)

↓

Cooling

↓

Cutting into attractive pieces (1.5 × 2.5 cm)

↓

Rapping into metallic paper

↓

Storage

Figure 7.2: Flow sheet for preparation of bael pulp

Ripe Bael Fruits

↓

Breaking of shell

↓

Removal of pulp with seed and fiber

↓

Mixing with water

↓

Passing through pulper

↓

Addition of preservative

↓

Pasteurization

↓

Storage

↓

Marketing

Pulp can also be obtained by heating the mixture of pulp and water a little by straining through a thick cloth.

8

Banana

Banana is the common name for herbaceous plants of the genus *Musa* and for the fruit they produce. Bananas come in a variety of sizes and colors when ripe, including yellow, purple, and red. Almost all modern edible parthenocarpic bananas come from the two wild species – *Musa acuminata* and *Musa balbisiana*. The scientific names of bananas are *Musa acuminata*, *Musa balbisiana* or hybrids *Musa acuminata × balbisiana*, depending on their genomic constitution. The old scientific names *Musa sapientum* and *Musa paradisiaca* are no longer used.

Banana is also used to describe Enset and Fe'i bananas, neither of which belongs to the aforementioned species. Enset bananas belong to the genus *Ensete* while the taxonomy of Fe'i-type cultivars is uncertain.

In popular culture and commerce, "banana" usually refers to soft, sweet "dessert" bananas. By contrast, *Musa* cultivars with firmer, starchier fruit are called plantains or "cooking bananas". The distinction is purely arbitrary and the terms 'plantain' and 'banana' are sometimes interchangeable depending on their usage. They are native to tropical South and Southeast Asia, and are likely to have been first domesticated in Papua New Guinea. Today, they are cultivated throughout the tropics. They are grown in at least 107 countries, primarily for their fruit, and to a lesser extent to make fiber, banana wine and as ornamental plants.

Comparison between the Two Wild Banana Ancestors in the Simmonds and Shepherd

Species	*Musa acuminata*	*Musa balbisiana*
Color of pseudostem	Black or grey-brown spots	Unmarked or slightly marked
Petiole canal	Erect edge, with scarred inferior leaves, not against the pseudostem	Closed edge, without leaves, against the pseudostem
Stalk	Covered with fine hair	Smooth

Species	*Musa acuminata*	*Musa balbisiana*
Pedicels	Short	Long
Ovum	Two regular rows in the locule	Four irregular rows in the locule
Elbow of the bract	Tall (< 0.28)	Short (> 0.30)
Bend of the bract	The bract wraps behind the opening	The bract raises without bending behind the opening
Form of the bract	Lance- or egg-shaped, tapering markedly after the bend	Broadly egg-shaped
Peak of the bract	Acute	Obtuse
Color of the bract	Dark red or yellow on the outside, opaque purple or yellow on the inside	Brown-purple on the outside, crimson on the inside
Discoloration	The inside of the bract is more bright toward the base	The inside of the bract is uniform
Scarification of the bract	Prominent	Not prominent
Free tepal of the male flower	Corrugated under the point	Rarely corrugated
Color of the male flower	White or cream	Pink
Color of the markings	Orange or bright yellow	Cream, yellow, or pale pink

Storage and Transport

Bananas must be transported over long distances from the tropics to world markets. To obtain maximum shelf life, harvest comes before the fruit is mature. The fruit requires careful handling, rapid transport to ports, cooling, and refrigerated shipping. The goal is to prevent the bananas from producing their natural ripening agent, ethylene. This technology allows storage and transport for 3–4 weeks at 13 °C (55 °F). On arrival, bananas are held at about 17 °C (63 °F) and treated with a low concentration of ethylene. After a few days, the fruit begins to ripen and is distributed for final sale. Unripe bananas cannot be held in home refrigerators because they suffer from the cold ripe bananas can be held for a few days at home. If the bananas you buy are too green, you can put them in a brown paper bag with an apple or tomato overnight to speed up the ripening process. They can be stored indefinitely frozen, then eaten like an ice pop or cooked as a banana mush.

Recent studies have suggested that carbon dioxide (which bananas produce) and ethylene absorbents extend fruit life even at high temperatures. This effect can be exploited by packing the fruit in a polyethylene bag and including an ethylene absorbent, *e.g.*, potassium permanganate, on an inert carrier. The bag is then sealed with a band or string. This treatment has been shown to more than double lifespan up to 3–4 weeks without the need for refrigeration.

Uses; Food and Cooking

Fruits

Bananas are a staple starch for many tropical populations. Depending upon cultivar and ripeness, the flesh can vary in taste from starchy to sweet, and texture from firm to mushy. Both skin and inner part can be eaten raw or cooked. Bananas' flavor is due, amongst other chemicals, to isoamyl acetate which is one of the main constituents of banana oil.

During the ripening process, bananas produce a plant hormone called ethylene, which indirectly affects the flavor. Among other things, ethylene stimulates the formation of amylase, an enzyme that breaks down starch into sugar, influencing the taste of bananas. The greener, less ripe bananas contain higher levels of starch and, consequently, have a "starchier" taste. On the other hand, yellow bananas taste sweeter due to higher sugar concentrations. Furthermore, ethylene signals the production of pectinase, an enzyme which breaks down the pectin between the cells of the banana, causing the banana to soften as it ripens.

Bananas are eaten deep fried, baked in their skin in a split bamboo, or steamed in glutinous rice wrapped in a banana leaf. Bananas can be made into jam. Banana pancakes are popular amongst backpackers and other travelers in South Asia and Southeast Asia. This has elicited the expression *Banana Pancake Trail* for those places in Asia that cater to this group of travelers. Banana chips are a snack produced from sliced dehydrated or fried banana or plantain, which have a dark brown color and an intense banana taste. Dried bananas are also ground to make banana flour. Extracting juice is difficult, because when a banana is compressed, it simply turns to pulp. Bananas feature prominently in Philippine cuisine, being part of traditional dishes and desserts like *maruya, turrón*, and *halo-halo*. Most of these dishes use the Saba or Cardaba banana cultivar. Pisang goreng, bananas fried with batter similar to the Filipino *maruya*, is a popular dessert in Malaysia, Singapore, and Indonesia. A similar dish is known in the United States as banana fritters.

Plantains are used in various stews and curries or cooked, baked or mashed in much the same way as potatoes. Seeded bananas (*Musa balbisiana*), one of the forerunners of the common domesticated banana, are sold in markets in Indonesia.

Flower

Banana hearts are used as a vegetable in South Asian and Southeast Asian cuisine, either raw or steamed with dips or cooked in soups, curries and fried foods. The flavor resembles that of artichoke. As with artichokes, both the fleshy part of the bracts and the heart are edible.

Nutrition and Research

Bananas are an excellent source of vitamin B_6 and contain moderate amounts of vitamin C, manganese and potassium. Along with other fruits and vegetables, consumption of bananas may be associated with a reduced risk of colorectal cancer and in women, breast cancer and renal cell carcinoma. Banana ingestion may affect dopamine production in people deficient in the amino acid tyrosine, a dopamine precursor present in bananas. In India, juice is extracted from the corm and used as a home remedy for jaundice, sometimes with the addition of honey, and for kidney stones. Individuals with a latex allergy may experience a reaction to bananas.

Banana, Raw, Edible Parts

Nutritional value per 100 g (3.5 oz)	
Energy	371 kJ (89 kcal)
Carbohydrates	22.84 g
Sugars	12.23 g
Dietary fiber	2.6 g
Fat	0.33 g
Protein	1.09 g
Vitamin A equiv.	3 μg
Thiamine (vit. B_1)	0.031 mg
Riboflavin (vit. B_2)	0.073 mg
Niacin (vit. B_3)	0.665 mg
Pantothenic acid (B_5)	0.334 mg
Vitamin B_6	0.4 mg
Folate (vit. B_9)	20 μg
Choline	9.8 mg
Vitamin C	8.7 mg
Calcium	5 mg
Iron	0.26 mg
Magnesium	27 mg
Manganese	0.3 mg
Phosphorus	22 mg
Potassium	358 mg
Zinc	0.15 mg

Figure 8.1: Flow sheet for preparation of banana chips

Fully Mature Banana fruits

↓

Cleaning/Washing

↓

Peeling/Trimming

↓

Slicing and Washing

↓

Dipping in solutions and blanching (citric acid 0.2 per cent, KMS 0.25 per cent, $CaCl_2$ 0.5 per cent, Salt 3 per cent, 2 Min)

↓

Removal excess water

↓

Drying

↓

Pan frying

↓

Removal excess oil

↓

Chips

↓

Packing

↓

Storage

Figure 8.2: Flow sheet for preparation of banana chips

Selection of raw (green) banana fruits

↓

Peeling banana fruits

↓

Cut into small chips

↓

Keep in 5 per cent NaCl solution + 0.25 per cent citric acid, 10 min

↓

Blanching in hot water 5 min

↓

Keep in 1 per cent KMS solution, 30 min

↓

Dry in tray drier at 50 – 55 °C, 48 h

↓

Packaging/Frying then packaging *vice versa*

↓

Storage

Figure 8.3: Flow sheet for preparation of banana toffee

Banana fruits fully ripe

↓

Peeling

↓

Homogenization of pulp

↓

Addition of sugar, fat, milk powder (750g, 100g, 100g/Kg Pulp)

↓

Heating upto 80 °Brix

↓

Addition of salt and citric acid (5g, 2g)

↓

Heating upto 82 °Brix

↓

Spreading in aluminium trays (1 cm thick; already smeared with fat)

↓

Cooling

↓

Cutting into attractive pieces (1.5 × 2.5 cm)

↓

Rapping into metallic paper

↓

Storage

Figure 8.4: Flow sheet for preparation of banana wine

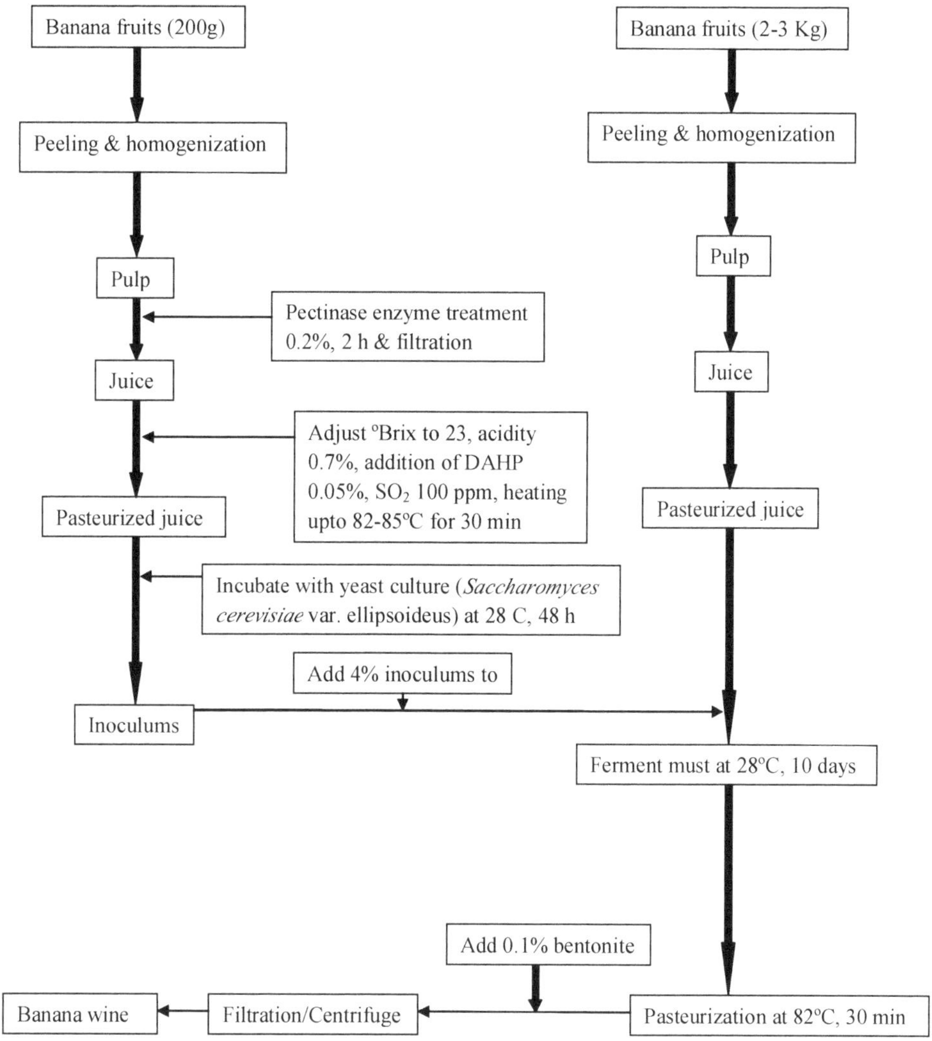

9

Barley

Barley is a major cereal grain, a member of the grass family. It serves as a major animal fodder, as base malt for beer and certain distilled beverages, and as a component of various health foods. It is used in soups and stews, and in barley bread of various cultures. In a 2007 ranking of cereal crops in the world, barley was fourth both in terms of quantity produced (136 million tons) and in area of cultivation (566,000 km^2).

Classification

In traditional classifications of barley these morphological differences have led to different forms of barley being classified as different species. Under these classifications two-rowed barley with shattering spikes (wild barley) is classified as *Hordeum spontaneum* K. Koch. Two-rowed barley with non-shattering spikes is classified as *H. distichum* L., six-rowed barley with non-shattering spikes as *H. vulgare* L. (or *H. hexastichum* L.), and six-rowed with shattering spikes as *H. agriocrithon* Åberg. The fact that these differences were driven by single-gene mutations, coupled with cytological and molecular evidence, has led most recent classifications to treat these forms as a single species, *H. vulgare* L.

Uses

Algicide

Barley straw, in England, is placed in mesh bags and floated in fish ponds or water gardens to help reduce algal growth without harming pond plants and animals. Barley straw has not been approved by the EPA for use as a pesticide and its effectiveness as an algicide in ponds has produced mixed results during university testing in the US and the UK.

Animal Feed

Half of the United States' barley production is used as an animal feed. Barley is an important feed grain in many areas of the world not typically suited for maize production, especially in northern climates - for example, northern and eastern Europe. Barley is the principal feed grain in Canada, Europe, and in the northern United States.[28] A finishing diet of barley is one of the defining characteristics of Western Canadian beef used in marketing campaigns.

Beverages

Alcoholic Beverages

A large part of the remainder is used for malting, for which barley is the best suited grain. It is a key ingredient in beer and whisky production. Two-row barley is traditionally used in German and English beers. Six-row barley was traditionally used in US beers, but both varieties are in common usage now. Distilled from green beer, whisky has been made primarily from barley in Ireland and Scotland, while other countries have utilized more diverse sources of alcohol; such as the more common corn, rye and wheat in the USA. In the USA, a grain type may be identified on a whisky label if that type of grain constitutes 51 per cent or more of the ingredients and certain other conditions are satisfied.

Barley wine was an alcoholic drink in the 18th century. It was prepared by boiling barley in water, then mixing the barley water with white wine and other ingredients like borage, lemon and sugar. In the 19th century a different barley wine was made prepared from recipes of ancient Greek origin.

Non-Alcoholic Beverages

Non-alcoholic drinks such as barley water and barley tea (called mugicha in Japan) have been made by boiling barley in water.

Food

Barley contains eight essential amino acids. According to a recent study, eating whole grain barley can regulate blood sugar (*i.e.* reduce blood glucose response to a meal) for up to 10 hours after consumption compared to white or even whole-grain wheat, which has a similar glycemic index. The effect was attributed to colonic fermentation of indigestible carbohydrates. Barley can also be used as a coffee substitute.

Hulled barley (or covered barley) is eaten after removing the inedible, fibrous outer hull. Once removed, it is called dehulled barley (or pot barley or scotch barley). Considered a whole grain, dehulled barley still has its bran and germ making it a nutritious and popular health food. Pearl barley (or pearled barley) is dehulled barley which has been steam processed further to remove the bran. It may be polished, a process known as "pearling". Dehulled or pearl barley may be processed into a variety of barley products, including flour, flakes similar to oatmeal, and grits.

Barley-meal, whole meal barley flour which is lighter than wheat meal but darker in color, is used in porridge and gruel in Scotland. Barley-meal gruel is known as

Sawiq in the Arab world.[39] With a long history of cultivation in the Middle East, barley is used in a wide range of traditional Assyrian, Arabic, Kurdish, Persian, and Assyrian foodstuffs including kashkak, kashk and murri. Barley soup is traditionally eaten during Ramadan in Saudi Arabia. It is also used in soups and stews in Eastern Europe. In Africa, where it is a traditional food plant, it has the potential to improve nutrition, boost food security, foster rural development and support sustainable land care.

The six row variety *bere* is cultivated in Orkney, Shetland, Caithness and the Western Isles in the Scottish Highlands and islands. The grain is used to make *beremeal*, used locally in bread, biscuits, and the traditional beremeal bannock. Like wheat and rye, barley contains gluten which makes it an unsuitable grain for consumption by those with celiac disease.

Health Benefits of Barley

Barley has been used for human consumption for the longest time. This is not only because it adds a unique delightful flavor to dishes, but because it adds nutritional value to the food we eat. Intake of barley is very likely to help you maintain your overall health in the long term. These whole grains are rich in protein, vitamins, minerals and amino acids essential for our health. More importantly, barley is one of the richest sources of both soluble and insoluble fiber. Insoluble fiber aids in proper excretion of waste products in the body, while soluble fiber (known as beta glucan) mixes with liquid, binds to fatty substances and allows them to leave the body. Barley is also seen to be rich in tocotrienols, which is an antioxidant which helps lessen risk for contracting heart disease and cholesterol problems.

Wheat and Barley

The health benefits of wheat and barley are very comparable, and some claim that they are one and the same. However, barley is rich in both soluble and insoluble fiber, while wheat is only rich in the latter. While both can help prevent "roughage" during excretion, barley also helps promote general heart health. All forms of barley contain almost all vitamins, minerals, amino acids and fiber that your body needs. However, the value of barley nutrition and its effect on your health depends largely on the type of barley that you consume. The figures may vary largely, but the main rule.

Kinds of Barley

Barley nutrition today comes in all forms possible. The grain, the leaves and the extract of barley are all used as food sources. Here are some of the most widely used forms of barley nutrition:

- *Hulled barley*: Barley in its hulled form contains the most amounts of nutrients. However, hulled barley takes a lot more time to soak when used with usual dishes.
- *Barley Grass*: Barley grasses, most especially on the young leaves, are excellent sources of vitamins, minerals, amino acids and antioxidants. Though it does not contain as much dietary fiber as can be found in barley

grains, it contains chlorophyll which can help detoxify your body from harmful toxins.

- ☆ *Barley Green Powder*: Due to the superior medicinal benefits of barley, many companies now produce barley powder which is made from pure barley grass. Most of them have delightful flavors and superior solubility that people take them as barley juice.
- ☆ *Barley Flour*: Flour made from barley can be used as substitute for wheat flour or even cake flour. The flour that is produced from whole barley is a lot more nutritious than flour from pearl barley mainly because the bran in the former is still left intact.
- ☆ *Barley Water*: The main claimed benefit of barley water is in supporting the kidneys especially in times of stress. It is also used as therapeutic water for those who have kidney and bladder ailments.
- ☆ *Pearl Barley*: Pearl barley is what you can find in groceries. These are hulled barley which has been processed to remove the bran.

As you can see, barley nutrition comes in all forms, so it is easier for us to take advantage of its nutritional benefits.

Raw barley

Nutritional value per 100 g (3.5 oz)	
Energy	1,474 kJ (352 kcal)
Carbohydrates	77.7 g
Sugars	0.8 g
Dietary fiber	15.6 g
Fat	1.2 g
Protein	9.9 g
Thiamine (vit. B_1)	0.2 mg
Riboflavin (vit. B_2)	0.1 mg
Niacin (vit. B_3)	4.6 mg
Pantothenic acid (B_5)	0.3 mg
Vitamin B_6	0.3 mg
Folate (vit. B_9)	23 μg
Vitamin C	0.0 mg
Calcium	29.0 mg
Iron	2.5 mg
Magnesium	79.0 mg
Phosphorus	221 mg
Potassium	280 mg
Zinc	2.1 mg

Figure 9.1: Flow sheet for preparation of barley beer

Mature barley grains

↓

Cleaning and grading

↓

Malting (soaking or steeping at 10-15 °C for 24 hr, germination at 16-25 °C for 5 to 7 days)

↓

Drying and powdering

↓

Mashing

↓

Fermentation with yeast (*Saccharomyces services*)

↓

Decantation/filtration

↓

Ageing/Maturation

↓

Finishing (CO_2 0.45-52 per cent)

↓

Sealing and labeling

↓

Storage

↓

Marketing

Figure 9.2: Flow sheet for preparation of barley water

Barley flour
↓
Addition of a little water
↓
Making thin paste
↓
Addition of remaining water
↓
Heating to gelatinize
↓
Cooling
↓
Filtering
↓
Addition to juice
↓
Addition of sugar
↓
Heating just to dissolve sugar
↓
Straining
↓
Addition of essence and preservative
↓
Bottling
↓
Capping
↓
Labeling
↓
Marketing

10

Ber

Ziziphus mauritiana, also known as Jujube, Chinee Apple, Indian plum, and permseret (Anguilla), is a tropical fruit tree species belonging to the family Rhamnaceae.

Z. mauritiana is a spiny, evergreen shrub or small tree up to 15 m high, with trunk 40 cm or more in diameter; spreading crown; stipular spines and many drooping branches. The fruit is of variable shape and size. It can be oval, obovate, oblong or round and that can be 1-2.5 in (2.5-6.25 cm) long, depending on the variety. The flesh is white and crisp. When slightly underipe, this fruit is a bit juicy and has a pleasant aroma. The fruit's skin is smooth, glossy, thin but tight.

It is most commonly found in the tropical and sub-tropical regions. Originally native to India it is now widely naturalized in tropical region from Africa to Afghanistan and China, and also through Malaysia and into Australia and in some Pacific regions. It can form dense stands and become invasive in some areas, including Fiji and Australia and has become a serious environmental weed in Northern Australia. It is a fast growing tree with a medium lifespan that can quickly reach up to 10–40 ft (3 to 12 m) tall.

Plants are capable of seed production once they reach a height of about 1 meter. Wild-growing plants in northern Australia may take 8 years to reach this size. In Australia, plants growing under natural conditions are capable of producing seeds once they reach a height of about 1m. Plants between 1 and 2m high produce, on average, less than five fruits per season. Large plants (>5m high) can produce 5000 or more fruits in a single season.

In India, some types ripen as early as October, others from mid-February to mid-March, others in March, or mid-March, to the end of April. In the Assiut Governorate, there are 2 crops a year, the main in early spring, the second in the fall. In India the trees flower in July to October and fruits are formed soon after. In February–March the fruits are mature and in some places a second crop is produced in the fall. Pickings are done by hand from ladders and about 110 lbs (50 kg) is harvested per day. The fruits remaining on the tree are shaken down. Only fully mature fruits are picked directly from the tree. They are transported in open bags to avoid fermentation.

Seedling trees bear 5,000 to 10,000 small fruits per year in India. Superior grafted trees may yield as many as 30,000 fruits. The best cultivar in India, with fruits normally averaging 30 to the lb (66 to the kg), yields 175 lbs (77 kg) annually. Special cultural treatment increases both fruit size and yield.

Uses

The fruit is eaten raw or pickled or used in beverages. It is quite nutritious and rich in vitamin C. It is second only to guava and much higher than citrus or apples. In India, the ripe fruits are mostly consumed raw, but are sometimes stewed. Slightly under ripe fruits are candied by a process of pricking, immersing in a salt solution. Ripe fruits are preserved by sun-drying and a powder is prepared for out-of-season purposes. It contains 20 to 30 per cent sugar, up to 2.5 per cent protein and 12.8 per cent carbohydrates. Fruits are also eaten in other forms, such as dried, candied, pickled, as juice, or as ber butter. In Ethiopia, the fruits are used to stupefy fish. The leaves are readily eaten by camels, cattle and goats and are considered nutritious.

In India and Queensland, the flowers are rated as a minor source of nectar for honeybees. The honey is light and of fair flavor. Ber timber is hard, strong, fine-grained, fine-textured, hard, tough, durable, and reddish in color. It is most often used to make, it has been used to line wells, to make legs for bedsteads, boat ribs, agricultural implements. The branches are used as framework in house construction and the wood makes good charcoal with a heat content of almost 4,900 kcal per kg. In addition, this species is used as firewood in many areas. In tropical Africa, the flexible branches are wrapped as retaining bands around conical thatched roofs of huts, and are twined together to form thorny corral walls to retain livestock.

The fruits are applied on cuts and ulcers; are employed in pulmonary ailments and fevers; and, mixed with salt and chili peppers, are given in indigestion and biliousness. The dried ripe fruit is a mild laxative. The seeds are sedative and are taken, sometimes with buttermilk, to halt nausea, vomiting, and abdominal pains in pregnancy. They check diarrhea, and are poulticed on wounds. Mixed with oil, they are rubbed on rheumatic areas. The leaves are applied as poultices and are helpful in liver troubles, asthma and fever and, together with catechu, are administered when an astringent is needed, as on wounds. The bitter, astringent bark decoction is taken to halt diarrhea and dysentery and relieve gingivitis. The bark paste is applied on sores. The root is purgative. A root decoction is given as a febrifuge, taenicide and emmenagogue, and the powdered root is dusted on wounds. Juice of the root bark is said to alleviate gout and rheumatism. Strong doses of the bark or root may be toxic. An infusion of the flowers serves as an eye lotion. The fatty-acid methyl ester of *Z. mauritiana* seed oil meets all of the major biodiesel requirements in the USA (ASTM D 6751-02, ASTM PS 121-99), Germany (DIN V 51606) and European Union (EN 14214). The average oil yield is 4.95 kg oil/tree or 1371 kg oil/hectare, and arid or semi-arid regions may be utilized due to its drought resistance.

Indian Jujube (Ber)–Fresh Fruit

Nutritional value per 100 g (3.5 oz)	
Energy	24.76 kJ (5.92 kcal)
Carbohydrates	17 g
Sugars	5.4-10.5 g
Dietary fiber	0.60 g
Fat	0.07 g
Protein	0.8 g
Water	81.6-83.0 g
Thiamine (vit. B_1)	0.02-0.024 mg
Riboflavin (vit. B_2)	0.02-0.038 mg
Niacin (vit. B_3)	0.7-0.873 mg
Calcium	25.6 mg
Iron	0.76-1.8 mg
Phosphorus	26.8 mg

Figure 10.1: Flow sheet for preparation of ber candy

Selection of ber fruits

↓

Washing

↓

Pricking

↓

Destoning

↓

Hot water blanching (5 min)

↓

Sulphur fumigation (2g/Kg for 2 h)

↓

Syruping

↓

Steeping in 40 per cent syrup + 1 per cent citric acid for 24 h

↓

Steeping in 50 per cent syrup + 1 per cent citric acid for 24 h

↓

Steeping in 60 per cent syrup + 1 per cent citric acid for 24 h

↓

Steeping in 70 per cent syrup + 1 per cent citric acid for 7 days

↓

Quickly washing

↓

Drying in shade (16-18 per cent moisture)

↓

Packaging

↓

Storage

Figure 10.2: Flow sheet for preparation of ber tutti-frutti

Selection of ber fruits
↓
Washing; removal of skin
↓
Pricking
↓
Destoning
↓
Cutting into small cubes
↓
Blanching (2-3 min)
↓
Syruping
↓
Steeping in 40 per cent syrup + 1 per cent citric acid + edible color for 24 h
↓
Steeping in 50 per cent syrup for 24 h
↓
Steeping in 60 per cent syrup for 24 h
↓
Steeping in 70 per cent syrup for 7 days
↓
Quickly washing
↓
Drying in shade (16-18 per cent moisture)
↓
Packaging
↓
Storage

Figure 10.3: Flow sheet for preparation of ber juice

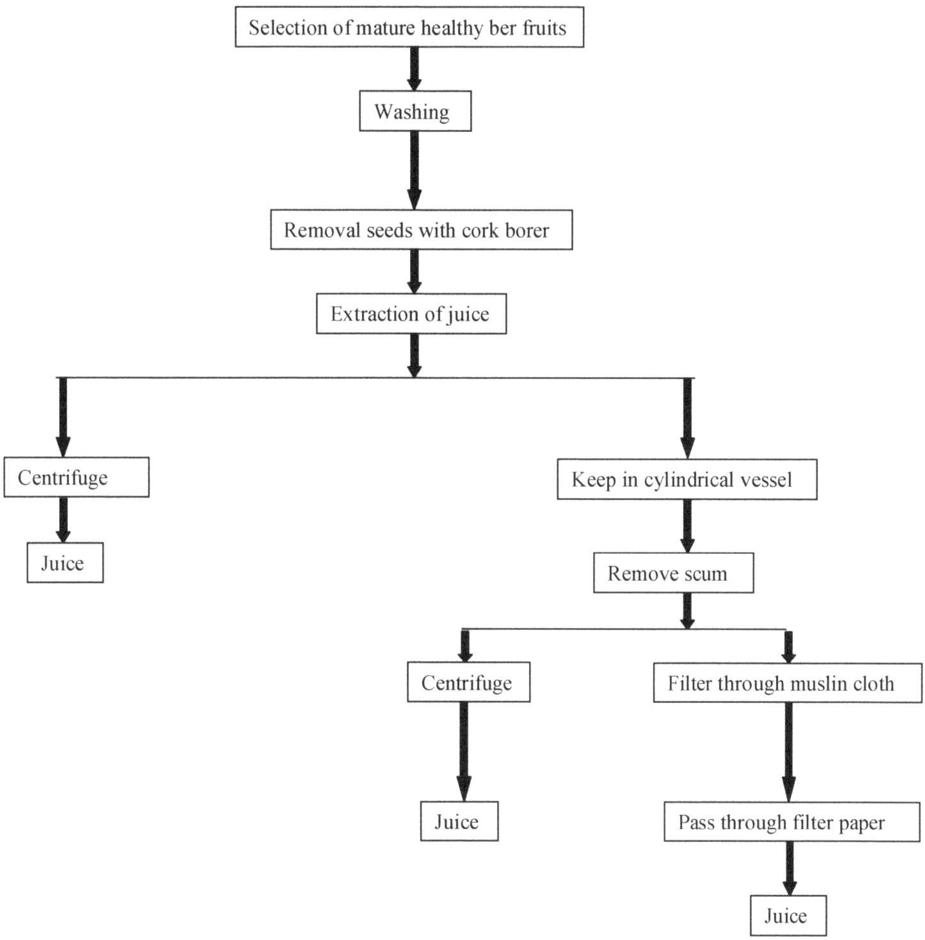

Figure 10.4: Flow sheet for preparation of dried ber

Selection of mature healthy ber fruits

↓

Pricking

↓

Removal of seeds

↓

Blanching

↓

Sulphiting (2g/kg, 2 h)

↓

Drying at 60 °C for 48 h

↓

Packing/Sealing

↓

Storage

The dried ber fruits may be cut into small pieces and used in bakery products. The dried fruits can also be soaked in sugar syrup for 24 hrs and cooked for 10 min to prepare ber candy.

Figure 10.5: Flow sheet for preparation of ber powder

Selection of mature healthy ber fruits

↓

Washing

↓

Removal of seeds

↓

Cutting into small pieces

↓

Blanching (2-3 min)

↓

Sulphiting (2g/kg 2 h)

↓

Drying

↓

Powdering

↓

Sieving (60 mesh)

↓

Packing

↓

Storage

Figure 10.6: Flow sheet for the preparation of ber chutney

Selection of mature yellow ber fruits

↓

Washing

↓

Grating into fine slices

↓ ← Add all recipe material

Heating mixture upto 68-69 °Brix

↓

Poured into sterilized bottles

↓

Cooling

↓

Seal air tight

↓

Storage

Sl.No.	*Material*	*Quantity*
1.	Ber small slices	1 Kg
2.	Sugar	1 Kg
3.	Salt	50 g
4.	Red chilli powder	20 g
5.	Onion chopped	60 g
6.	Garlic chopped	15 g
7.	Cardamom powder	15 g
8.	Cinnamon	15 g
9.	Vinegar	180 ml

Figure 10.7: Flow sheet for the preparation of ber murabba

Selection of mature yellow ber fruits

↓

Washing

↓

Grating into fine slices

↓ ← Add all recipe material

Heating mixture upto 68-69 $^{\circ}$Brix

↓

Poured into sterilized bottles

↓

Seal air tight

↓

Storage

Sl.No.	*Material*	*Quantity*
1.	Ber small slices/shreds	1 Kg
2.	Sugar	1 Kg
3.	Salt	2 g
4.	Citric acid	2 g
5.	Vinegar	25 ml
6.	Essence	As required

Figure 10.8: Flow sheet for the preparation of ber pickles

Selection of mature yellow ber fruits

↓

Washing

↓

Removal of seeds

↓

Cut into quarters

↓

Slices immersed in 5 per cent citric acid for 12 h

↓

Draining salt water

↓

Keep for drying for 2 h

↓

Mixed with turmeric and salt powder

↓

Heating of oil

↓ ← Add all recipe material

Mixed with ber slices

↓ ← Add sodium benzoate

Transferred into sterilized glass jars

↓

Add boiled and cooled oil

↓

Keep for curing, 15 days

↓

Storage

Sl.No.	*Material*	*Quantity*	*Sl.No.*	*Material*	*Quantity*
1.	Ber slices	1.5 kg	6.	Red chillies	50g
2.	Salt	250g	7.	Asafoetida	60g
3.	Edible oil	250g	8.	Turmeric	25g
4.	Fenugreek	25g	9.	Sodium benzoate	0.1 g
5.	Mustard course ground	100g			

Figure 10.9: Flow sheet for preparation of ber wine

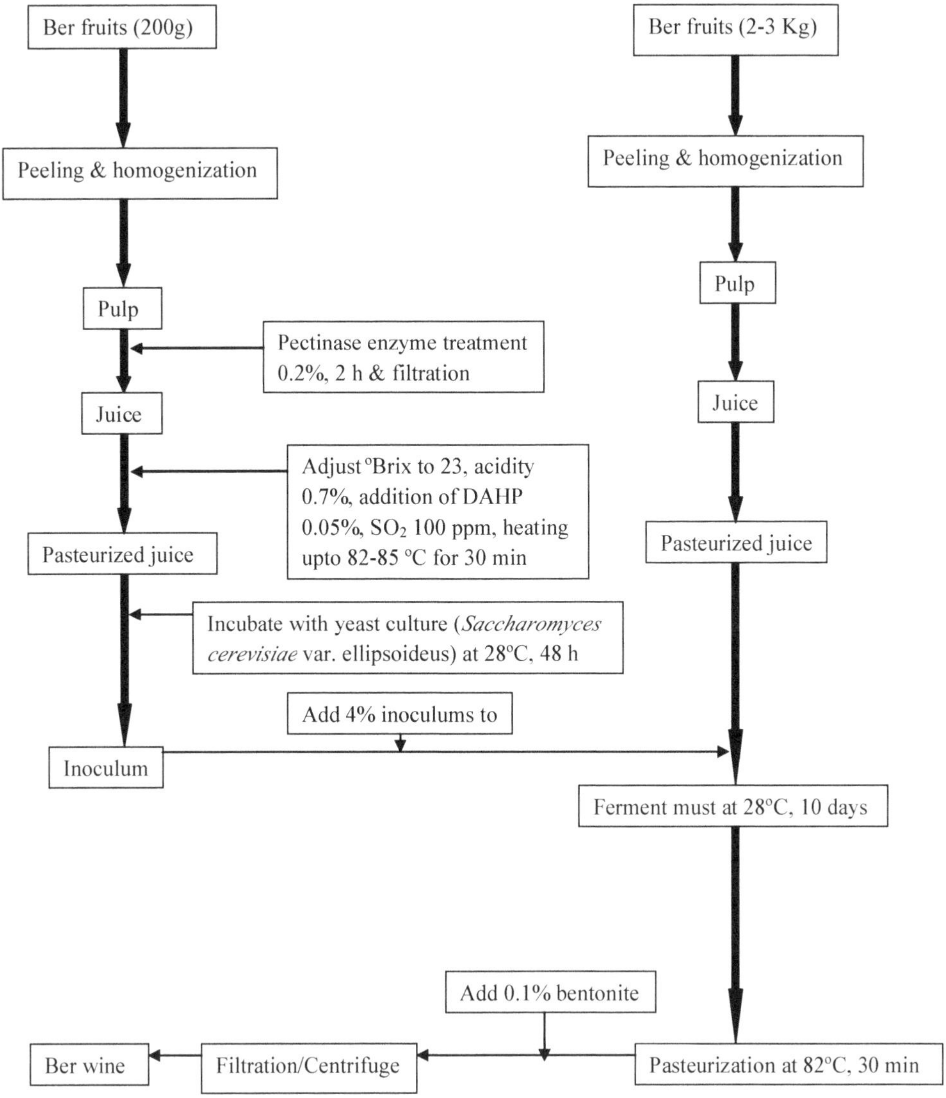

11

Cashewnut

The cashew is a tree in the family Anacardiaceae. Its English name derives from the Portuguese name for the fruit of the cashew tree, *caju*, which in turn derives from the indigenous Tupi name, *acajú*. It is now widely grown in tropical climates for its cashewnuts and cashew apples. The tree is small and evergreen, growing to 10-12m (~32 ft) tall, with a short, often irregularly shaped trunk. The leaves are spirally arranged, leathery textured, elliptic to obovate, 4 to 22 cm long and 2 to 15 cm broad, with a smooth margin. The flowers are produced in a panicle or corymb up to 26 cm long, each flower small, pale green at first then turning reddish, with five slender, acute petals 7 to 15 mm long. The largest cashew tree in the world covers an area of about 7,500 square meters (81,000 sq ft).

The true fruit of the cashew tree is a kidney or boxing-glove shaped drupe that grows at the end of the cashew apple. The drupe develops first on the tree, and then the pedicel expands into the cashew apple. Within the true fruit is a single seed, the cashewnut. Although a nut in the culinary sense, in the botanical sense the nut of the cashew is a seed. The seed is surrounded by a double shell containing an allergenic phenolic resin, anacardic acid, a potent skin irritant chemically related to the more well known allergenic oil urushiol which is also a toxin found in the related poison ivy. Properly roasting cashews destroys the toxin, but it must be done outdoors as the smoke (not unlike that from burning poison ivy) contains urushiol droplets which can cause severe, sometimes life-threatening, reactions by irritating the lungs. People who are allergic to cashew urushiols may also react to mango or pistachio which is also in the Anacardiaceae family. Some people are allergic to cashewnuts, but cashews are a less frequent allergen than nuts or peanuts. While native to Northern South America, the Portuguese took the cashew plant to Goa, India, between the years of 1560 and 1565. From there it spread throughout Southeast Asia and eventually Africa.

Uses [Medicinal and Industry]

The cashew nutshell liquid (**CNSL**), a byproduct of processing cashew, is mostly composed of anacardic acids. These acids have been used effectively against tooth abscesses due to their lethality to a wide range of Gram-positive bacteria. Many parts of the plant are used by the Patamona of Guyana medicinally. The bark is scraped and soaked overnight or boiled as an antidiarrheal; it also yields a gum used in varnish. Seeds are ground into powders used for antivenom for snake bites. The nut oil is used topically as an antifungal and for healing cracked heels. Anacardic acid is also used in the chemical industry for the production of cardanol, which is used for resins, coatings, and frictional materials.

Culinary

The cashewnut is a popular snack, and its rich flavor means that it is often eaten on its own, lightly salted or sugared, or covered in chocolate.

Cashew is very commonly used in Indian cuisine. The nut can be used whole for garnishing sweets or curries, or ground into a paste that forms a base of sauces for curries (*e.g. Korma*), or some sweets. It is also used in powdered form in the preparation of several Indian sweets and desserts. In Goan cuisine both roasted and raw kernels are used whole for making curries and sweets. The cashew apple is eaten raw or used in curries. The cashewnut can also be harvested in its tender form, when the shell has not hardened and is green in color. The shell is soft and can be cut with a knife and the kernel extracted, but it is still corrosive at this stage, so gloves are required. The kernel can be soaked in turmeric water to get rid of the corrosive material before use. This is mostly found in Kerala cuisine, typically in avial, a dish that contains several vegetables, grated coconut, turmeric and green chillies. Cashewnuts also appear in Thai cuisine and Chinese cuisine, generally in whole form. In Malaysia, the young leaves are eaten raw in a salad or with Sambal belacan (shrimp paste with chilli and lime).

In Brazil, the cashew fruit juice is popular all across the country. In Panama, the cashew fruit is cooked with water and sugar for a prolonged period of time to make a sweet, brown, paste-like dessert called "dunce de marañón". Marañón is one of the Spanish names for cashew. In the Philippines, cashew is a known product of Antipolo, and is eaten with suman. Pampanga also has a sweet dessert called turrones de casuy which is cashew marzipan wrapped in white wafer. In Indonesia, roasted and salted cashewnut is called *mente*, while the cashew apple is called *jambu monyet* (literally means monkey rose apple).

Alcohol

In Goa, India, the cashew apple (the accessory fruit) is mashed; the juice is extracted and kept for fermentation for 2–3 days. Fermented juice then undergoes a double distillation process. The resulting beverage is called feni. In the southern region of Mtwara, Tanzania, the cashew apple (*bibo* in Swahili) is dried and saved.

Later it is reconstituted with water and fermented, then distilled to make strong liquor often referred to by the generic name, gongo.

In Mozambique, it is very common among the cashew farmers to make strong liquor from the cashew apple which is called "agua ardente" (burning water). According to *An Account of the Island of Ceylon* written by Robert Percivalan alcohol had been distilled in the early twentieth century from the juice of the fruit, and had been manufactured in the West Indies. Apparently the Dutch considered it superior to brandy as a "liqueur."

Nutrition

The fats and oils in cashewnuts are 54 per cent monounsaturated fat (18:1), 18 per cent polyunsaturated fat (18:2), and 16 per cent saturated fat (9 per cent palmitic acid (16:0) and 7 per cent stearic acid (18:0)).

Cashewnuts, Raw

Nutritional value per 100 g (3.5 oz)	
Energy	2,314 kJ (553 kcal)
Carbohydrates	30.19 g
Sugars	5.91 g
Dietary fiber	3.3 g
Fat	43.85 g
Protein	18.22 g
Thiamine (vit. B_1)	0.42 mg
Riboflavin (vit. B_2)	0.06 mg
Niacin (vit. B_3)	1.06 mg
Pantothenic acid (B_5)	0.86 mg
Vitamin B_6	0.42 mg
Folate (vit. B_9)	25 μg
Vitamin C	0.5 mg
Calcium	37 mg
Iron	6.68 mg
Magnesium	292 mg
Manganese	1.66 mg
Phosphorus	593 mg
Potassium	660 mg
Sodium	12 mg
Zinc	5.78 mg

Figure 11.1: Flow sheet for preparation of processed cashewnut

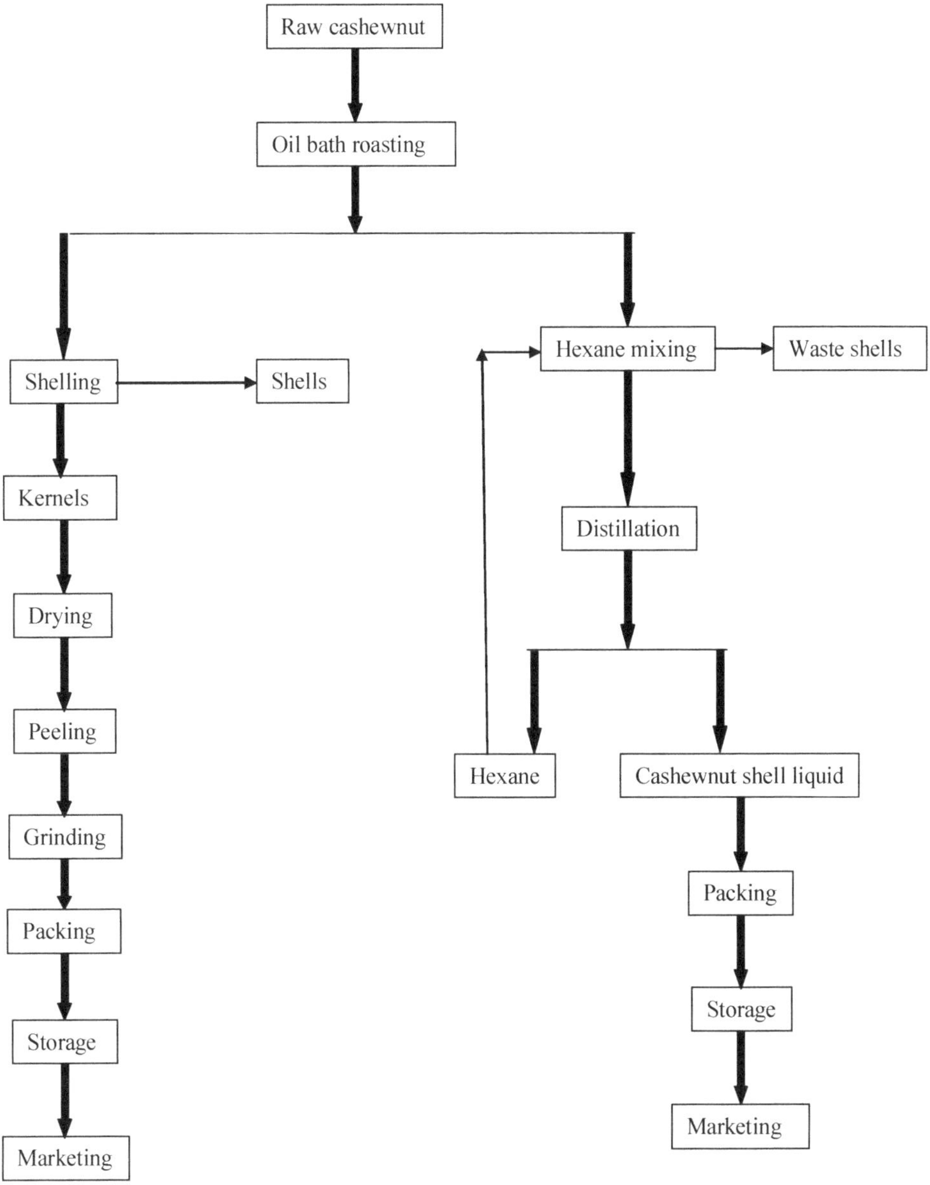

12

Citrus Fruits

1. Lemon

The lemon is both a small evergreen tree (*Citrus × limon*, often given as *C. limon*) native to Asia, and the tree's ellipsoidal yellow fruit. The fruit is used for culinary and non-culinary purposes throughout the world – primarily for its juice, though the pulp and rind (zest) are also used, mainly in cooking and baking. Lemon juice is about 5 per cent to 6 per cent (approximately 0.3 M) citric acid, which gives lemons a sour taste, and a pH of 2–3. Many lemon flavored drinks and foods are available, including lemonade and sherbet lemons. The distinctive sour taste of lemon juice makes it a key ingredient in many dishes across the world.

The exact origin of the lemon has remained a mystery, though it is widely presumed that lemons first grew in India, northern Burma, and China. In South and South East Asia, it was known for its antiseptic properties and it was used as an antidote for various poisons. Lemons entered Europe (near southern Italy) no later than the 1st century CE, during the time of Ancient Rome. However, they were not widely cultivated. It was later introduced to Persia and then to Iraq and Egypt around CE 700. The lemon was first recorded in literature in a 10th century Arabic treatise on farming, and was also used as an ornamental plant in early Islamic gardens. It was distributed widely throughout the Arab world and the Mediterranean region between 1000 and 1150. The genetic origin of the lemon, however, was reported to be hybrid between sour orange and citron.

Differences Between Orange and Lemon

Whilst in some cases it is generally accepted to be clear (sweet/sour) or (orange/yellow) in reality it is not. Various examples occur, which do not agree, either with the popular definition of oranges and lemons, or with that given by botanists. For instance the k/ratta citrus of India has an orange leaf, an orange exterior, an orange pulp, yet its fruit is prominently mammillate (*i.e.* covered with rounded protuberances), and its flower is tinged purple. Again, the acid-less citrus, the* mitha-nimboo, has a lemon-yellow exterior, a white pulp, and it is mammillate; its leaves are like those of a lemon, yet its flowers are pure white. The jhambiri proper has orange-like leaves, the flowers are slightly tinged with purple, the pulp, varies from white to pale yellow, or orange, and the fruit is mammillate, and of two kinds externally, either of an orange color, or of a lemon-yellow, and possibly also of a fawn color.

Varieties

- ✰ *Bush lemon tree*: This naturalized lemon grows wild in subtropical Australia. They are very hardy, and have a thick skin with a true lemon flavor; the zest is good for cooking. It grows to about 4m in a sunny position.
- ✰ *Eureka*: Because it grows year-round and abundantly, this is the common supermarket lemon.
- ✰ *Femminello St. Teresa, or Sorrento*: Native to Italy, this fruit's zest is high in lemon oils. It is the variety traditionally used in the making of limoncello.
- ✰ *Jhambiri C. jhambiri* (Tan): The Rough lemon has a lemon yellow exterior and a very sour pulp. It is widely used as a rootstock in S. Asia.
- ✰ *Lisbon*: A good quality bitter lemon with high juice and acid levels, the fruits of Lisbon are very similar to Eureka. The vigorous and productive trees are very thorny, particularly when young.
- ✰ *Meyer*: This is a cross between a lemon and possibly an orange or a mandarin, and was named for Frank N. Meyer, who first discovered it in 1908. Thin-skinned and slightly less acidic than the Lisbon and Eureka lemons, Meyer lemons require more care when shipping and are not widely grown on a commercial basis. Meyer lemons have a much thinner rind, and often mature to a yellow-orange color. Meyer lemons are slightly more frost-tolerant than other lemons.
- ✰ *Ponderosa*: The tree is very hardy and can handle frosts; the fruit are thick-skinned and very large. Likely a citron-lemon hybrid.
- ✰ *Variegated Pink*: A varietal of the eureka or Lisbon cultivars with variegated patterns in the foliage and the rinds of immature green fruit. Upon maturing to yellow, the variegated pattern recedes in the fruit rind. The flesh and juice are pink or pinkish-orange instead of yellow.
- ✰ *Verna*: A Spanish variety of unknown origin.
- ✰ *Yen Ben*: An Australasian cultivar.
- ✰ *Yuzu*: Cultivated in Japan and Korea for centuries, yuzu have a flavor akin to a mixture of meyer lemon and white grapefruit. Yuzu is likely a wild hybrid between an ichang papeda and a sour mandarin, and is a close relative of sudachi and kaffir limes. Yuzu rival citranges and kumquats as the most cold-tolerant citrus.

Culinary Uses

Lemon juice, rind, and zest are used in a wide variety of culinary applications:

- Lemon juice is used to make lemonade, soft drinks, cocktails, and marinades for both fish, where its acid neutralizes amines in fish by converting them into non-volatile ammonium salts, and meat, where the acid partially hydrolyzes tough collagen fibers, tenderizing the meat, but the low pH denatures the proteins, causing them to dry out when cooked. Lemon juice is also used as a short-term preservative on certain foods that tend to oxidize and turn brown after being sliced, such as apples, bananas and avocados, where its acid denatures the enzymes which cause browning and degradation. Lemon juice and rind are used to make marmalade and lemon liqueur.
- Lemon slices and lemon rind are used a garnish for both food and drinks.
- Lemon zest, the grated outer rind of the fruit, is used to add flavor to baked goods, puddings, rice and other dishes.
- Pickled lemons are a Moroccan delicacy.

Non Culinary Uses

Aromatherapy, First Aid and Medicine

- In one of the most comprehensive scientific investigations done yet, researchers at The Ohio State University revealed lemon oil aroma used in aromatherapy does not influence the human immune system, but may enhance mood.
- The low pH of juice makes it antibacterial.

Commercial Use

- Lemons were the primary commercial source of citric acid prior to the development of fermentation-based processes.
- A halved lemon is used as a finger moistener for those counting large amounts of bills, such as tellers and cashiers.

Household Use

- The peel oil is used as a wood cleaner and polish, where the solvent property of *d*-limonene is employed to dissolve old wax, fingerprints, and grime.
- A halved lemon dipped in salt or baking powder can be used to brighten copper cookware. The acid dissolves the tarnish and the abrasives assist the cleaning.
- As a sanitary kitchen deodorizer the juice can deodorize, remove grease, bleach stains, and disinfect; when mixed with baking soda, it can remove stains from plastic food storage containers.

Insecticide

- The *d*-limonene in lemon oil is used as a nontoxic insecticide treatment. See orange oil.

Nutritive Value

The average lemon contains approximately 3 tablespoons of juice. Allowing lemons to come to room temperature before squeezing (or heating briefly in a microwave) makes the juice easier to extract. Lemons left unrefrigerated for long periods of time are susceptible to mold.

Lemon, Raw, Without Peel

Nutritional value per 100 g (3.5 oz)	
Energy	121 kJ (29 kcal)
Carbohydrates	9.32 g
Sugars	2.50 g
Dietary fiber	2.8 g
Fat	0.30 g
Protein	1.10 g
Thiamine (vit. B_1)	0.040 mg
Riboflavin (vit. B_2)	0.020 mg
Niacin (vit. B_3)	0.100 mg
Pantothenic acid (B_5)	0.190 mg
Vitamin B_6	0.080 mg
Folate (vit. B_9)	11 μg
Vitamin C	53.0 mg
Calcium	26 mg
Iron	0.60 mg
Magnesium	8 mg
Phosphorus	16 mg
Potassium	138 mg
Zinc	0.06 mg

2. Lime

Lime is a term referring to a number of different citrus fruits, both species and hybrids, which are typically round, green to yellow in color, 3–6 cm in diameter, and containing sour and acidic pulp. Limes are a good source of vitamin C. Limes are often used to accent the flavors of foods and beverages. Limes are grown all year round and are usually smaller and more sour than lemons.

Uses

In cooking, lime is valued both for the acidity of its juice and the floral aroma of its zest. It is a very common ingredient in authentic Mexican, Southwestern United States, Vietnamese and Thai dishes. It is also used for its pickling properties in ceviche. The use of dried limes (called black lime or *loomi*) as a flavouring is typical of Persian cuisine and Iraqi cuisine, as well as in Gulf-style *baharat* (a spice mixture that is also called *kabsa* or *kebsa*). Lime is an essential ingredient of any cuisine from India, and many varieties of pickles are made, *e.g.* sweetened lime pickle, salted pickle, and lime chutney. Lime leaves are also an herb in South, East, and Southeast Asia. Lime is frequently used to add flavour to cold and hot drinks, including water, tonic and other cocktails.

Other Uses

To prevent scurvy during the 19th century, British sailors were issued a daily allowance of citrus, such as lemon, and later switched to lime, which was not as effective at preventing scurvy but was easier to obtain on Britain's Caribbean colonies. It was later discovered that the greater effectiveness of lemons derived from the 4-fold higher quantities of vitamin C lemon juice contains compared to the West Indian limes used by the British.

Lime juice is the juice of **limes** (citrus). It may be squeezed from fresh limes, or purchased in bottles in both unsweetened and sweetened varieties. Lime juice is used to make limeade, and as an ingredient (typically as sour mix) in many cocktails. Lime extracts and essential oils are frequently used in perfumes, cleaning products, and aromatherapy. In India, the lime is used in Tantra for removing evil spirits. It is also combined with Indian chilis to make a protective charm to repel the evil eye. Furthermore, it was believed that hanging limes over sick people cured them of the illness by repelling evil spirits lurking inside the body.

Health Effects

When the skin is exposed to ultraviolet light after lime juice contact, a reaction known as phytophotodermatitis can occur, which can cause darkening of the skin, swelling or blistering. The agent responsible for this is psoralen. Lime contains 24 milligrams less vitamin C per 100 grams than the lemon.

Lime, Raw (Edible Parts)

Nutritional value per 100 g (3.5 oz)	
Energy	126 kJ (30 kcal)
Carbohydrates	11 g
Sugars	1.7 g
Dietary fiber	3 g
Fat	0.2 g
Protein	0.7 g
Water	88 g
Vitamin C	29 mg

3. Orange

An orange–specifically, the sweet orange–is the citrus *Citrus x sinensis* (*Citrus sinensis* (L.) Osbeck) and its fruit. It is the most commonly grown tree fruit in the world. The orange is a hybrid of ancient cultivated origin, possibly between pomelo

(*Citrus maxima*) and mandarin (*Citrus reticulata*). It is an evergreen flowering tree generally growing to 9–10 m in height (although very old specimens have reached 15 m). The leaves are arranged alternately, are ovate in shape with crenulate margins and are 4–10 cm long. The orange fruit is a hesperidium, a type of berry. Orange trees are widely cultivated in tropical and subtropical climates for the delicious sweet fruit, which is peeled or cut (to avoid the bitter rind) and eaten whole, or processed to extract orange juice, and also for the fragrant peel. In 2008, 68.5 million tons of oranges were grown worldwide, primarily in Brazil and the US states California and Florida. Oranges probably originated in Southeast Asia and were cultivated in China by 2500 BC. The fruit of *Citrus sinensis* is called *sweet orange* to distinguish it from *Citrus aurantium*, the bitter orange. The name is thought to derive ultimately from the Sanskrit for the orange tree, with its final form developing after passing through numerous intermediate languages.

Terminology and Scope

All citrus trees are of the single genus, *Citrus*, and remain almost entirely interfertile; that is, there is only one "superspecies" which includes grapefruits, lemons, limes, oranges, and numerous other types and hybrids. Nevertheless, names have been given to the various members of the genus. The name "orange" applies primarily to the sweet orange, *Citrus sinensis*, which accounts for about 70 per cent of world citrus production. This article is limited to *Citrus sinensis* and its hybrids.

Other citrus species known as oranges include:

- ✰ The bitter orange, *Citrus aurantium*, also known as Seville orange, sour orange (especially when used as rootstock for a sweet orange tree), bigarade orange, and marmalade orange.
- ✰ The bergamot orange, *Citrus bergamia Risso*, which is grown primarily in Italy and used primarily for the peel, which flavours Earl Grey tea.
- ✰ The mandarin orange *Citrus reticulata*, which itself has an enormous number of cultivars (most notably the satsuma *(C. unshiu)*, tangerine *(Citrus × tangerina)* and clementine *(C. clementina)*. In some cultivars the mandarin resembles the sweet orange and is difficult to distinguish from it, but it is generally smaller and/or oblate rather than round in shape, easier to peel, and less acid.
- ✰ The trifoliate orange (*Poncirus trifoliata*) is sometimes included in the genus and classified as an orange (*Citrus trifoliata*). It is often used as rootstock for sweet orange trees, especially as a hybrid with other *Citrus* cultivars. The trifoliate orange is a thorny shrub or small tree grown primarily for its foliage and flowers, or as a barrier hedge; however, it bears a downy fruit resembling a small citrus fruit, from which marmalade is sometimes made. It is native to northern China and Korea, and is also known as "hardy orange" (because it can withstand sub-freezing temperatures) or "Chinese bitter orange".

Taxonomy of the orange (and citrus in general) presents difficulties; the interfertility of citrus has resulted in numerous hybrids, bud unions, and cultivars; taxonomy is often controversial, confusing, or inconsistent. The fruit of a member of the genus *Citrus* is considered a hesperidium, a kind of modified berry, because it has numerous seeds, is fleshy and soft, derives from a single ovary, and is covered by a rind created by a leathery thickening of the ovary wall. An orange seed is called a "pip". The white thread-like material attached to the inside of the peel is called pith. Although the sweet orange will grow to different sizes and colors according to local conditions, it most commonly has ten *carpels*, or segments, inside. Unripe fruit is green. The pebbled exterior of ripe fruit can be bright orange to yellow-orange, but often retains a considerable amount of the green color of unripe fruit. Orange trees are generally grafted; the bottom part of the tree, including the roots and trunk, is called the rootstock, while the fruit-producing top part of the tree is called budwood (when talking about the process of grafting) or scion (when talking about the variety of orange). *Citrus sinensis* (L.) Osbeck is broken down into four groups with distinct characteristics: Common oranges, blood oranges, navels, and acidless oranges.

Nutritional Value

Orange Raw

Nutritional value per 100 g (3.5 oz)	
Energy	192 kJ (46 kcal)
Carbohydrates	11.54 g
Sugars	9.14 g
Dietary fiber	2.4 g
Fat	0.21 g
Protein	0.70 g
Thiamine (vit. B_1)	0.100 mg
Riboflavin (vit. B_2)	0.040 mg
Niacin (vit. B_3)	0.400 mg
Pantothenic acid (B_5)	0.250 mg
Vitamin B_6	0.051 mg
Folate (vit. B_9)	17 μg
Vitamin C	45 mg
Calcium	43 mg
Iron	0.09 mg
Magnesium	10 mg
Phosphorus	12 mg
Potassium	169 mg
Zinc	0.08 mg

Figure 12.1: Flow sheet for preparation of citrus juice

Mandarin and Sweet oranges
↓
Washing
↓
Peeling by hand
↓
Separation and cleaning of segments
↓
Juice extraction (screw type extractor)
↓
Straining
↓
Heating at 80 °C, 15 min
↓
Bottling or canning
↓
Crown corking or can sealing
↓
Pasteurization
↓
Cooling
↓
Storage
↓
Marketing

Canned juice of sweet oranges such as Malta and Mosambi is not bitter. The juice from mandarin orange, however often develops a characteristics bitter taste on heating the can or bottle. This bitterness is due to the limonin present in the juice and is not acceptable.

The bitterness can be reduced considerably by dipping the segments in 2 per cent boiling NaOH for 2-3 minutes to remove their adhering outer covering and fibrous material which cause bitterness, before extraction of juice. Addition of 5-6 per cent sugar also helps in reducing the bitter taste. Other important precautions for reducing the bitterness are:

1. Selection of fully ripe fruits only.
2. Hot water treatment prior to peeling or cutting of fruits and
3. Avoiding excess pressure during extraction.

To improve its quality and preserve the vitamin content, the juice is generally de-aerated and flash pasteurization. Preservation of pure orange juice with its entire natural flavor is still a technical problem.

Figure 12.2: Flow sheet for preparation of lemon/lime juice

Fully mature fruits
↓
Washing
↓
Cutting into halves
↓
Juice extraction (use juice extractor)
↓
Straining of juice
↓
Pure juice
↓
Pasteurization
↓
Can/bottle filling
↓
Sealing
↓
Storage
↓
Marketing

Figure 12.3: Flow sheet for removal of astringency from orange juice

Fully Mature Oranges
↓
Peeling
↓
Dipping segments in hot 2 per cent NaOH solution for 2-3 min
↓
Dipping in 0.5 per cent citric acid solution to neutralize alkali
↓
Use for juice extraction

Figure 12.4: Flow sheet for preparation of lime pickle in salt

Fully Mature Lime
↓
Washing
↓
Cutting into 4 pieces
↓
Squeezing out juice from ¼ amount of fruit
↓
Mixing spices and salt with juice
↓
Mixing with lime pieces
↓
Filling in jars
↓
Covering with lid
↓
Keeping in sun for 4-6 days
↓
Storage at ambient temperature

Sl.No.	*Item*	*Quantity*	*Sl.No.*	*Item*	*Quantity*
1.	Lime	1 kg	5.	Cumin	10 g
2.	Salt	200 g	6.	Cardamom	10 g
3.	Red chilli powder	15 g	7.	Black pepper powder	10 g
4.	Cinnamon	10 g	8.	Clove headless	5 No

Figure 12.5: Flow sheet for preparation of lime pickle in oil

Fully Mature Lime

↓

Washing

↓

Cutting into 4 pieces

↓

Squeezing out juice from ¼ amount of fruit

↓

Mixing spices and salt with juice

↓

Mixing with lime pieces

↓

Filling in jars

↓

Covering with lid

↓

Keeping in sun for 4-6 days

↓

Addition of oil after heating & cooling it

↓

Storage at ambient temperature

Sl.No.	*Item*	*Quantity*	*Sl.No.*	*Item*	*Quantity*
1.	Lime	1 kg	6.	Cumin	10 g
2.	Salt	120 g	7.	Cardamom	10 g
3.	Red chilli powder	10 g	8.	Black pepper powder	10 g
4.	Cinnamon powder	10 g	9.	Asafoetida	2 g
5.	Turmeric powder	10 g	10.	Mustard oil	500 ml

Figure 12.6: Flow sheet for preparation of marmalade

Fully Ripe Fruits

↓

Washing

↓

Peeling outer yellow portion thinly

↓

Cutting yellow portion into shreds (1.9-2.5 cm long & 0.8-0.12 cm thick)

↓

Cutting of 0.3-0.45 cm thick slices of peeled fruit of crushing into pulp in a greater

↓

Boiling (2-3 time water; 40-60 min)

↓

Straining the extract

↓

Testing for pectin content

↓

Addition of sugar if required

↓

Cooking to 103-105ºC

↓

Addition of soft prepared shreds (62g/Kg extract)

↓

Boiling till jellying point

↓

Testing for end-point

↓

Cooling

↓

Addition of Flavouring

↓

Filling in sterilized bottles

↓

Sealing

↓

Storage

↓

Marketing

13

Coconut

The coconut palm, *Cocos nucifera*, is a member of the family Arecaceae (palm family). It is the only accepted species in the genus *Cocos*. The term coconut can refer to the entire coconut palm, the seed, or the fruit, which is not a botanical nut. The spelling cocoanut is an old-fashioned form of the word. The term is derived from 16th century Portuguese and Spanish *cocos*, meaning "grinning face", from the three small holes on the coconut shell that resemble human facial features. Found across much of the tropic and subtropical area, the coconut is known for its great versatility as seen in the many domestic, commercial, and industrial uses of its different parts. Coconuts are part of the daily diet of many people. Its endosperm is known as the edible "flesh" of the coconut; when dried it is called copra. The oil and milk derived from it are commonly used in cooking and frying; coconut oil is also widely used in soaps and cosmetics. The clear liquid coconut water within is a refreshing drink and can be processed to create alcohol. The husks and leaves can be used as material to make a variety of products for furnishing and decorating. It also has cultural and religious significance in many societies that use it.

Fruit

Botanically the coconut fruit is a drupe, not a true nut. Like other fruits it has three layers: exocarp, mesocarp, and endocarp. The exocarp and mesocarp make up the *husk* of the coconut. Coconuts sold in the shops of non-tropical countries often have had the exocarp (outermost layer) removed. The mesocarp is composed of fibers called coir which have many traditional and commercial uses. The shell has three germination pores (stoma) or *eyes* that are clearly visible on its outside surface once the husk is removed. A full-sized coconut weighs about 1.44 kilograms (3.2 lb). It takes around 6000 full-grown coconuts to produce a tonne of copra.

Seed

Within the shell is a single seed. When the seed germinates, the root (radical) of its embryo pushes out through one of the eyes of the shell. The outermost layer of the seed, the *testa*, adheres to the inside of the shell. In a mature coconut, a thick albuminous endosperm adheres to the inside of the testa. This endosperm or *meat* is the white and fleshy edible part of the coconut. Although coconut meat contains less fat than many oilseeds and seeds such as almonds, it is noted for its high amount of medium-chain saturated fat. About 90 per cent of the fat found in coconut meat is saturated, a proportion exceeding that of foods such as lard, butter, and tallow. There has been some debate as to whether or not the saturated fat in coconuts is healthier than other forms of saturated fat (see coconut oil). Like most nut meats, coconut meat contains less sugar and more protein than popular fruits such as bananas, apples and oranges. It is relatively high in minerals such as iron, phosphorus and zinc.

The endosperm surrounds a hollow interior space, filled with air and often a liquid referred to as coconut water (distinct from coconut milk). Immature coconuts are more likely to contain coconut water and less meat. They are often sold with a small portion of the husk cut away to allow access to the coconut water. Young coconuts used for coconut water are called tender coconuts. The water of a tender coconut is liquid endosperm. It is sweet (mild) with an aerated feel when cut fresh. Depending on its size a tender contains 300 to 1,000 ml of coconut water.

Culinary Use

The various parts of the coconut have a number of culinary uses. The nut provides oil for frying, cooking, and making margarine. The white, fleshy part of the seed–the coconut meat–is edible and used fresh or dried in cooking especially in confections and desserts like macaroons. Desiccated coconut can be used as an ingredient or to produce coconut milk which is frequently added to curry dishes and other savory viands. Coconut flour has also been developed for use in baking and to combat malnutrition. Coconut chips have been sold in tourist regions like Hawaii and the Caribbean. Coconut butter is often used to describe solidified coconut oil, but has also been adopted as a name by certain specialty products made out of coconut milk solids or puréed coconut meat and oil.

Coconut Water

Coconut water contains sugar, fiber, proteins, antioxidants, vitamins and minerals, and provides an isotonic electrolyte balance. It is consumed as a refreshing drink throughout the humid tropics and is gaining popularity as isotonic sports drink. Mature fruits have significantly less liquid than young immature coconuts, barring spoilage. Coconut water can be fermented to produce coconut vinegar.

Coconut Milk

Coconut milk, not to be confused with coconut water, is obtained primarily by extracting juice by pressing the grated coconut's white kernel or by passing hot water or milk through grated coconut, which extracts the oil and aromatic compounds. It has a fat content around 17. When refrigerated and left to set, coconut cream will rise to the top and separate from the milk. The milk can be used to produce virgin coconut oil by controlled heating and removal of the oil fraction.

Toddy and Nectar

The sap derived from incising the flower clusters of the coconut is drunk as neera also known as toddy or *tuba* (Philippines), *tuak* (Indonesia and Malaysia) or *karewe* (fresh and not fermented, collected twice a day, for breakfast and dinner) in Kiribati. When left to ferment on its own it becomes palm wine. Palm wine is distilled to produce arrack. In the Philippines this alcoholic drink is called *lambanog* or "coconut vodka".

The sap can be reduced by boiling to create a sweet syrup or candy such as *te kamamai* in Kiribati or *dhiyaa hakuru* and *Addu bondi* in Maldives. It can be reduced further to yield coconut sugar also referred to as palm sugar or jaggery. A young well-maintained tree can produce around 300 liters (66 imp gal; 79 U.S. gal) of toddy per year while a forty year old tree may yield around 400 liters (88 imp gal; 110 US gal).

Heart of Palm and Coconut Sprout

Apical buds of adult plants are edible, and are known as "palm cabbage" or heart of palm. They are considered a rare delicacy, as harvesting the buds kills the palms. Hearts of palm are eaten in salads, sometimes called "millionaire's salad". Newly germinated coconuts contain an edible fluff of marshmallow-like consistency called coconut sprout, produced as the endosperm nourishes the developing embryo.

India

In Kerala, the most common way of cooking vegetables is to add grated coconut and then steam them with spices after frying in a little oil. People from Kerala also make chutney, which involves grinding the coconut with salt, chillies, and whole spices. The *uruttu chammanthi* as the granulated chutney is known is eaten with rice or kanji (rice gruel). It is also invariably the main side dish served with idli, vadai, and dosai. Coconut ground with spices is also mixed in sambar and other various lunch dishes for extra taste. Dishes that are garnished with grated coconut are generally referred to as *poduthol* in North Malabar and *thoran* in rest of Kerala. *Puttu* is a culinary delicacy of Kerala and Tamil Nadu, in which layers of coconut alternate with layers of powdered rice, all of which fit into a bamboo stalk. In recent times this has been replaced with steel or aluminium tubes, which is then steamed over a pot. Coconut is regularly broken in the middle class families in Tamil Nadu for food. Coconut meat can be eaten as a snack sweetened with jaggery or molasses.

Medicinal Uses

Coconuts may help benign prostatic hyperplasia. In rats, virgin coconut oil reduced total cholesterol, triglycerides, phospholipids, LDL, and VLDL cholesterol levels and increased HDL cholesterol in serum and tissues. The hexane fraction of coconut peel may contain novel anticancer compounds. Young coconut juice has estrogen-like characteristics. Inside a coconut is a cavity filled with coconut water, which is sterile until opened. It mixes easily with blood, and was used during World War II in emergency transfusions. It can also serve as an emergency short-term intravenous hydration fluid. This is possible because the coconut water has a high level of sugar and other salts that makes it possible to be used in the bloodstream, much like the modern lactated ringer solution or a dextrose/water solution as an IV. Coconut is also commonly used as a traditional remedy in Pakistan to treat bites from rats. In Brazil coconut is known as *coco-da-Bahia* or *coqueiro-da-India*. The tea from the husk fiber is widely used to treat several inflammatory disorders.

Coconut, Meat, Raw

Nutritional value per 100 g (3.5 oz)	
Energy	1,481 kJ (354 kcal)
Carbohydrates	15.23 g
Sugars	6.23 g
Dietary fiber	9.0 g
Fat	33.49 g
Saturated	29.70 g
Monounsaturated	1.43 g
Polyunsaturated	0.37 g
Protein	3.3 g
Thiamine (vit. B_1)	0.066 mg
Riboflavin (vit. B_2)	0.02 mg
Niacin (vit. B_3)	0.54 mg
Pantothenic acid (B_5)	0.300 mg
Vitamin B_6	0.054 mg
Folate (vit. B_9)	26 μg
Vitamin C	3.3 mg
Calcium	14 mg
Iron	2.43 mg
Magnesium	32 mg
Phosphorus	113 mg
Potassium	356 mg
Zinc	1.1 mg

Coconut Shell Ash Compound

Compound	*Per cent*
K_2O	45.01
Na_2O	15.42
CaO	6.26
MgO	1.32
$Fe_2O_3 + Al_2O_3$	1.39
P_2O_5	4.64
SO_3	5.75
SiO_2	4.64

Coconut Shell Compound (Dry Basis)

Compound	*Per cent*
Cellulose	33.61
Lignin	36.51
Pentosans	29.27
Ash	0.61

Figure 13.1: Flow sheet for preparation of coconut burfee

Coconut fruits fully mature

↓

Peeling/removing outer shell

↓

Homogenization of pulp

↓

Addition of sugar, fat, milk powder (1000g, 100g, 100g/Kg Pulp)

↓

Heating upto 78 °Brix

↓

Addition of salt and citric acid (5g, 2g)

↓

Heating upto 80 °Brix

↓

Spreading in aluminium trays (1 cm thick; already smeared with fat)

↓

Cooling

↓

Cutting into attractive pieces (1.5 × 2.5 cm)

↓

Rapping into metallic paper

↓

Storage

Figure 13.2: Flow sheet for preparation of wet milling of coconut

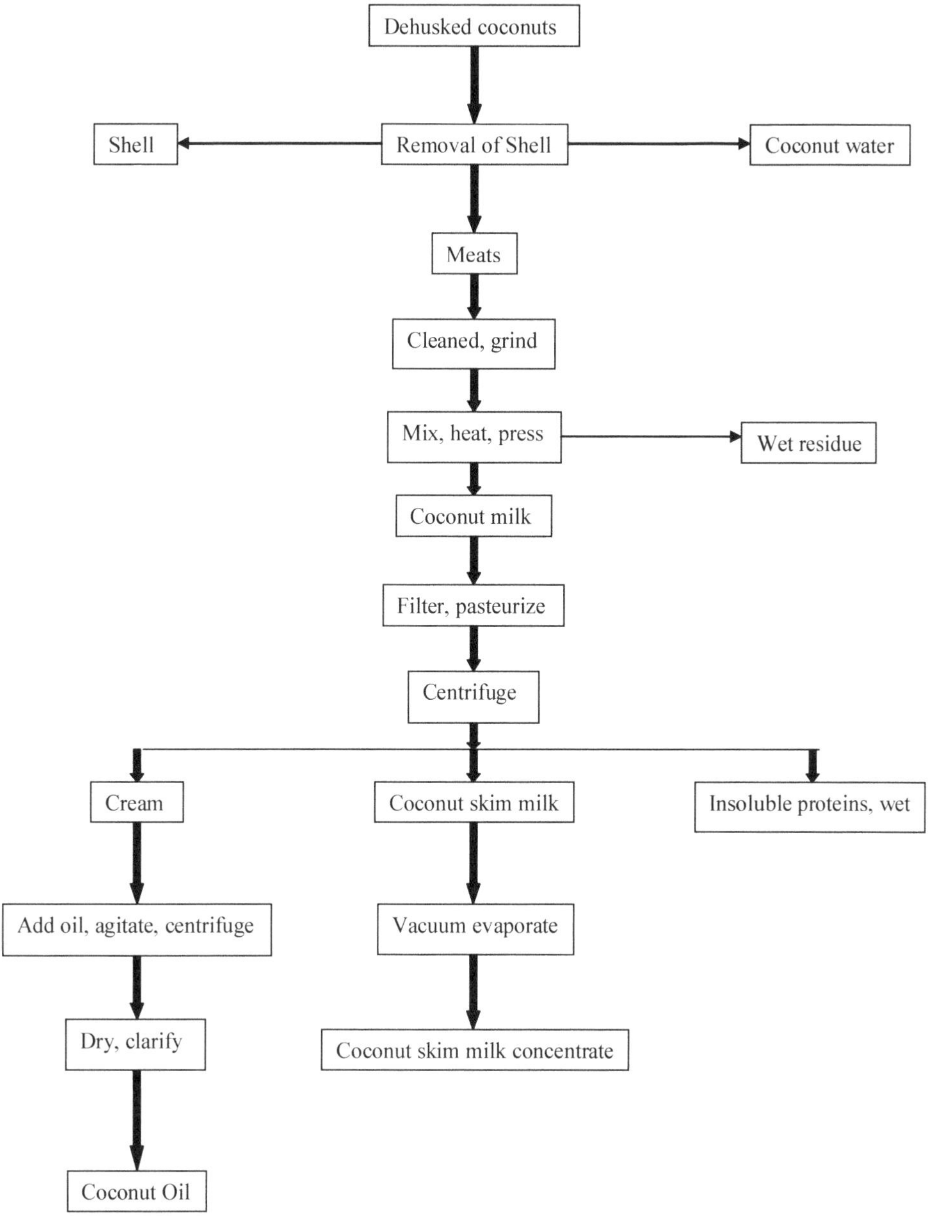

14

Custard Apple

Annona squamosa (also called sugar-pineapple or sweetsop) is a species of *Annona* native to the tropical Americas and widely grown in Colombia, El Salvador, India, Pakistan and the Philippines. Its exact native range is unknown due to extensive cultivation, but thought to be in the Caribbean; the species was described from Jamaica.

It is a semi-evergreen shrub or small tree reaching 6–8 meters (20–26 ft) tall. The leaves are alternate, simple, oblong-lanceolate, 5–17 cm (2.0–6.7 in) long and 2–5 centimeters (0.79–2.0 in) broad. The flowers are produced in clusters of 3-4, each flower 1.5–3 cm (0.59–1.2 in) across, with three large petals and three minute ones, yellow-green spotted purple at the base.

The fruit is usually round; slightly pine cone-like, 6–10 cm (2.4–3.9 in) diameter and weighing 100–230 g (3.5–8.1 oz), with a scaly or lumpy skin. There are variations in shape and size. The fruit flesh is sweet, white to light yellow, and resembles and tastes like custard. The edible portion coats the seeds generously; a bit like the gooey portion of a tomato seed. Sugar-apple has a very distinct, sweet-smelling fragrance. The texture of the flesh that coats the seeds is a bit like the center of a very ripe guava (excluding the seeds). It is slightly grainy, a bit slippery, very sweet and very soft. The seeds are scattered through the fruit flesh; the seed coats are blackish-brown, 12–18 mm (0.47–0.71 in) long, and hard and shiny.

There are also new varieties being developed in Taiwan. There is a pineapple sugar-apple, which is similar in sweetness but has a very different taste. Like the name suggests, it tastes like pineapple. The arrangement of seeds is in spaced rows, with the fruit's flesh filling most of the fruit and making grooves for the seeds, instead of the flesh only occurring around the seeds. Diterpenoid alkaloid atisine is the main component of the root. Others include atidine, histisine, hetisine, hetidine, heterophyllisine, heterophylline, heterlophylline, isoatisine, dihydroatisine, and hetisinone benzoyl heteratisine and citronella oil.

It is quite a prolific bearer, and it will produce fruit in as little as two to three years. A tree five years old may produce as many as 50 sugar apples. Poor fruit production has been reported in Florida because there are few natural pollinators (honeybees have a difficult time penetrating the tightly closed female flowers); however, hand pollination with a natural fiber brush is effective in increasing yield. Natural pollinators include beetles (coleoptera) of the families Nitidulidae, Staphylinidae, Chrysomelidae, Curculionidae and Scarabeidae. In the Philippines, the fruit is commonly eaten by the Philippine fruit bat (*kabag* or *kabog*), which then spreads the seeds from island to island. A Filipino company produces sugar apple wine. It is a host plant for larvae of the butterfly *Graphium agamemnon* (tailed jay). It is used by some societies in India to prepare a hair tonic. The seeds are also ground and applied to hair to get rid of lice; however, it must be kept away from the eyes as it is highly irritant and can cause blindness. Heat-extracted oil from the seeds has been employed against agricultural pests. Studies have shown the ether extract of the seeds to have no residual toxicity after 2 days. High concentrations are potent for 2 days and weaken steadily, all activity being lost after 8 days. In Mexico, the leaves are rubbed on floors and put in hen's nests to repel lice.

Benefits of Custard Apple

Custard Apple is a sweet fruit, which is widely grown across the world. The fruit is compound in nature, measuring 3¼ to 6½ inches (8-16 cm) in diameter. The shape of the fruit may be symmetrically heart-shaped, lopsided, or irregular. It may be nearly round or oblate, with a deep or shallow depression at the base. The skin or covering of custard apple is thin but tough, which is usually yellow or brownish when ripe. Beneath the thin covering, a thick, cream-white layer of custard like, granular, flesh is present, which surrounds seeds. The flavor of the fruit is sweet and agreeable. Besides, custard apple has many health and nutritional benefits. Given below is the nutrition value of the fruit.

Health and Nutrition Benefits of Eating Custard Apple

- Custard apple is a storehouse of Vitamin C, which is an anti-oxidant and helps in neutralizing free radicals.
- Vitamin A present in the fruit is good for hair, eyes and healthy skin.
- Custard apple contains magnesium, which plays vital role in relaxing muscles and protecting heart against diseases.
- Potassium and Vitamin B6 are also present in custard apple.
- Copper is yet another useful ingredient of custard apple.
- It is a rich source of dietary fiber, which helps indigestion.
- As it contains low fat levels, it is good for maintaining optimum health.
- The paste of the flesh of the fruit can be used for local application on ulcers, abscesses and boils.
- The fruit, in its unripe form, can be dried, crushed and used for treating diarrhea and dysentery.

- ☆ Custard apple serves as an expectorant, stimulant, coolant and haematinic and is even useful in treating anemia.
- ☆ The seeds of the fruit have insecticidal and abortifacient properties.

Nutritional Value of Custard Apple

Given below is nutritional value per 100 gm of custard apple

Nutrient	*Amount*
Carbohydrate	23.71 gm
Protein	1 to 4.3 gm
Natural Sugar	14 to 18 gm
Fiber	1 to 3.2 g
Total Soluble Solids (Brix)	22.3 per cent
Energy	76 to 96 cal (310 to 420 kJ)
Total Acidity	0.4 mg
Calcium	17 to 22 mg
Sodium	4 to 14 mg
Potassium	250 to 578 mg
Magnesium	32 to 88 mg
Copper	2.4 mg
Iron	0.7 mg
Zinc	0.2 to 2.7mg
Vitamin A	1 mg
Thiamine (B1)	0.05 to 0.08 mg
Riboflavin (B2)	0.08 to 0.1 mg
Niacin (B3)	0.5 to 0.8 mg
Vitamin C	22 to 43 mg

Figure 14.1: Flow sheet for preparation of custard apple pulp

Fully Ripe Custard apple
↓
Washing and Cleaning
↓
Sorting
↓
Pulping
↓
Heating
↓
Preservative addition
↓
Mixing
↓
Can filling and sealing
↓
Pasteurization
↓
Cooling and Labeling
↓
Storage
↓
Marketing

Figure 14.2: Flow sheet for preparation of custard apple: Sapota jams

Custard apple : Sapota fruits (50:50)

↓

Cleaning/Washing

↓

Peeling

↓

Extraction of pulp

↓

Addition of sugar, 750g/kg

↓

Heating upto 65 °Brix

↓

Addition of citric acid, 3g

↓

Heating upto 68.5 °Brix

↓

Filling in sterilized glass jars

↓

Cooling

↓

Sealing

↓

Storage

Figure 14.3: Flow sheet for preparation of custard apple toffee

Custard apple pulp

↓

Addition of sugar, fat, skim milk powder

↓

Heating upto 80 °Brix

↓

Addition of salt and citric acid

↓

Spreading on aluminium trays (already smeared with fat/Oil)

↓

Cooling

↓

Cutting into attractive pieces

↓

Rap into metallic paper

↓

Storage

15

Fig

The Common fig (*Ficus carica*) is a deciduous tree growing to heights of up to 6 m (19 ft) in the genus Ficus from the family Moraceae known as Common fig tree. It is a temperate species native to the Middle East.

Ficus carica is a monoecious, deciduous tree or a large shrub. It is native to the Middle East. It was later cultivated from Afghanistan to Portugal, and from the 15th century onwards was grown in areas including Northern Europe and the New World. In the 16th century, Cardinal Pole introduced fig trees to Lambeth Palace. Like the mulberry tree, it has a substance called latex which is extracted for industrial purposes in the related *Ficus elastica*. It grows to a height of 6.9–10 meters (23–33 ft) tall, with smooth grey bark. *Ficus carica* is well known for its fragrant leaves that are large and lobed. The leaves are 12–25 centimeters (4.7–9.8 in) long and 10–18 centimeters (3.9–7.1 in) across, and deeply lobed with three or five lobes. *Ficus carica* have pyriform Sicon infructescences, the fleshy fruit fig, with inner unisexual flowers. The fruit is 3–5 centimeters (1.2–2.0 in) long, with a green skin, sometimes ripening towards purple or brown. *Ficus carica* has milky sap (laticifer). The sap of the fig's green parts is an irritant to human skin.

The Common fig tree has been cultivated since ancient times and grows wild in dry and sunny areas, with deep and fresh soil, also in rocky areas, from sea level to 1,700 meters. Prefers light and medium soils, requires well-drained soil and can grow in nutritionally poor soil. Like all fig trees, *Ficus carica* requires wasp pollination of a particular species of wasp to produce seeds. The plant can tolerate estacional drought, and the Middle Eastern and Mediterranean climate is especially suitable for the plant. Situated in favorable habitat old specimens when mature, can reach a considerable size and form a large dense shade tree. Its aggressive root system precludes its use in many urban areas of cities: in walls, cracks and buildings but in nature helps the plant to take root in the most inhospitable areas. Common fig tree is mostly a phreatophyte that live in areas with standing or running water, grows well in the valleys of the rivers and ravines saving no water, having strong need of water that extracted from the ground. The deep-rooted plant search groundwater, in aquifers, ravines, or cracks in the rocks. The fig tree, with the water, cools the environment in hot places, creating a fresh and pleasant habitat for many animals that take shelter in its shade in the times of intense heat.

The flower is not visible, as it blooms inside the infructescence. Although commonly referred to as a fruit, the fig is actually the infructescence or sicon of the tree, known as a false fruit or multiple fruit, in which the flowers and seeds are borne. It is a hollow-ended stem containing many flowers. The small orifice (ostiole) visible on the middle of the fruit is a narrow passage, which allows to the specialized fig wasp, to enter the fruit and pollinate the flower, where after the fruit grows seeds. *See Ficus: Fig pollination and fig fruit. Ficus carica* is dispersed by birds and mammals that scatter their seeds in droppings. It is an important food source for much of the fauna in some areas, and it owes its expansion. The common fig tree also sprouts from the root and stolon issues. The genus Dorstenia, which, like the fig, is in the mulberry family (Moraceae), exhibits similar tiny flowers arranged on a receptacle but in this case the receptacle is a more or less flat, open surface. It is a structure intermediate between the fig and the mulberry, where the fig corresponds in structure to an invaginated or inside-out mulberry.

The infrutescence is pollinated by a symbiosis with a kind of fig wasp. The fertilized female wasp enters the fig through the sicon, which is a tiny hole in the crown (the ostiole). She crawls on the inflorescence inside the fig and pollinates some of the female flowers. She lays her eggs inside some of the flowers and dies. After weeks of development in their galls, the male wasps emerge before females through holes they produce by chewing the galls. The male wasps then fertilize the females by depositing semen in the hole in the gall. The males later return to the females and enlarge the holes to facilitate the females to emerge. Then some males enlarge holes in the scion, which enables females to disperse after collecting pollen from the developed male flowers. Females have a short time (<48 hours) to find another fig tree with receptive siconios to spread the pollen and assist the tree in reproduction.

Two crops of figs are potentially produced each year. The first or breva crop develops in the spring on last year's shoot growth. In contrast, the main fig crop develops on the current year's shoot growth and ripens in the late summer or fall. The main crop is generally superior in both quantity and quality than the breva crop. However, some cultivars produce good breva crops (*e.g.*, Black Mission, Croisic, and Ventura).

There are basically three varieties of common figs:

- ☆ Caducous (or Smyrna) figs require pollination by the fig wasp and caprifigs to develop crops. Some cultivars are Calimyrna, Marabout, and Zidi.
- ☆ Persistent (or Common) figs do not need pollination; fruit develop through parthenocarpic means. This is the variety of fig most commonly grown by home gardeners. Adriatic, Black Mission, Brown Turkey, Brunswick, and Celeste are some representative cultivars.

- ☆ Intermediate (or San Pedro) figs do not need pollination to set the breva crop, but do need pollination, at least in some regions, for the main crop. Examples are Lampeira, King, and San Pedro.

Figs plants are easy to propagate through several methods. Propagation using seeds is not the preferred method since vegetative methods existing are quicker and more reliable, that is, they do not yield the inedible caprifigs. However, those desiring to can plant seeds of dried figs with moist sphagnum moss or other media in a zip lock bag and expect germination in a few weeks to several months. The tiny plants can be transplanted out little by little once the leaves open, and despite the tiny initial size can grow to 1 foot (30 cm) or more than one year from planting seeds.

For propagation in the mid-summer months, air layer new growth in August (mid-summer) or insert hardened off 15–25 cm (6-10 inches) shoots into moist perlite or a sandy soil mix, keeping the cuttings shaded until new growth begins; then gradually moves them into full sun. An alternative propagation method is bending over a taller branch, scratching the bark to reveal the green inner bark, then pinning the scratched area tightly to the ground. Within a few weeks, roots will develop and the branch can be clipped from the mother plant and transplanted where desired. For spring propagation, before the tree starts growth, cut 15–25 cm (6-10 inches) shoots that have healthy buds at their ends, and set into a moist perlite and/or sandy soil mix located in the shade. Once the cuttings start to produce leaves, bury them up to the bottom leaf to give the plant a good start in the desired location.

Figs can also be found in continental climate with hot summer, as far north as Hungary and Moravia, and can be harvested up to four times per year. Thousands of cultivars, most named, have been developed or come into existence as human migration brought the fig to many places outside its natural range. It has been an important food crop for thousands of years, and was also thought to be highly beneficial in the diet.

The edible fig is one of the first plants that were cultivated by humans. Nine sub fossil figs of a parthenocarpic type dating to about 9400–9200 BC were found in the early Neolithic village Gilgal I (in the Jordan Valley, 13 km north of Jericho). The find predates the domestication of wheat, barley, and legumes, and may thus be the first known instance of agriculture. It is proposed that they may have been planted and cultivated intentionally, one thousand years before the next crops were domesticated (wheat and rye).

Figs were also a common food source for the Romans. Cato the Elder, in his *De Agri Cultura*, lists several strains of figs grown at the time he wrote his handbook: the Mariscan, African, Herculanean, Saguntine, and the black Tellanian (*De agri cultura*). The fruits were used, among other things, to fatten geese for the production of a precursor of foie gras. Figs can be eaten fresh or dried, and used in jam-making. Most commercial production is in dried or otherwise processed forms, since the ripe fruit does not transport well, and once picked does not keep well.

Nutrition

Figs are one of the highest plant sources of calcium and fiber. According to USDA data for the Mission variety, dried figs are richest in fiber, copper, manganese, magnesium, potassium, calcium, and vitamin K, relative to human needs. They have smaller amounts of many other nutrients. Figs have a laxative effect and contain many antioxidants. They are good source of flavonoids and polyphenols including gallic acid, chlorogenic acid, syringic acid, (+)-catechin, (–)-epicatechin and rutin. In one study, a 40-gram portion of dried figs (two medium size figs) produced a significant increase in plasma antioxidant capacity.

Fig, Dried, Uncooked

Nutritional value per 100 g (3.5 oz)	
Energy	1,041 kJ (249 kcal)
Carbohydrates	63.87 g
Sugars	47.92 g
Dietary fiber	9.8 g
Fat	0.93 g
Protein	3.30 g
Thiamine (vit. B_1)	0.085 mg
Riboflavin (vit. B_2)	0.082 mg
Niacin (vit. B_3)	0.619 mg
Pantothenic acid (B_5)	0.434 mg
Vitamin B_6	0.106 mg
Folate (vit. B_9)	9 μg
Vitamin C	1.2 mg
Calcium	162 mg
Iron	2.03 mg
Magnesium	68 mg
Phosphorus	67 mg
Potassium	680 mg
Zinc	0.55 mg

Figure 15.1: Flow sheet for preparation of fig toffee

Fig fruits fully ripe

↓

Washing and Cutting

↓

Homogenization of pulp

↓

Addition of sugar, fat, milk powder (750g, 100g, 100g/Kg Pulp)

↓

Heating upto 80 °Brix

↓

Addition of salt and citric acid (5g, 2g)

↓

Heating upto 82 °Brix

↓

Spreading in aluminium trays (1 cm thick; already smeared with fat)

↓

Cooling

↓

Cutting into attractive pieces (1.5 × 2.5 cm)

↓

Rapping into metallic paper

↓

Storage

Figure 15.2: Flow sheet for preparation of dried fig by steaming

Fig fruits (mature)

↓

Cleaning and washing

↓

Steaming (90°C, 10lbs, 5 min)

↓

Sulphuring (2 g/kg 30 min except Danna variety)

↓

Syruping
(50°Brix syrup for 24 hrs, to get 26-30°Brix; use glucose + fructose 50:50)

↓

Wiping

↓

Drying (at 50-55°C for 18-20 hrs)

↓

Dried fig (18-20 per cent moisture)

↓

Packaging

↓

Storage

Figure 15.3: Flow sheet for preparation of dried fig

Selection of mature fig fruits

↓

Washing

↓

Dipping in 5000 ppm ethaphon solution for 3 min

↓

Blanching 2-4 min

↓

Sulphur fumigation 2g/kg, 2hrs

↓

Drying in tray drier

↓

Packaging

↓

Storage

16

Grape

A grape is a non-climacteric fruit, specifically a berry, that grows on the perennial and deciduous woody vines of the genus *Vitis*. Grapes can be eaten raw or they can be used for making jam, juice, jelly, vinegar, wine, grape seed extracts, raisins, molasses and grape seed oil. Grapes are also used in some kinds of confectionery. Grapes are typically an ellipsoid shape resembling a prolate spheroid.

The cultivation of the domesticated grape began 6,000-8,000 years ago in the Near East. Yeast, one of the earliest domesticated microorganisms, occurs naturally on the skins of grapes, leading to the innovation of alcoholic drinks such as wine. First traces of red wine is seen in ancient Armenia where apparently, to date, the oldest winery was found, dating to around 4,000 BC. By the 9th century AD the city of Shiraz was known to produce some of the finest wines in the Middle East. Thus it has been proposed that Syrah red wine is named after Shiraz a city in Persia where the grape was used to make Shirazi wine. Ancient Egyptian hieroglyphics record the cultivation of purple grapes, and history attests to the ancient Greeks, Phoenicians and Romans growing purple grapes for both eating and wine production. Later, the growing of grapes spread to Europe, North Africa, and eventually North America.

Native purple grapes belonging to the *Vitis* genus proliferated in the wild across North America, and were a part of the diet of many North American Native Americans, but were considered by European colonists to be unsuitable for wine. The first Old World *Vitis vinifera* purple grapes were cultivated in California.

Grapes are a type of fruit that grow in clusters of 15 to 300, and can be crimson, black, dark blue, yellow, green, orange, and pink. "White" grapes are actually green in color, and are evolutionarily derived from the purple grape. Mutations in two regulatory genes of white grapes turn off production of anthocyanins which are responsible for the color of purple grapes. Anthocyanins and other pigment chemicals of the larger family of polyphenols in purple grapes are responsible for the varying shades of purple in red wines.

Most grapes come from cultivars of *Vitis vinifera*, the European grapevine native to the Mediterranean and Central Asia. Minor amounts of fruit and wine come from American and Asian species such as:

- *Vitis labrusca*, the North American table and grape juice grapevines (including the concord cultivar), sometimes used for wine, are native to the Eastern United States and Canada.
- *Vitis riparia*, a wild vine of North America, is sometimes used for winemaking and for jam. It is native to the entire Eastern U.S. and north to Quebec.
- *Vitis rotundifolia*, the muscadines, used for jams and wine, are native to the Southeastern United States from Delaware to the Gulf of Mexico.
- *Vitis amurensis* is the most important Asian species.

Table and Wine Grapes

Commercially cultivated grapes can usually be classified as either table or wine grapes, based on their intended method of consumption: eaten raw (table grapes) or used to make wine (wine grapes). While almost all of them belong to the same species, *Vitis vinifera*, table and wine grapes have significant differences, brought about through selective breeding. Table grape cultivars tend to have large, seedless fruit (see below) with relatively thin skin. Wine grapes are smaller, usually seeded, and have relatively thick skins (a desirable characteristic in winemaking, since much of the aroma in wine comes from the skin). Wine grapes also tend to be very sweet: they are harvested at the time when their juice is approximately 24 per cent sugar by weight. By comparison, commercially produced "100 per cent grape juice", made from table grapes is usually around 15 per cent sugar by weight.

Seedless Grapes

Although grape seeds contain many nutrients, some consumers choose seedless grapes; seedless cultivars now make up the overwhelming majority of table grape plantings. Because grapevines are vegetatively propagated by cuttings, the lack of seeds does not present a problem for reproduction. It is, however, an issue for breeders, who must either use a seeded variety as the female parent or rescue embryos early in development using tissue culture techniques.

There are several sources of the seedlessness trait, and essentially all commercial cultivators get it from one of three sources: Thompson Seedless, Russian Seedless, and Black Monukka, all being cultivars of *Vitis vinifera*. There are currently more than a dozen varieties of seedless grapes. Several, such as Einset Seedless, Reliance and Venus, have been specifically cultivated for hardiness and quality in the relatively cold climates of northeastern United States and southern Ontario.

An offset to the improved eating quality of seedlessness is the loss of potential health benefits provided by the enriched phytochemical content of grape seeds (see Health claims, below).

Raisins, Currants and Sultanas

In most of Europe, dried grapes are referred to as "raisins" or the local equivalent. In the UK, three different varieties are recognized, forcing the EU to use the term "Dried vine fruit" in official documents.

A *raisin* is any dried grape. While *raisin* is a French loanword, the word in French refers to the fresh fruit; *grappe* (from which the English *grape* is derived) refers to the bunch (as in *une grappe de raisins*).

A *currant* is a dried Zante Black Corinth grape, the name being a corruption of the French *raisin de Corinthe* (Corinth grape). *Currant* has also come to refer to the blackcurrant and redcurrant, two berries unrelated to grapes.

A *sultana* was originally a raisin made from Sultana grapes of Turkish origin (known as Thompson Seedless in the United States), but the word is now applied to raisins made from either white grapes, or red grapes which are bleached to resemble the traditional sultana.

Health Claims

French Paradox

Comparing diets among Western countries, researchers have discovered that although the French tend to eat higher levels of animal fat, surprisingly the incidence of heart disease remains low in France. This phenomenon has been termed the French Paradox, and is thought to occur from protective benefits of regularly consuming red wine. Apart from potential benefits of alcohol itself, including reduced platelet aggregation and vasodilation, polyphenols (*e.g.*, resveratrol) mainly in the grape skin provide other suspected health benefits, such as:

- ✰ Alteration of molecular mechanisms in blood vessels, reducing susceptibility to vascular damage
- ✰ Decreased activity of angiotensin, a systemic hormone causing blood vessel constriction that would elevate blood pressure
- ✰ Increased production of the vasodilator hormone, nitric oxide (endothelium-derived relaxing factor)

Although adoption of wine consumption is not recommended by some health authorities, a significant volume of research indicates moderate consumption, such as one glass of red wine a day for women and two for men, may confer health benefits. Emerging evidence is that wine polyphenols like resveratrol provide physiological benefit whereas alcohol itself may have protective effects on the cardiovascular system.

Resveratrol

Grape phytochemicals such as resveratrol (a polyphenol antioxidant), have been positively linked to inhibiting any cancer, heart disease, degenerative nerve disease, viral infections and mechanisms of Alzheimer's disease.

Protection of the genome through antioxidant actions may be a general function of resveratrol. In laboratory studies, resveratrol bears a significant transcriptional overlap with the beneficial effects of calorie restriction in heart, skeletal muscle and brain. Both dietary interventions inhibit gene expression associated with heart and skeletal muscle aging, and prevent age-related heart failure.

Resveratrol is the subject of several human clinical trials, among which the most advanced is a one year dietary regimen in a Phase III study of elderly patients with Alzheimer's disease.

Synthesized by many plants, resveratrol apparently serves antifungal and other defensive properties. Dietary resveratrol has been shown to modulate the metabolism of lipids and to inhibit oxidation of low-density lipoproteins and aggregation of platelets.

Resveratrol is found in wide amounts among grape varieties, primarily in their skins and seeds which, in muscadine grapes, have about one hundred times higher concentration than pulp. Fresh grape skin contains about 50 to 100 micrograms of resveratrol per gram.

Anthocyanins and Other Phenolics

Anthocyanins tend to be the main polyphenolics in purple grapes whereas flavan-3-ols (*i.e.* catechins) are the more abundant phenolic in white varieties. Total phenolic content, a laboratory index of antioxidant strength, is higher in purple varieties due almost entirely to anthocyanin density in purple grape skin compared to absence of anthocyanins in white grape skin. It is these anthocyanins that are attracting the efforts of scientists to define their properties for human health. Phenolic content of grape skin varies with cultivar, soil composition, climate, geographic origin, and cultivation practices or exposure to diseases, such as fungal infections.

Red wine may offer health benefits more so than white because potentially beneficial compounds are present in grape skin, and only red wine is fermented with skins. The amount of fermentation time a wine spends in contact with grape skins is an important determinant of its resveratrol content. Ordinary non-muscadine red wine contains between 0.2 and 5.8 mg/L, depending on the grape variety, because it is fermented with the skins, allowing the wine to absorb the resveratrol. By contrast, a white wine contains lower phenolic contents because it is fermented after removal of skins. Wines produced from muscadine grapes may contain more than 40 mg/L, an exceptional phenolic content. In muscadine skins, ellagic acid, myricetin, quercetin, kaempferol, and trans-resveratrol are major phenolics. Contrary to previous results, ellagic acid and not resveratrol is the major phenolic in muscadine grapes. The flavonols syringetin, syringetin 3-O-galactoside, laricitrin and laricitrin 3-O-galactoside are also found in purple grape but absent in white grape.

Seed Constituents

Since the 1980s, biochemical and medical studies have demonstrated significant antioxidant properties of grape seed oligomeric proanthocyanidins. Together with tannins, polyphenols and polyunsaturated fatty acids, these seed constituents display inhibitory activities against several experimental disease models, including cancer, heart failure and other disorders of oxidative stress. Grape seed oil from crushed seeds is used in cosmeceuticals and skincare products for many perceived health benefits. Grape seed oil is notable for its high contents of tocopherols (vitamin E), phytosterols, and polyunsaturated fatty acids such as linoleic acid, oleic acid and alpha-linolenic acid.

Concord grape juice

Commercial juice products from Concord grapes have been applied in medical research studies, showing potential benefits against the onset stage of cancer, platelet aggregation and other risk factors of atherosclerosis, loss of physical performance and mental acuity during aging and hypertension in humans.

Grapes, purple or green

Nutritional value per 100 g (3.5 oz)	
Energy	288 kJ (69 kcal)
Carbohydrates	18.1 g
Sugars	15.48 g
Dietary fiber	0.9 g
Fat	0.0 g
Protein	0.72 g
Thiamine (vit. B_1)	0.069 mg
Riboflavin (vit. B_2)	0.07 mg
Niacin (vit. B_3)	0.188 mg
Pantothenic acid (B_5)	0.05 mg
Vitamin B_6	0.086 mg
Folate (vit. B_9)	2 μg
Vitamin B_{12}	0 μg
Vitamin C	10.8 mg
Vitamin K	22 μg
Calcium	10 mg
Iron	0.36 mg
Magnesium	7 mg
Manganese	0.071 mg
Phosphorus	20 mg
Potassium	191 mg
Sodium	3.02 mg
Zinc	0.07 mg

Figure 16.1: Flow sheet for preparation of grape juice

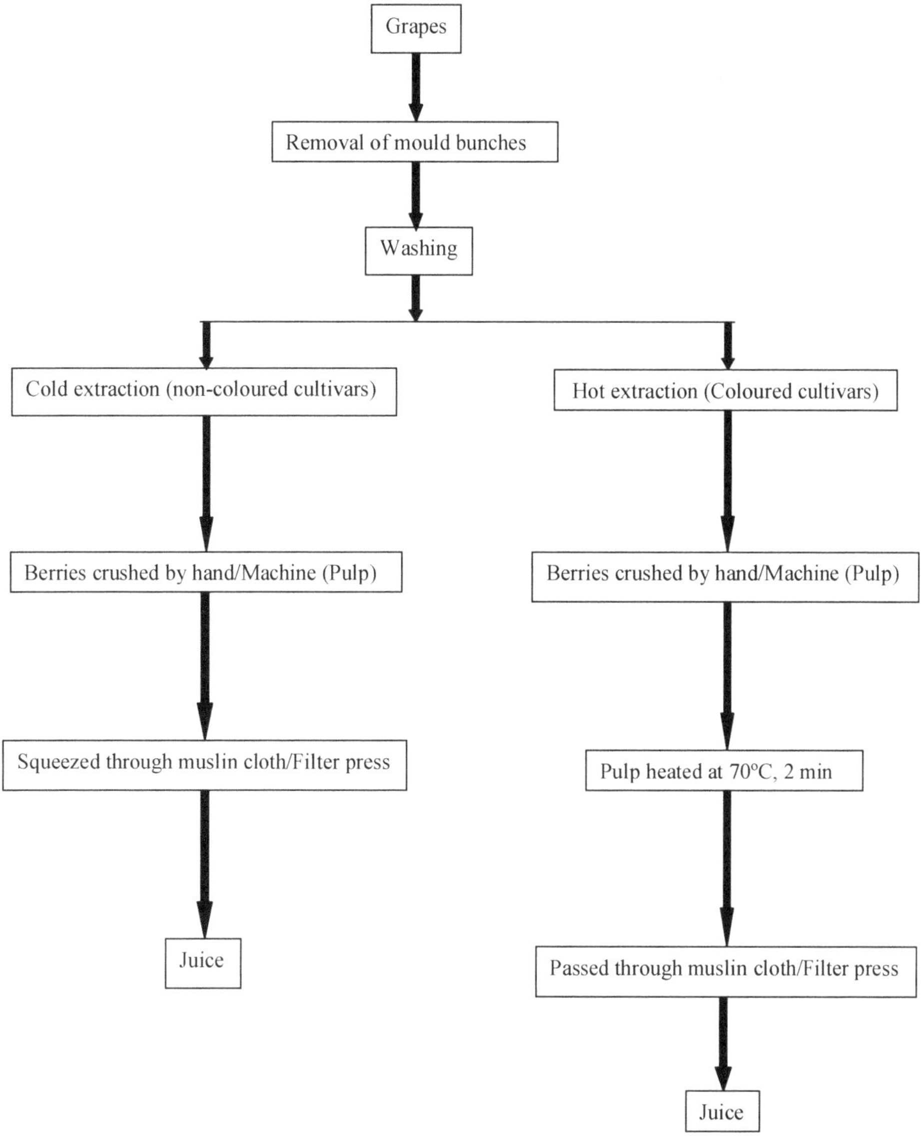

Figure 16.2: Flow sheet for preparation of grape pomace toffee

Grape pomace (1 Kg)

↓

Add sugar (1000g), fat (100g), and skim milk powder (50g)

↓

Heating upto 80 °Brix

↓

Addition of salt (5g) and citric acid (2g)

↓

Heating upto 82 °Brix

↓

Spread on aluminium trays (already smeared with fat/oil)

↓

Cooling

↓

Cutting into small attractive pieces

↓

Rapping into metallic paper

↓

Storage

Figure 16.3: Process for preparation of grape raisins by golden bleach method

Grape bunches

↓

Sorting of diseased, cracked, malformed and discolored berries

↓

Dipping in 0.3 per cent NaOH solution at 100°C for 3 seconds

↓

Rinsing in running tap water

↓

Sulphuring @ 2 g/kg for 2 hrs

↓

Drying

↓

Raisins

↓

Storage

Figure 16.4: Process for preparation of grape raisins by soda oil dip method

Grape bunches

↓

Sorting of diseased, cracked, malformed and discolored berries

↓

Dipping in a solution containing 4.5 per cent K_2CO_3, 0.5 per cent Na_2CO_3 and 1.0 per cent preyal dipping oil for 5 min

↓

Drying

↓

Raisins

↓

Storage

Figure 16.5: Process for preparation of grape raisins by sugar syruping method

Grape bunches

↓

Sorting of diseased, cracked, malformed and discolored berries

↓

Dipping in 0.3 per cent NaOH solution at 100°C for 3 seconds

↓

Rinsing in running tap water

↓

Sulphuring @ 1 g/kg for 2 hrs

↓

Sugar syruping for 48 hrs

↓

Washing under tap water

↓

Drying at 50-55°C

↓

Raisins

↓

Storage

Figure 16.6: Flow sheet for preparation of grape wine

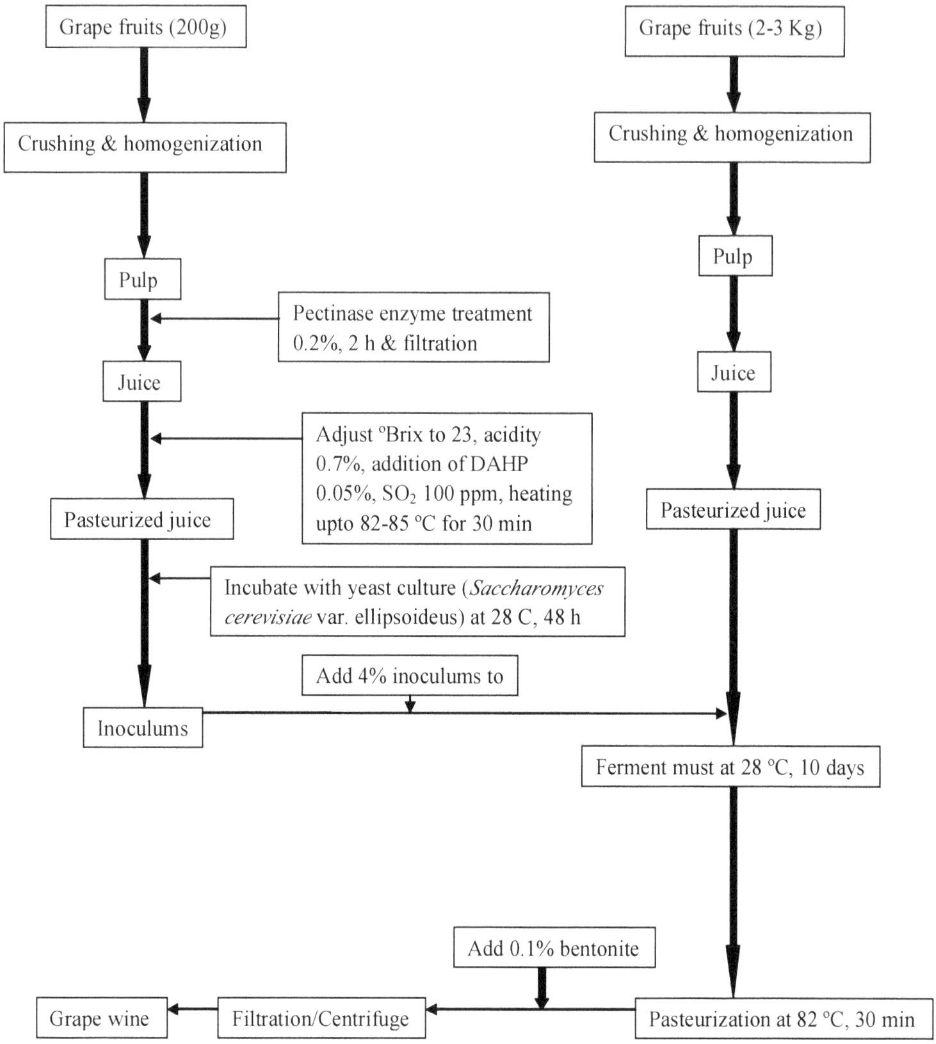

Grape Wine

The unmodified word *wine* signifies the juice of grapes fermented by wine yeast and appropriately (and legally) finished into an alcoholic beverage. Small amounts of wine are made from other fruits and, particularly by amateur wine-makers, from other food materials such as honey. These products require modifiers for the term wine, *e.g.*, honey wine *(mead)*, cherry or blueberry wine, etc. Rice's starchy carbohydrates have to undergo conversion into yeast-fermentable gars prior and concurrent to the alcoholic fermentation. Thus, it has more in common with beer, and its product should be referred to as *sake* instead of rice wine. Fruits other than wine grapes typically are too low in sugar and too tart to make satisfactory wine directly. They must be *ameliorated* (within legal limits) by addition of sugar and dilution or neutralization of the fruits acids.

As commercial food products, wines and the industry that produces them are unique in many ways. Whereas a few very large producers account for the majority of U.S. production volume, there are thousands of small producers who may grow their grapes, convert them into wine, bottle it, and sell or even ship much of it directly to consumers. This high degree of vertical integration often applies even in the largest of the winemaking operations, which may own vineyards and wineries, as well as wine bottle and closure manufacturers. However, the degree of vertical integration is limited by the enforcement of the so-called three-tier system that dates back to the post-Prohibition era which separates producers, distributors (wholesalers) and retailers, such as wine shops or restaurants. This system has been challenged legally in recent years, in particular by small wineries which wish to ship their wines directly to consumers across the U.S. Direct shipment of wine is still considered a felony in certain states all of which have legal power over the local alcoholic beverage industries. Whereas wineries can be extensive operations employing specialized equipment, they may be successful, even today (2012), if small, modestly equipped, and focused on regional types and styles of wines. Owing to a short season of ripe fruit (usually only about two to six weeks depending on grape varieties grown) and the very limited possibility of adapting a winery to process other foods, much winery equipment has only a short and intensive period of annual use. This and the high cost of vineyard land is often coincides located in prime real estate markets makes the capital investment into the wine business costly and limits short-term returns on such investments.

While standardization is usual in most (food) products, diversity and variation is sought and admired in wines. No other product has a similarly extensive aficionado's literature, including dedicated and influential magazines, wine and food writers' newspaper columns, winery and wine club newsletters and websites, etc, as well as specific scientific and technical journals on *viticulture* (grape growing) and *enology* (wine science and technology). Diversity is sought not only between classes of wine, *e.g.*, sparkling wines vs. Sherries, but more significantly among wines of the very same class.

Most of these differences are related to the grape variety, to the topography, geology (soil structure, *e.g.*, drainage and water holding capacity) and meso/micro-climates of the vineyard and especially to the seasonal weather pattern of a particular year or *vintage*. Even if these are essentially similar, individual wines have cachets and nuances that are not always attributable to specific technical differences, but which nevertheless distinguish them to the discriminating consumer. Different winemakers and sometimes different lots made by the same winemaker capitalize upon these subtle differences in style to maintain consumer interest. For example, each red table wine labeled as a single grape varietal, of the same vintage and from the same area, is almost certain to differ in style among producers from that winegrowing area. One winemaker may emphasize later, riper picking of the grapes, another one storage in new oak barrels, a third longer *pomace* (skins and seeds) contact with the wine, etc. To retain satisfied customers, consistency in objectively hard-to-assess quality is sought, but differences among styles as much as the winery's images are important in sustaining interest in the "art of winemaking," and may partially explain why a bottle of wine from the same vintage and same growing region may sell for less than $1 or more than $1,000.

Wines retain an aspect of mystic power and connections to art and culture.

Detailed connoisseurship and wine appreciation are occasionally deplored as a form of pretentious snobbery. They are, nevertheless, important in themselves and can prove entertaining to understand without being intimidating. Pure technology cannot fully displace or even explain the romance of wine. Today most of the world's wine is produced by modern technology and is improved thereby, but winemaking is still something of an art, as far as the attainment of its ultimate quality is concerned. Considerable experience and study are required to appreciate fully all the kinds of wine available. It is very important to realize that the inherent qualities of a wine are based solely on an individual consumer's preference, experience and personal associations and do not require absolute ratings by self-proclaimed wine experts.

Wine Classification

Types and Styles

In classifying wines, many parameters might be used: fruit (species, variety, or condition), composition (color, alcohol, sugar, acid, or carbon dioxide), fermentation procedure (carbonic maceration, alcoholic, aerobic yeast film, malolactic), regulations (taxes, etc), geography, climate, weather (vintage). The last three are considered important in determining prices and reputations for individual wines, but are not very helpful in wine group classification. All of the above legally permitted variations, different producers, and marketing descriptions contribute to an almost infinite number of kinds of wines.

An effort to categorize all the significant classes of wine in the general form of a dichotomous key follows. A number of all-or-none alternatives are useful here, *e.g.*, does or does not the wine have added flavors, added alcohol, red color, significant sweetness, appreciable carbon dioxide, oxidized character, or recognizable flavor from the grape variety. It must be borne in mind that there is considerable leeway among styles within each type of wine. Not every wine fits neatly into such a classification. Regulations, taxation levels, geographical specifics, and tradition impose some limits on wines available in or from a specific locale.

Most wines with less than 14 per cent alcohol are classed as table wines because they are usually consumed with meals. Note that as used here, premium wines are included. (In the countries of the E.D., the classification "table wine" is usually a region's lowest quality category and consists of inexpensive everyday wines.) Sparkling wines are included in this group because producing the sparkle and retaining it during consumption of a bottle necessitates a modest alcohol level. The fortified group of wines have had distilled grape spirits added in order to reach, usually, at least about 18 per cent ethanol. Such dessert and aperitif wines are intended to be consumed after or before a meal. Sweet and otherwise strongly flavored wines have gravitated here because they are more stable and thus can be consumed over several sittings. Other pertinent notes follow the list.

1. "Natural" wines, <14 per cent alcohol. Their nature and keeping qualities have traditionally depended heavily on complete fermentation and protection from air.
   A. "Still" wines without obvious carbon dioxide.
      1. Red wines containing anthocyanin pigments
      2. White wines not containing anthocyanin pigments
      3. Blush or Rose wines made either by brief skin contact from red grapes, or by blending of white and red wines. Blush wine styles tend to be lighter in color than roses, although there is no official definition.
   B. "Sparkling" wines with obvious carbon dioxide. Color predominantly white or rose, occasionally red.
      1. From second fermentation of added sugar; usual gauge pressure about 2.026-6.078 × 10^3 hPa (2-6 bar)
      2. Wines with residual excess carbon dioxide from first alcoholic fermentation, not from re-fermentation of added sugar; usual gauge pressure about 0.2026-2.026 × 10^3 hPa (0.2-2.0 bar)
      3. Artificially carbonated wines.
2. Wines with 14-17 per cent alcohol.
   A. Traditionally, miscellaneous, white and red, usually sweet types, mainly with proprietary names. Note that is nowadays not unusual to find natural red still wines, especially from California, in the 14-15 per cent alcohol range when produced from very ripe, often slightly dehydrated fruit.
   B. Special types.
      1. Sacramental, kosher, ecclesiastical, etc wines: usually sweet, *e.g.*, *vino santo,* with dessert-type names produces for special markets
      2. Blending wines to increase the alcohol (or other component) or to impart a special character to other wines
      3. Wines re-fermented with aerobic yeast: flor (Spanish Fino or Manzanilla) Sherries, California submerged-culture dry Sherry.
3. Fortified Wines, 18-21 per cent alcohol. Their nature and keeping qualities depend heavily on the addition and quality of distilled wine spirits.
   A. Without added food flavoring materials.
   B. With added (natural) food flavors or plant extracts.

Figure 16.7: Flow sheet for preparation of White wine

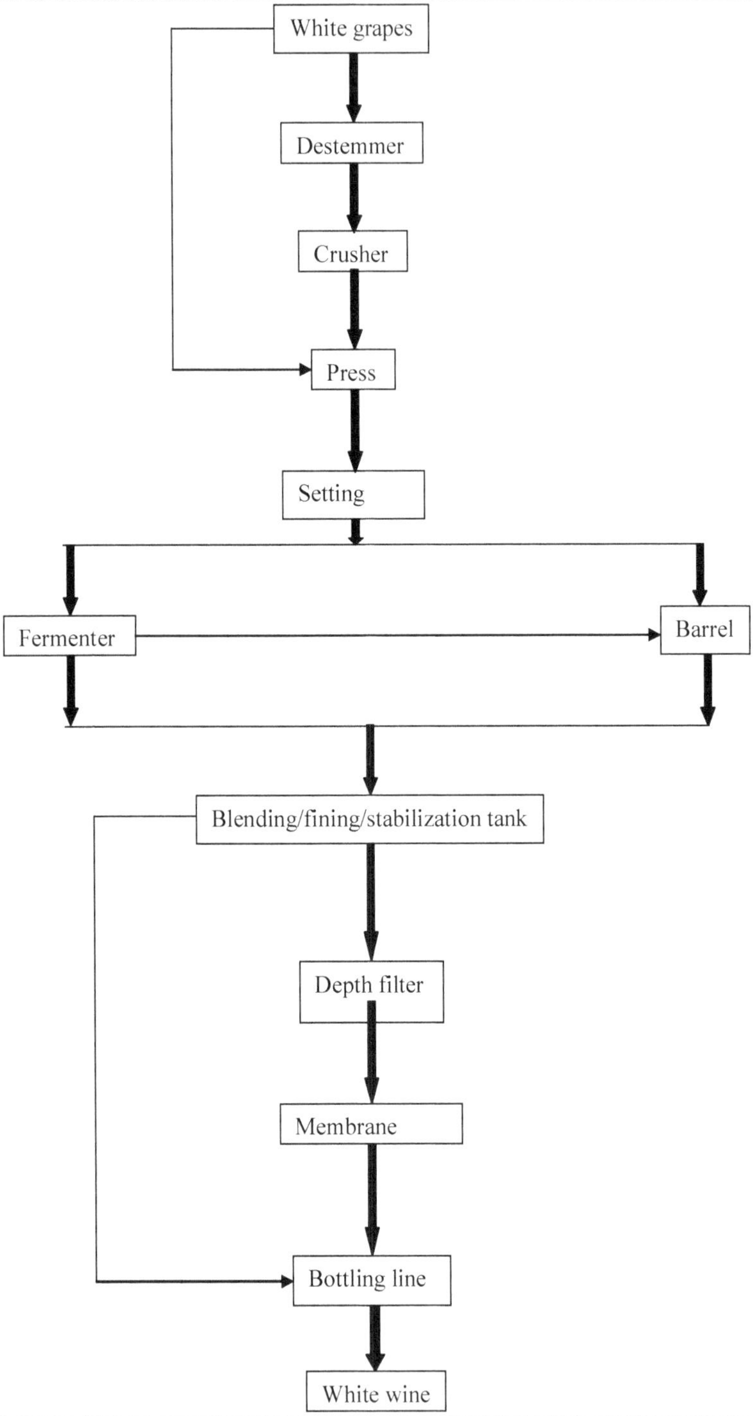

Figure 16.8: Flow sheet for preparation of Rose wine

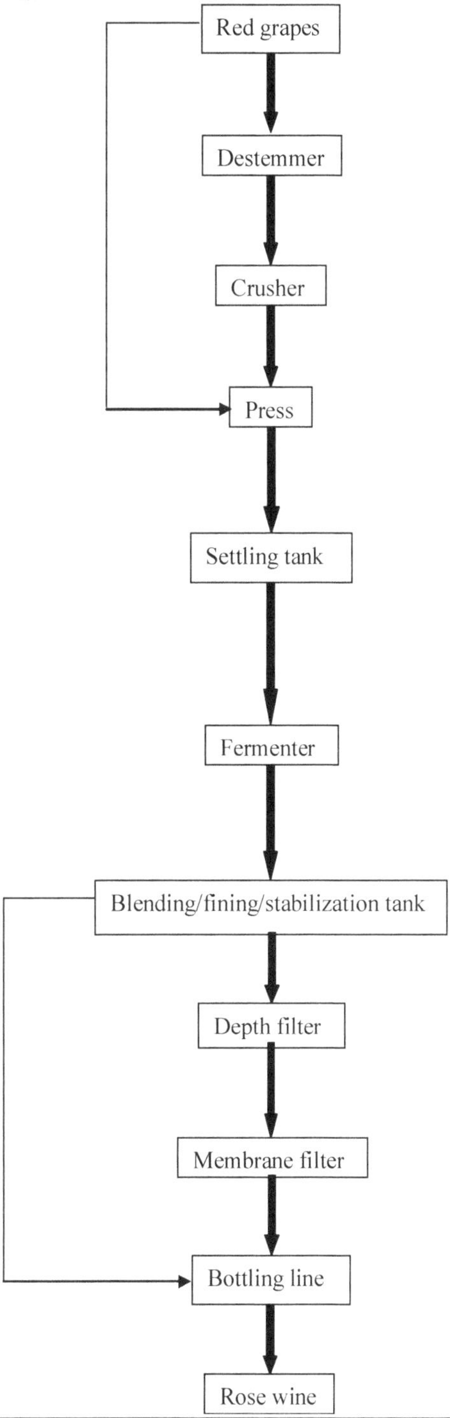

Figure 16.9: Flow sheet for preparation of sparkling wine

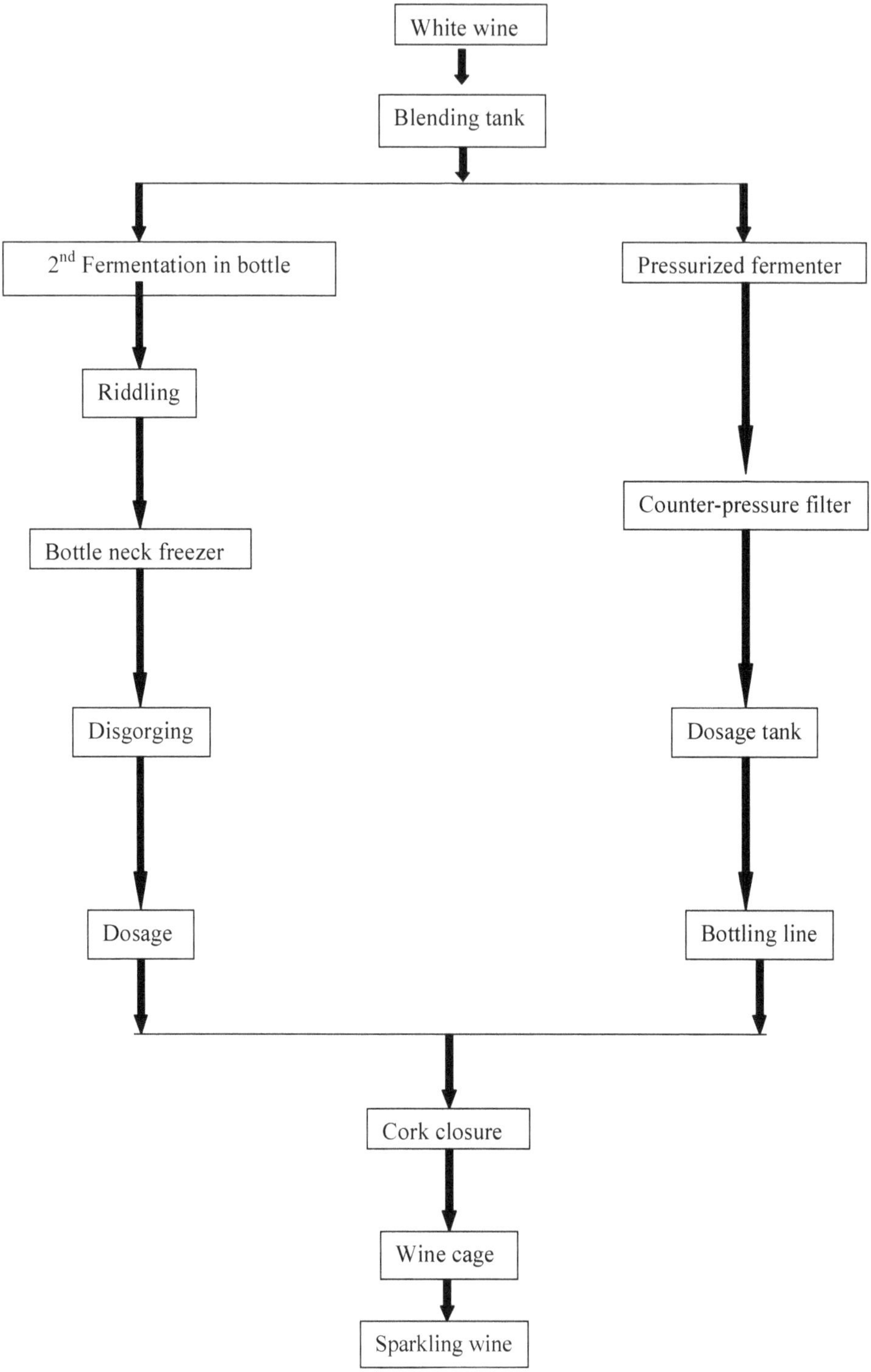

Figure 16.10: Flow sheet for preparation of Nouveau style wine

Red grapes

↓

Destemmer

↓

Maceration tank + CO_2

↓

Press

↓

Fermenter

↓

Stabilization/Filing/Blending tank

↓

Depth filter

↓

Membrane filter

↓

Bottling line

↓

Nouveau style wine

Figure 16.11: Flow sheet for preparation of ice wine

Partially frozen grapes

↓

Press

↓

Settling tank

↓

Fermenter

↓

Stabilization/fining tank

↓

Depth filter

↓

Membrane filter

↓

Bottling line

↓

Ice wine

Figure 16.12: Flow sheet for preparation of Port style wine

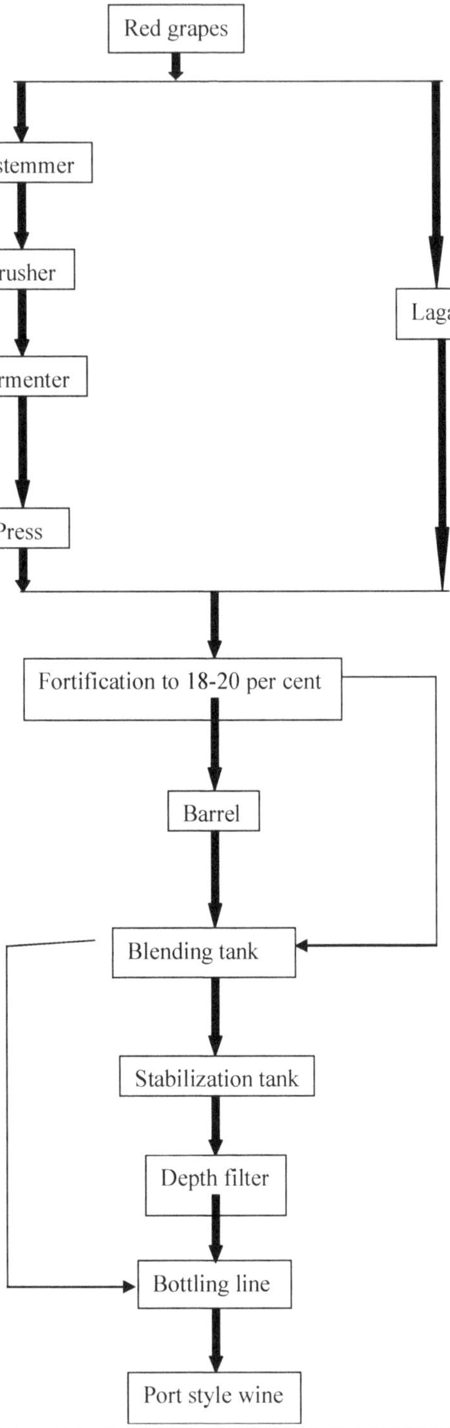

Figure 16.13: Flow sheet for preparation of Flor Sherry style wine

Wine grapes

↓

Destemmer/crusher

↓

Press

↓

Settling tank

↓

Fermenter

↓

Fortification to 15-15.5 per cent

↓

Barrel Solera w/Flor

↓

Blending/stabilization tank

↓

Depth filter

↓

Bottling tank

↓

Bottling line

↓

Flor Sherry style wine

Figure 16.14: Flow sheet for preparation of Red wine

Red grapes

↓

Destemmer

↓

Crusher

↓

Fermenter

↓

Press

↓

Barrel

↓

Blending/Fining/Stabilization tank

↓

Depth filter

↓

Membrane filter

↓

Bottling line

↓

Red wine

17

Guava

Guavas are plants in the myrtle family (Myrtaceae) genus *Psidium* (meaning "pomegranate" in Latin), which contains about 100 species of tropical shrubs and small trees. They are native to Mexico, Central America, and northern South America. Guavas are now cultivated and naturalized throughout the tropics and subtropics in Southeast Asia, Hawaii, the Caribbean, Florida and Africa.

Guava Fruit

Guava fruit, usually 4 to 12 cm long, are round or oval depending on the species. The outer skin may be rough, often with a bitter taste, or soft and sweet. Varying between species, the skin can be any thickness, is usually green before maturity, but becomes yellow, maroon, or green when ripe. Guava fruit generally have a pronounced and typical fragrance, similar to lemon rind but less sharp. Guava pulp may be sweet or sour, off-white ("white" guavas) to deep pink ("red" guavas), with the seeds in the central pulp of variable number and hardness, depending on species.

Culinary Uses

In Hawaii, guava fruit is eaten with soy sauce and vinegar. Occasionally, a pinch of sugar and black pepper are added to the soy sauce and vinegar mixture. The guava fruit is cut up and dipped into the sauce. In Pakistan and India, guava fruit is often eaten raw, typically cut into quarters with a pinch of salt and pepper and sometimes cayenne powder/masala. Street vendors often sell guava fruit for a couple of rupees each. In the Philippines, ripe guava is used in cooking sinigang.

The fruit is also often prepared as a dessert, in fruit salads. In Asia, fresh guava slices are often dipped in preserved prune powder or salt. In India it is often sprinkled with red rock salt, which is very tart. Because of the high level of pectin, guavas are extensively used to make candies, preserves, jellies, jams, marmalades (*Brazilian goiabada*), and also for juices and aguas frescas. Guava juice is very popular in Hawaii, Cuba, Puerto Rico, Colombia, Venezuela, Egypt, Mexico, and South Africa. "Red" guavas can be used as the base of salted products such as sauces, substituting for tomatoes, especially for those sensitive to the latter's acidity. In Asia, a drink is made from an infusion of guava fruits and leaves. In Brazil, the infusion made with guava tree leaves (chá-de-goiabeira, *i.e.* "tea" of guava tree leaves) is considered medicinal.

Nutritional Value

Guavas are rich in dietary fiber, vitamins A and C, folic acid, and the dietary minerals, potassium, copper and manganese. Having a generally broad, low-calorie profile of essential nutrients, a single common guava (*P. guajava*) fruit contains about four times the amount of vitamin C as an orange.

However, nutrient content varies across guava cultivars. Although the strawberry guava (*P. littorale* var. *cattleianum*), notably containing 90 mg of vitamin C per serving, has about 25 per cent of the amount found in more common varieties, its total vitamin C content in one serving still provides 100 per cent of the Dietary Reference Intake for adult males. Guavas contain both carotenoids and polyphenols like (+)-gallocatechin, guaijaverin, leucocyanidin and amritoside – the major classes of antioxidant pigments – giving them relatively high potential antioxidant value among plant foods. As these pigments produce the fruit skin and flesh color, guavas that are red-orange have more pigment content as polyphenol, carotenoid and pro-vitamin A, retinoid sources than yellow-green ones.

Potential Medical Uses

Since the 1950s, guavas – particularly the leaves – have been a subject for diverse research in chemical identity of their constituents, pharmacological properties and history in folk medicine. Most research, however, has been conducted on apple guava (*P. guajava*), with other species remaining undefined. From preliminary medical research in laboratory models, extracts from apple guava leaves or bark are implicated in therapeutic mechanisms against cancer, bacterial infections, inflammation and pain. Essential oils from guava leaves display anti-cancer activity *in-vitro*.

Guava leaves are used in folk medicine as a remedy for diarrhea and, as well as the bark, for their supposed antimicrobial properties and as an astringent. Guava leaves or bark are used in traditional treatments against diabetes. In Trinidad, a tea made from young leaves is used for diarrhea, dysentery and fever.

Common Guava, per 165 g of individual fruit portion

Calories	112
Moisture	133 g
Dietary Fiber	8.9 g
Protein	4.2 g
Fat	1.6 g
Ash	2.3 g
Carbohydrates	23.6 g
Calcium	30 mg
Phosphorus	66 mg
Iron	0.4 mg
Potassium	688 mg
Copper	0.4 mg
Beta-carotene (Vitamin A)	1030 IU
Ascorbic acid (Vitamin C)	377 mg

Thiamin (Vitamin B_1)	0.1 mg
Riboflavin (Vitamin B_2)	0.1 mg
Niacin (Vitamin B_3)	1.8 mg
Folic acid	81 mcg

Figure 17.1: Flow sheet for preparation of guava jelly

Fully mature guava fruits
↓
Washing and Cleaning
↓
Cutting in to small pieces
↓
Addition of water (1:1) and 2 g citric acid/kg
↓
Heating for 30 min
↓
Straining the pulp
↓
Again heating same pulp with ½ quantity of water
↓
Second time straining the pulp
↓
Pulp extract keep for sedimentation 12 hrs
↓
Take supernant
↓
Pectin test (add pectin/sugar as per requirement)
↓
Boiling to get jelly test
↓
Filling pasteurized hot bottles
↓
Waxing and Sealing
↓
Pasteurization
↓
Labeling and Storage
↓
Marketing

Figure 17.2: Flow sheet for preparation of Guava juice

Fully mature/ripe guava fruits
↓
Washing
↓
Removing of seeds
(Pass through 60 mesh stainless steel sieve)
↓
Pulping using screw type juicer
↓
Pectinase enzyme treatment (0.2 per cent) incubation at 30 °C, 4 hrs
↓
Filtration/Centrifugation
↓
Juice
↓
Pasteurization
↓
Packaging
↓
Storage

Figure 17.3: Flow sheet for preparation of RTS from guava and papaya juice

Guava : Papaya Fruit juice
↓
Blending (70:30)
↓
Addition of water, sugar and citric acid
↓
Filtration
↓
Checking °Brix 15, acidity 0.25 per cent
↓
Filling into pasteurized bottles
↓
Pasteurization
↓
Cooling and labeling
↓
Storage

Figure 17.4: Flow sheet for preparation of guava pulp

Fully Ripe Guava Fruits
↓
Washing
↓
Cutting into pieces
↓
Mixing with water (1:1)
↓
Passing through pulper
↓
Separation of Pulp and Seeds
↓
Pulp
↓
Addition of preservative
↓
Can/Pouch filling
↓
Pasteurization/Cold storage
↓
Marketing

Pulp can also be obtained by grinding the pieces finely with water in a grinder by straining through thick cloth.

Figure 17.5: Flow sheet for preparation of guava toffee

Fully ripe guava fruits

↓

Washing

↓

Slicing

↓

Rubbing on sieve/strain through muslin cloth or pulping with screw type machine

↓

Remove seeds, fibers

↓

Addition of sugar, fat, skim milk powder (Pulp 1.0 kg, 1000g sugar, 100g SMP)

↓

Heating upto 80 °Brix

↓

Addition of salt 5g

↓

Heating upto 82 °Brix

↓

Spreading in aluminium trays already smeared with oil/fat

↓

Cooling

↓

Cutting in to desired pieces

↓

Rapping into metallic paper

↓

Storage

Figure 17.6: Flow sheet for preparation of semi-dehydrated guava

Slightly Ripe Firm fruit of Guava
↓
Washing under tap water
↓
Peeling by hand peeler
↓
Cutting in eight equal pieces
↓
Blanching (96 °C), 5 min
↓
Draining and cooling
↓
Syruping with 70 °Brix sugar syrup, 24 h
↓
Slightly washing
↓
Draining (1/2 h)
↓
Wiping with wet sponge
↓
Drying in tray dryer (60 °C)
↓
Dehydrated product (10.5 per cent Moisture content)
↓
Packaging
↓
Storage

Figure 17.7: Flow sheet for preparation of guava wine

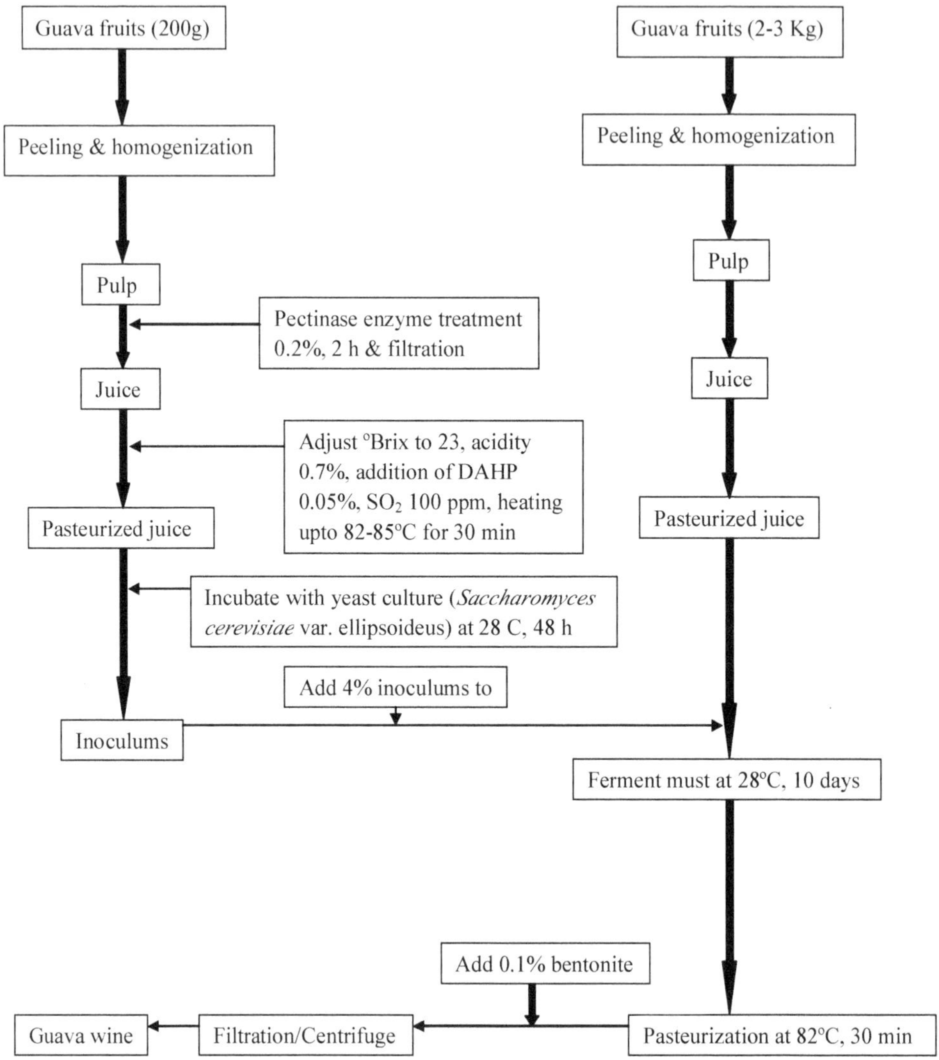

18

Jackfruit

Jackfruit is the name of the fruit of 'jackfruit' tree, which belongs to the mulberry family. It is native to southwestern India, Bangladesh, Philippines, Sri Lanka and possibly, east of the Malay Peninsula. Mainly found to be cultivated in tropical conditions, jackfruit is said to be the largest tree-borne fruit in the world, with its diameter being at least 25 cm. on the other extreme, there can be jackfruits measuring as much as 36 kg in weight, 90 cm in length and 50 cm in diameter. The best part of the fruit is that even its seeds have been found to have a high nutritional value. If you want to know more about the health and nutrition benefits of eating jackfruit, browse further through internet or see encyclopaedia.

Although many who eat jackfruit love it for its delicious taste, quite a few are unaware of the health benefits of jackfruit. Jackfruit contains Vitamin A, Vitamin C, thiamin, riboflavin, calcium, potassium, iron, sodium, zinc and niacin among many other nutrients. Jackfruit also benefits one's health as it has a low caloric content: 100 grams of jackfruit only contains 94 calories.

Jackfruit is a rich source of potassium with 303 milligram found in 100 grams of jackfruit. Studies show that food rich in potassium helps to lower blood pressure. Therefore, jackfruit has the health benefit of providing relief for anyone who suffers from high blood pressure. Another benefit of eating jackfruit is that it is a good source of Vitamin C. The human body does not make Vitamin C naturally so we must eat food that contains Vitamin C to reap its health benefits. The health benefits of Vitamin C are that it is an antioxidant that protects the body against free radicals, Vitamin C holds the cells in our bodies together, and Vitamin C strengthens our immune systems and keeps our gums healthy.

Jackfruit contains phytonutrients: lignans, isoflavones and saponins which have health benefits that are wide ranging. These phytonutrients have anti-cancer, antihypertensive, anti-ulcer and anti-ageing properties. The phytonutrients found in jackfruit, therefore, can prevent cancer cells from forming in the body, can lower blood pressure, can fight against stomach ulcers and can slow down the degeneration of cells that make the skin look young and vital. Jackfruit root has been found to help those suffering from asthma. Boiling the root of the jackfruit and extracting its high nutritional content has been found to control asthma. Jackfruit also benefits the body in that jackfruit root can also be used to treat skin problems. The extract of jackfruit root is also believed to be able to help cure diarrhea and fever.

Nutritional Value of Jackfruit

Given below is the amount of nutrients present in 100 gm of jackfruit:

Nutrient	*Amount*
Total Carbohydrates	24 g
Dietary Fiber	1 g
Protein	1 g
Total Fat	0.3 mg
Saturated Fat	0.63 per cent
Monounsaturated Fat	0.44 per cent
Polyunsaturated Fat	0.086 mg
Energy	94 Calories
Vitamin A	297 IU
Vitamin C	6.7 mg
Thiamin	0.03 mg
Riboflavin	0.11 mg
Niacin	0.4 mg
Vitamin B6	0.108 mg
Foliate	14 μg
Calcium	34 g
Sodium	3 mg
Iron	0.6 mg
Magnesium	37 g
Phosphorus	36 g
Potassium	303 g
Sodium	2 mg
Zinc	0.42 g
Copper	0.187 mg
Manganese	0.197 mg
Selenium	0.6 μg

Health and Nutrition Benefits of Eating Jackfruit

- Being rich in potassium, jackfruit has been found to be helpful in the lowering of blood pressure.
- The extract of Jackfruit root is believed to help cure fever as well as diarrhea.
- Jackfruit contains phytonutrients, with health benefits ranging from anti-cancer to antihypertensive.
- The root of this fruit has been found to be beneficial for those suffering from asthma.
- Jackfruit proves to be a very good source of vitamin C, which is known for its high antioxidant properties.
- The fruit contains isoflavones, antioxidants, and phytonutrients, all of which are credited for their cancer-fighting properties.
- Jackfruit is known to contain anti-ulcer properties and is also good for those suffering from indigestion.
- Boasting of anti-ageing properties, the fruit can help slow down the degeneration of cells and make the skin look young and supple.
- Jackfruit serves as a good supply of proteins, carbohydrates and vitamins, for the human body.
- It is believed that the fruit can help prevent and treat tension and nervousness.
- Since it contains few calories and a very small amount of fat, jackfruit is good for those trying to lose weight.
- If you are suffering from constipation, regular consumption of the fruit will surely prove beneficial.
- The root of jackfruit is said to be good for the treatment of a number of skin problems.

Figure 18.1: Flow sheet for preparation of jackfruit pickle

Jack fruit (small mature)
↓
Washing
↓
Peeling
↓
Cutting into suitable sized pieces
↓
Steeping in 10 per cent salt solution for a week
↓
Drain off brine
↓
Washing pieces with water
↓
Frying all ingredients in oil (except vinegar, jiggery, sugar)
↓
Mixing with pieces and frying for 5 min
↓
Cooling
↓
Filling in jar
↓
Keeping in sun for a week
↓
Making paste of sugar/jiggery and vinegar
↓
Adding to pieces
↓
Sealing and labeling
↓
Storage and Marketing

Sl.No.	*Item*	*Quantity*	*Sl.No.*	*Item*	*Quantity*
1.	Jackfruit	1 kg	7.	Cardamom (large)	10 g
2.	Salt	100 g	8.	Aniseed powder	10 g
3.	Onion chopped	50 g	9.	Jaggery/Sugar	250 g
4.	Clove headless	8 Nos	10.	Black pepper	10 g
5.	Red chilli powder	15 g	11.	Vinegar	150 ml
6.	Cumin	10 g	12.	Mustard oil	350 ml

19

Jamun

Jambul (*Syzygium cumini*) is an evergreen tropical tree in the flowering plant family Myrtaceae. Jambul is native to Bangladesh, India, Nepal, Pakistan, Sri Lanka, and Indonesia. The name of the fruit is sometimes mistranslated as blackberry, which is a different fruit in an unrelated family.

Jambul is also known as jambu/jambula/jamboola, Java plum, jamun, jaam/kalojaam, jamblang, jambolan, black plum, Damson plum, Ducats plum, Jambolan plum, or Portuguese plum. Malabar plum may also refer to other species of *Syzygium*. In Tamil, this fruit is called Naaval Pazham or Navva Pazham.

The tree was introduced to Florida, USA in 1911 by the USDA, and is also now commonly grown in Suriname and Trinidad and Tobago. In Brazil, where it was introduced from India during Portuguese colonization, it has dispersed spontaneously in the wild in some places, as its fruits are eagerly sought by various native birds such as thrushes, tanagers and the Great Kiskadee. This species is considered an invasive in Hawaii, USA.

A fairly fast growing species, it can reach heights of up to 30 m and can live more than 100 years. Its dense foliage provides shade and is grown just for its ornamental value. The wood is strong and is water resistant. Because of this it is used in railway sleepers and to install motors in wells. It is sometimes used to make cheap furniture and village dwellings though it is relatively hard to work on.

Jambul trees start flowering from March to April. The flowers of jambul are fragrant and small, about 5 mm in diameter. The fruits develop by May or June and resemble large berries. The fruit is oblong, ovoid, starts green and turns pink to shining crimson black as it matures. A variant of the tree produces white colored fruit. The fruit has a combination of sweet, mildly sour and astringent flavour and tends to color the tongue purple. The seed is also used in various alternative healing systems like Ayurveda (to control diabetes, for example. Unani and Chinese medicine for digestive ailments. The leaves and bark are used for controlling blood pressure and gingivitis. Wine and vinegar are also made from the fruit. It has a high source in vitamin A and vitamin C.

According to Hindu tradition, Rama subsisted on the fruit in the forest for 14 years during his exile from Ayodhya. Because of this, many Hindus regard jambul as a 'fruit of the gods,' especially in Gujarat, India, where it is known locally as jamboon. Lord Krishna has been described as having skin the color of Jamun. In Hindu mythology several protagonists have been described as having the color of jambul.

Jamun Plum, Raw (NDB NO:09145)

Nutritional value per 100 g (3.5 oz)	
Energy	251 kJ (60 kcal)
Carbohydrates	15.56 g
Fat	0.23 g
Protein	0.72 g
Water	83.13 g
Vitamin A	3 IU
Thiamine (vit. B_1)	0.006 mg
Riboflavin (vit. B_2)	0.012 mg
Niacin (vit. B_3)	0.260 mg
Pantothenic acid (B_5)	0.160 mg
Vitamin B_6	0.038 mg
Vitamin C	14.3 mg
Calcium	19 mg
Iron	0.19 mg
Magnesium	15 mg
Phosphorus	17 mg
Potassium	79 mg
Sodium	14 mg

Figure 19.1: Flow sheet for processing of jamun fruits into various products

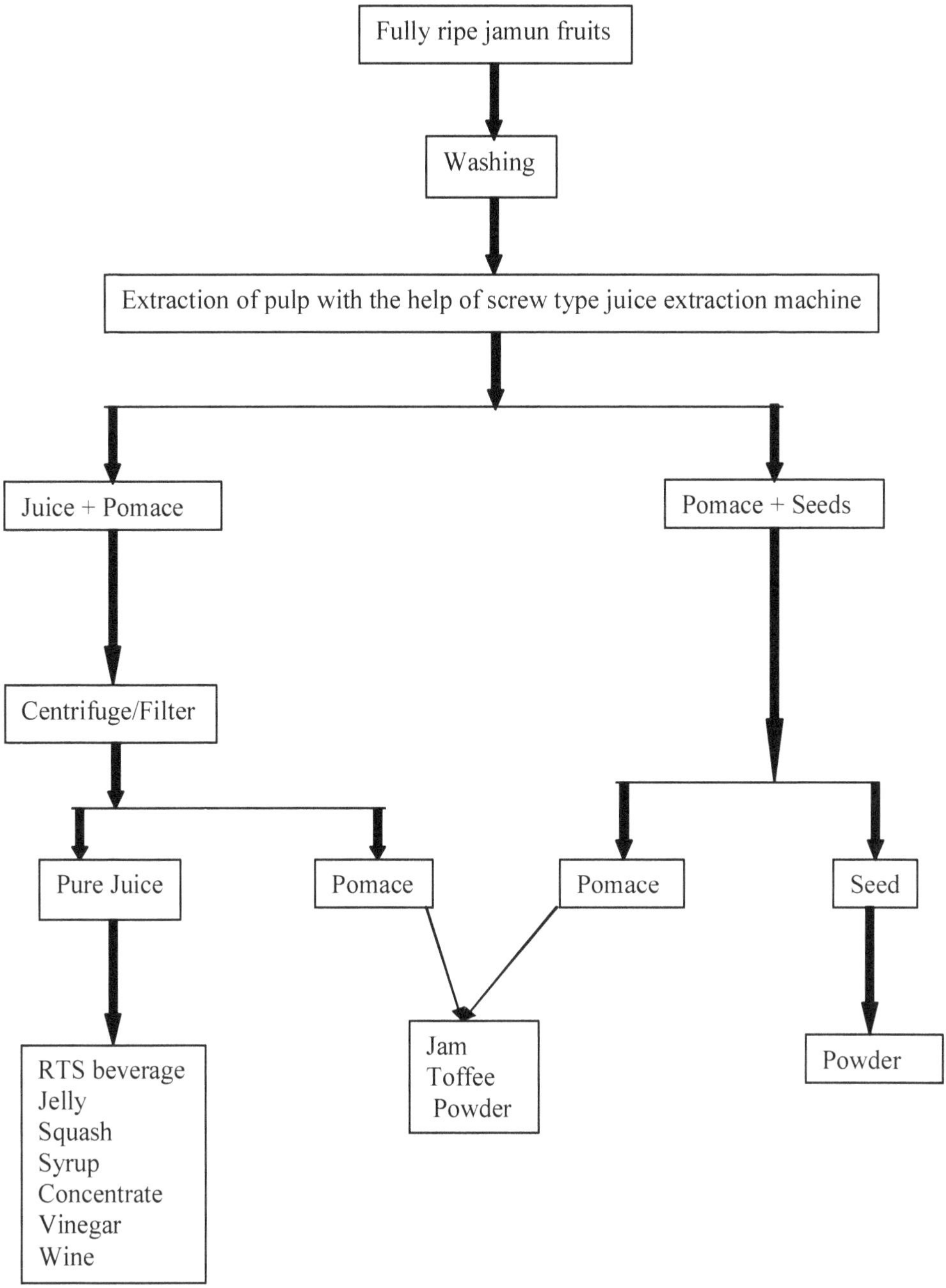

Figure 19.2: Flow sheet for processing of jamun seed and pomace powder

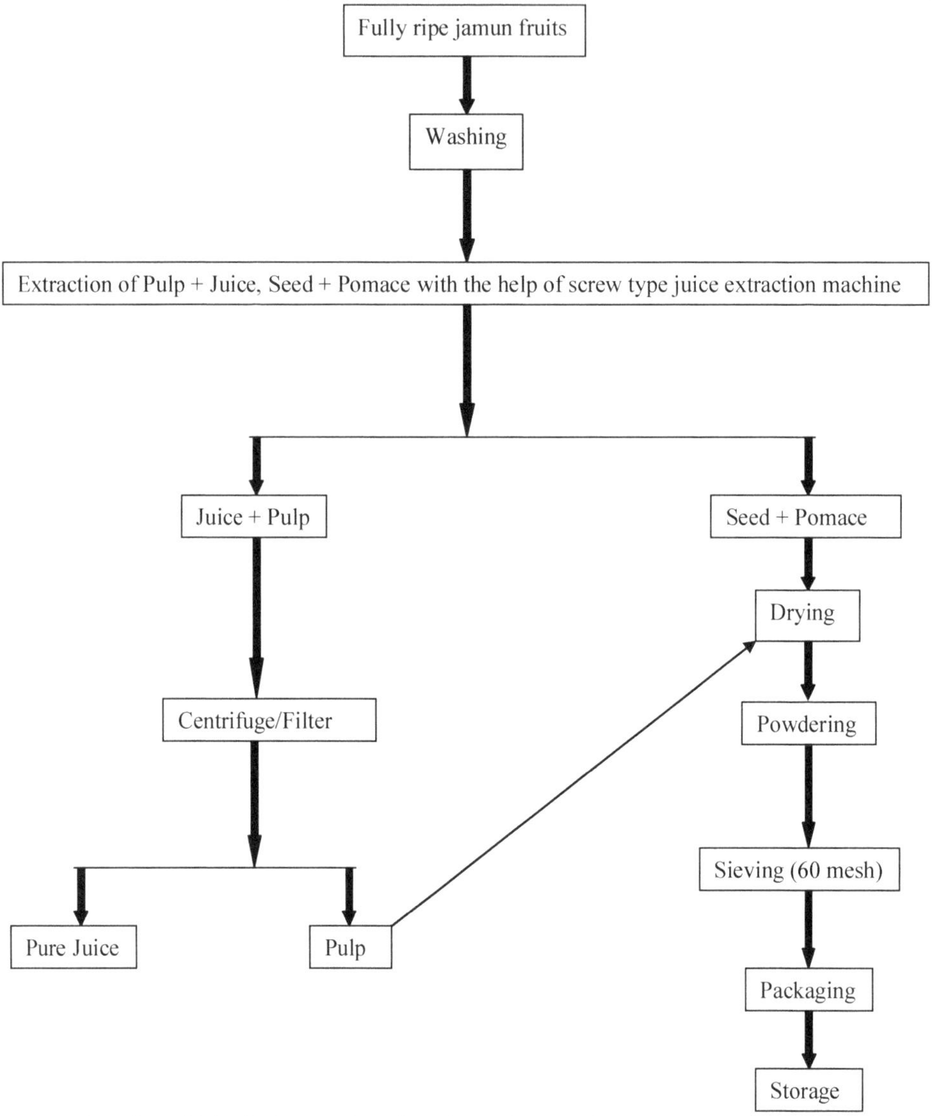

Figure 19.3: Flow sheet for preparation of jamun juice

Jamun fruits
↓
Washing
↓
Passing through grating machine → Remove seeds
↓
Heating pulp to 60 °C, 3 min
↓
Pass through basket press → Remove pomace
↓
Collection of juice
↓
Heating of juice to 85 °C, 2 min
↓
Centrifuge/Filter
↓
Cooling
↓
Addition of sodium benzoate
↓
Packaging
↓
Storage

Figure 19.4: Flow sheet for preparation of jamun toffee

Jamun juice extracted Pulp/pomace

↓

Add corn flour, skim milk powder, sugar, fat etc.

↓

Heating upto 80 °Brix

↓

Add citric acid and salt

↓

Heating upto 82 °Brix

↓

Spread in the tray upto ½ to 1′ thickness (tray already smeared with fat/oil

↓

Cool to room temperature

↓

Cut into attractive pieces

↓

Packaging into metallic paper

↓

Storage

Figure 19.5: Flow sheet for preparation of jamun wine

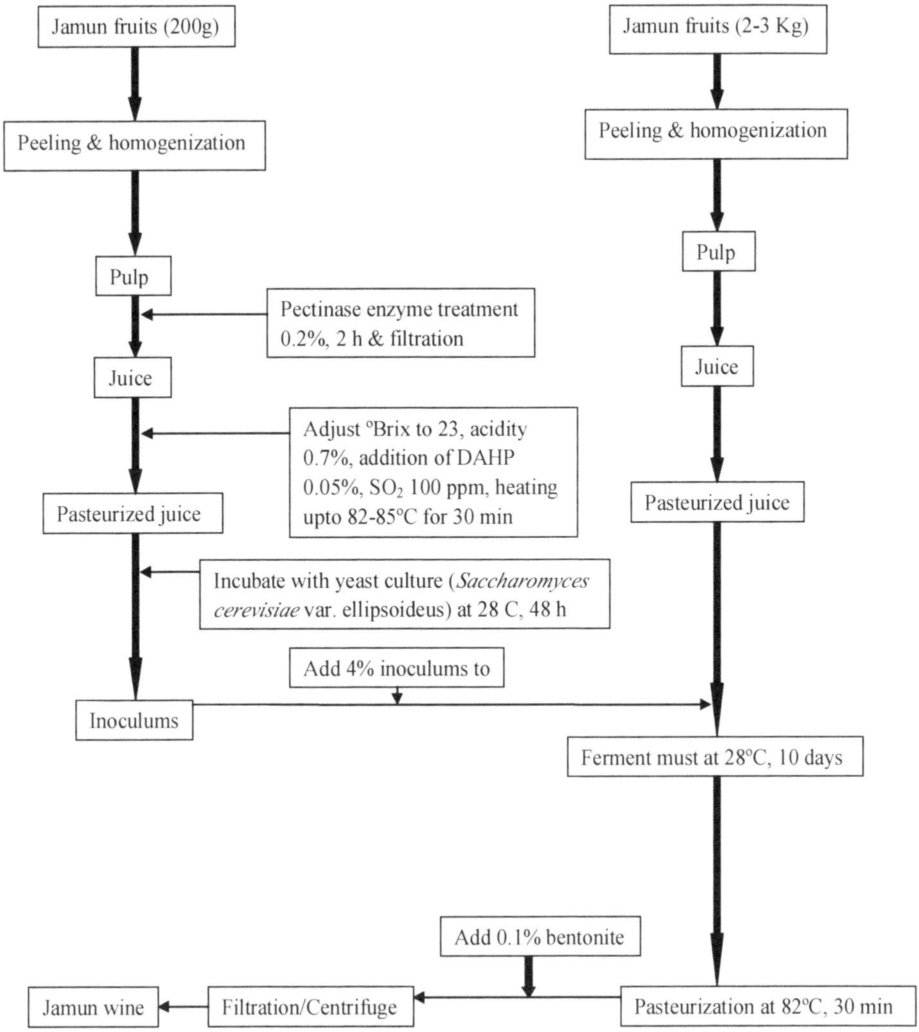

20

Karonda-Plum Phalsa

1. Karonda

Karonda (*Carissa carandus* L.) is a fruit of dry areas and flourishes well on lands with high temperatures. At present it is grown on a limited scale in Rajasthan, Gujarat and Uttar Pradesh. It is a fruit of Indian origin and under the changing world trade scenario it can be exploited on a commercial scale as a fruit for the processing industries.

The karonda fruit is a rich source of iron and contains a fair amount of Vitamin C and, therefore, is very useful for cure of anemia and has antiscorbutic (counteracting scurvy) properties. Mature fruit contains high amount of pectin and, therefore, besides being used for making pickle, it can be exploited for making jelly, jam, squash, syrup and chutney, which are of great demand in the international market.

Karonda plants can be grown for raising beautiful juvenile hedges and because of the presence of auxiliary spines it can be a very good bio-fence. Karonda can also be planted on the boundaries and bunds of fields, keeping space in between for other crops. The leaves of the plant are easily biodegradable and enrich the soil with a lot of organic carbon and other mineral nutrients. Due to the presence of dense foliage, the evaporation rate from the soil underneath a karonda plantation is extremely slow and, therefore, it is useful for moisture-stress conditions. The roots of the plant are heavily branched and make it suitable for stabilizing eroding slopes.

A very hardy, drought-tolerant plant that thrives well in a wide range of soils, it is commercially propagated through seed in August and September, though inarching and budding can also be practiced for vegetative propagation. Cuttings may also succeed. Planting is done with the first shower of monsoon at a distance of 1.5 m. There are no well-established varieties of Karonda, but depending on the color, different strains are grouped as grape green, green with purple blush, white with pink blush and maroon. Pink blushed varieties are richer in iron than green and maroon ones.

Plants raised from seed come into bearing two years after planting. Flowering starts in March and the fruit ripens from July to September in North India. The fruit is ready for harvest after 100-110 days of fruit-set, when they develop natural color with a specific gravity around 1.02. Two to three pickings are required for complete harvest of 3-4 kg fruit per plant. The fruit harvested at maturity can be stored for 5-7 days at room temperature. If harvested at a ripe stage, it has a very short shelf life and can be stored only up to two days at normal temperature. Long-duration storage of the fruit can be done in SO_2 (KMS) @ 2000ppm.

In the lower hills of Himachal Pradesh, vast chunks of marginal and wastelands, which are unsuitable for other crops of commercial importance, karonda can be a good alternative, particularly for poor farmers. The maintenance cost of plantations is low in comparison to other fruits.

Karonda is a berry-sized fruit that is commonly used as a condiment or additive to Indian pickles and spices. It is a very hardy, drought-tolerant plant that thrives well in a wide range of soils.

Uses

The karonda fruit is a rich source of iron and contains a fair amount of Vitamin C. It is antiscorbutic and very useful for cure of anemia. Mature fruit contains high amount of pectin and, therefore, besides being used for making pickle, it can be exploited for making jelly, jam, squash, syrup and chutney.

Nutrition Facts

Karonda, Serving Size 1cup slices (150.0g)	
Calories	62
Total Fat	1.3 g
Cholesterol	0 mg
Sodium	3 mg
Potassium	260 mg
Total Carbohydrate	13.63 g
Protein	0.5 g
Vitamin A	2 μg
Thiamin	0.04 mg
Riboflavin	0.06 mg
Niacin	0.2 mg
Vitamin B_{12}	0 μg
Vitamin C	38 mg
Calcium	11 mg
Iron	1.31 mg
Magnesium	16 mg
Phosphorus	7 mg
Copper	0.21 mg

2. Plum

A plum or gage is a stone fruit tree in the genus *Prunus*, subgenus *Prunus*. The subgenus is distinguished from other subgenera (peaches, cherries, bird cherries, etc.) in the shoots having a terminal bud and solitary side buds (not clustered), the flowers in groups of one to five together on short stems, and the fruit having a groove running down one side and a smooth stone.

Mature plum fruit may have a dusty-white coating that gives them a glaucous appearance; this is easily rubbed off. This is an epicuticular wax coating and is known as "wax bloom". Dried plum fruits are called dried plums or prunes, although prunes are a distinct type of plum, and may have antedated the fruits now commonly known as plums.

The subgenus is divided into three sections:

1 Sect. *Prunus* (Old World plums)- leaves in bud rolled inwards; flowers 1-3 together; fruit smooth, often wax-bloomed
   - *P. cerasifera* (cherry plum)
   - *P. cocomilia* (Italian plum)
   - *P. consociiflora*
   - *P. domestica* (species of most "plums" and "prunes" sold as such)
   - *P. insititia* (damsons)
   - *P. simonii*
   - *P. spinosa* (blackthorn or sloe)

2. Sect. *Prunocerasus* (New World plums) - leaves in bud folded inwards; flowers 3-5 together; fruit smooth, often wax-bloomed
   - *P. alleghaniensis* (Allegheny plum)
   - *P. americana* (American plum)
   - *P. angustifolia* (Chickasaw plum)
   - *P. hortulana*
   - *P. maritima* (beach plum)
   - *P. mexicana* (Mexican plum)
   - *P. nigra* (Canada plum, Black plum)
   - *P.* × *orthosepala* (*P. americana* × *P. angustifolia*)
   - *P. subcordata* (Klamath, Oregon, or Sierra plum)

3. Sect. *Armeniaca* (apricots) - leaves in bud rolled inwards; flowers very short-stalked; fruit velvety; treated as a distinct subgenus by some authors
   - *P. armeniaca* (apricot)
   - *P. brigantina*
   - *P. mume* (Chinese plum, Japanese apricot)
   - *P. sibirica*

Uses

Plum fruit tastes sweet and/or tart; the skin may be particularly tart. It is juicy and can be eaten fresh or used in jam-making or other recipes. Plum juice can be fermented into plum wine; when distilled, this produces a brandy known in Eastern Europe as *Slivovitz, Rakia, Þuică* or *Pálinka*. In central England, a cider-like alcoholic beverage known as plum jerkum is made from plums.

Dried plums (or prunes) are also sweet and juicy and contain several antioxidants. Plums and prunes are known for their laxative effect. This effect has been attributed to various compounds present in the fruits, such as dietary fiber, sorbitol, and isatin. Prunes and prune juice are often used to help regulate the functioning of the digestive system. Dried prune marketers in the United States have, in recent years, begun marketing their product as "dried plums". This is due to "prune" having negative connotations connected with elderly people suffering from constipation.

Dried, salted plums are used as a snack, sometimes known as *saladito* or *salao*. Various flavors of dried plum are available at Chinese grocers and specialty stores worldwide. They tend to be much drier than the standard prune. Cream, ginsing, spicy, and salty are among the common varieties. Licorice is generally used to intensify the flavor of these plums and is used to make salty plum drinks and toppings for shaved ice or *baobing*.

Pickled plums are another type of preserve available in Asia and international specialty stores. The Japanese variety, called *umeboshi*, is often used for rice balls, called *onigiri* or *omusubi*. The *ume*, from which *umeboshi* are made, is more closely related, however, to the apricot than to the plum.

As with many other members of the rose family, plum seeds contain cyanogenic glycosides, including amygdalin. These substances are capable of decomposing into a sugar molecule and hydrogen cyanide gas. While plum seeds are not the most toxic within the rose family (the bitter almond is the most toxic, large doses of these chemicals from any source are hazardous to human health. Prune kernel oil is made from the fleshy inner part of the pit of the plum. Plums come in a wide variety of colors and sizes. Some are much firmer-fleshed than others, and some have yellow, white, green or red flesh, with equally varying skin color.

Plums (without pit) *Prunus* spp.

Nutritional value per 100 g (3.5 oz)	
Energy	192 kJ (46 kcal)
Carbohydrates	11.4 g
Sugars	9.9 g
Dietary fiber	1.4 g
Fat	0.28 g
Protein	0.70 g
Vitamin A	345 IU
Vitamin C	9.5 mg
Phosphorus	16 mg
Potassium	157 mg

Plums, Dried (Prunes), Uncooked

Nutritional value per 100 g (3.5 oz)	
Energy	1,006 kJ (240 kcal)
Carbohydrates	63.88 g
Sugars	38.13 g
Dietary fiber	7.1 g
Fat	0.38 g
Protein	2.18 g
Vitamin A	781 IU
Vitamin C	0.6 mg
Phosphorus	69 mg
Potassium	732 mg

Plum is a fruit that is related to the family of peaches and cherries. It is one of those fruits that are rich in dietary fiber, which in turn proves effective in improving the digestive system. There are thousands of varieties of plums that are available throughout the world, ranging in colors like red, blue-black, purple, yellow, green or amber. Plums are believed to have originated in Asia and since then, they have been grown all over the world. The fresh and juicy taste of the plums makes it refreshing to eat them on a hot summer day! Dried plums, called prunes, are also quite delicious.

Nutritive Value of Plum as per USDA

Nutrients	*Amount*
Basic	
Protein	0.52 g
Carbohydrates	8.59 g
Water	56.23 g
Ash	0.26 g
Calories	
Total Calories	36.30
Calories from Fat	3.68
Calories from Saturated Fat	0.29
Fiber	
Dietary Fiber	0.99 g
Soluble Fiber	0.55 g
Insoluble Fiber	0.44 g
Sugar	
Total Sugar	6.66 g
Monosaccharides	2.97 g
Disaccharides	1.98 g
Other Carbs	0.94 g

Nutrients	*Amount*
Fats	
Total Fat	0.41 g
Saturated Fat	0.03 g
Mono Fat	0.27 g
Poly Fat	0.09 g
Vitamins	
Vitamin A	213.18 IU
Vitamin A	
Carotenoid	21.12 RE
Beta Carotene	126.72 μg
Thiamin	0.03 mg
Riboflavin	0.06 mg
Niacin	0.33 mg
Niacin equiv	0.33 mg
Vitamin B_6	0.05 mg
Vitamin C	6.27 mg
Vitamin E alpha equiv	0.40 mg
Vitamin E	0.59 IU
Vitamin E	0.57 mg
Foliate	1.45 mg
Pantothenic Acid	0.12 mg
Minerals	
Calcium	2.64 mg
Copper	0.03 mg
Iron	0.07 mg
Magnesium	4.62 mg
Manganese	0.03 mg
Phosphorous	6.60 mg
Potassium	113.52 mg
Selenium	0.33 μg
Zinc	0.07 mg
Mono Fats	
18:1 Oleic	0.26 g
Poly Fats	
18:2 Linoleic	0.09 g
Other Fats	
Omega 6 Fatty Acids	0.09 g

Nutrients	*Amount*
Amino Acids	
Alanine	0.02 g
Arginine	0.01 g
Aspartate	0.16 g
Glutamate	0.02 g
Glysine	0.01 g
Histidine	0.01 g
Isoleucine	0.01 g
Leucine	0.01 g
Lysine	0.01 g
Phenylalanine	0.01 g
Proline	0.02 g
Serine	0.01 g
Threonine	0.01 g
Valine	0.01 g
Other Nutrients	
Sugar Alcohol	0.93 g
Sorbitol	0.93 g
Organic Acids	827.64 mg
Citric Acid	22.44 mg
Malic Acid	805.20 mg

Nutrition Benefits of Eating Plums

- ✰ Plums, even their dried form known as 'prunes', are high in unique phytonutrients called neochlorogenic and chlorogenic acid. Their function is that of an antioxidant and is of much benefit to the body.
- ✰ Since plums are rich in antioxidants, they provide protection from superoxide anion radical and also prevent damage to our neurons and fats that form a part of our cell membranes.
- ✰ Consumption of plums helps in the production and absorption of iron in the body, thus leading to better blood circulation, which further leads to the growth of healthy tissues.
- ✰ Regular consumption of plums can prevent mascular degeneration and any other infection of the eye, in the long run. Your eyes will be healthy and strong for long time and you can also retain a sharp eye-sight.
- ✰ Researchers have found that plums have anti-cancer agents that may help prevent the growth of cancerous cells and tumors in the body.
- ✰ Eating plums also reduces your chances of contracting a heart disease in the long run. Plums have certain cleansing agents that keep the blood pure and prevent complications of the heart.

- Plums have high content of Vitamin C, which means that they help protect the body against health conditions like asthma, colon cancer, osteoarthritis and rheumatoid arthritis.
- Plum juice concentrate is effective in preventing and reducing human influenza A.

Health Benefits of Plums

- Delicious, fleshy, succulent plums are low in calories and contain no saturated fats; but contain numerous health promoting compounds, minerals and vitamins.
- Certain health benefiting compounds present in the plum fruits, such as dietary fiber, sorbitol, and isatin are known to help regulate the functioning of the digestive system and thereby used in constipation conditions.
- Fresh plums are an excellent source of vitamin C, which is also a powerful natural antioxidant. Consumption of foods rich in vitamin C helps body develop resistance against infectious agents, counter inflammation and scavenge harmful free radicals.
- Fresh plums, especially yellow Mirabelle type, are very good source of vitamin A and beta carotene. Vitamin A is essential for vision. It is also required for maintaining healthy mucus membranes and skin. Consumption of natural fruits rich in vitamin A known to protect from lung and oral cavity cancers.
- The fruit is also good in health promoting flavonoid poly phenolic antioxidants such as lutein, cryptoxanthin and zeaxanthin in significant amounts. These compounds help act as protective scavengers against oxygen-derived free radicals and reactive oxygen species (ROS) that play a role in aging and various disease process. Zeaxanthin, an important dietary carotenoid selectively absorbed into the retinal macula lutea where it is thought to provide antioxidant and protective light-filtering functions.
- Plums are rich in minerals like potassium, fluoride and iron. Iron is required for red blood cell formation. Potassium in an important component of cell and body fluids that helps controlling heart rate and blood pressure.
- Rich in B-complex group of vitamins such as niacin, vitamin B_6 and pantothenic acid. These vitamins are acting as cofactors help body metabolize carbohydrates, proteins and fats. They also provide about 5 per cent RDA levels of vitamin K. Vitamin K is essential for many clotting factors function in the blood as well as in bone metabolism and helps reduce Alzheimer's disease in the elderly.

3. Phalsa

Blackcurrant, *Ribes nigrum*, (Phalsa or Falsa) is a species of *Ribes* berry native to central and northern Europe and northern Asia, and is a perennial.

It is a small shrub, growing to 1–2 m tall. The leaves are alternate, simple, 3–5 cm long and broad, and palmately lobed with five lobes, with a serrated margin. The flowers are 4–6 mm diameter, with five reddish-green to brownish petals; they are produced in racemes 5–10 cm long.

When not in fruit, the plant looks similar to the redcurrant shrub, distinguished by a strong fragrance from leaves and stems. The fruit is an edible berry 1 cm diameter, very dark purple in color, almost black, with a glossy skin and a persistent calyx at the apex, and containing several seeds dense in nutrients. An established bush can produce up to 5 kilograms of berries during summer.

Nutrients and Phytochemicals

The fruit has extraordinarily high vitamin C content (302 per cent of the Daily Value per 100 g, table), good levels of potassium, phosphorus, iron and vitamin B_5, and a broad range of other essential nutrients.

Other phytochemicals in the fruit (polyphenols/anthocyanins) have been demonstrated in laboratory experiments with potential to inhibit inflammation mechanisms suspected to be at the origin of heart disease, cancer, microbial infections or neurological disorders like Alzheimer's disease. Major anthocyanins in blackcurrant pomace are delphinidin-3-O-glucoside, delphinidin-3-O-rutinoside, cyanidin-3-O-glucoside, and cyanidin-3-O-rutinoside which are retained in the juice concentrate among other yet unidentified polyphenols.

Blackcurrant seed oil is also rich in many nutrients, especially vitamin E and several unsaturated fatty acids including alpha-linolenic acid and gamma-linolenic acid. In a human pilot study, ingestion of blackcurrant seed oil by mothers reduced atopic dermatitis in their breast-fed newborns that were supplemented with the oil over two years.

Culinary Uses

In the UK, blackcurrant cordial is often mixed with cider to make a drink called "Cider and Black". The addition of lager results in "Diesel" or "Snakebite and Black" available at pubs. Adding a small amount of blackcurrant juice to Guinness is preferred by some to heighten the taste of the popular stout. Macerated blackcurrants are also the primary ingredient in the apéritif *crème de cassis*. Japan imports $3.6 million of New Zealand blackcurrants for uses as dietary supplements, snacks, functional food products and as quick-frozen (IQF) produce for culinary production as jams, jellies or preserves. In Russia, blackcurrant leaves may be used for flavouring tea or preserves. Sweetened vodka may also be infused with blackcurrant leaves or berries, making a deep yellowish-green beverage with a sharp flavour and astringent taste.

Besides being juiced and used in jellies, syrups, and cordials, blackcurrants are also used in cooking because their astringency creates flavour in many sauces, meat dishes, and desserts.

It was once thought that currants needed to be "topped and tailed" (the flower remnants and the stalks removed) before cooking. This is not the case, though, as these parts are easily assimilated during the cooking process. If one prefers, the whole blackcurrant stem and fruit can be frozen, then shaken vigorously. The tops and tails will break off, and the fruit can then be easily separated. Ribena, a non-carbonated soft drink flavored with blackcurrants, takes its name from *Ribes*.

Confectionery

Blackcurrant berries have a distinctive sweet and sharp taste popular in jam, juice, ice cream, and liqueur (see Ribena). They are a common ingredient of *Rødgrød*, a popular kissel-like dessert in North German and Danish cuisines. In the UK, Europe and Commonwealth countries, some types of confectionery include a blackcurrant flavour, and in Belgium and the Netherlands, *cassis* is a flavoured currant soft drink. In the United States, blackcurrant flavour is rather rare in candies and jellies compared to UK sweets. In the United States, grape flavour is often used in brands of candy where blackcurrant would appear in Europe. Blackcurrant liqueur mixed with white wine is called Kir or Kir Royale when mixed with Champagne.

Phalsa/Currants, European Black, raw

Nutritional value per 100 g (3.5 oz)	
Energy	264 kJ (63 kcal)
Carbohydrates	15.4 g
Fat	0.4 g
Protein	1.4 g
Thiamine (vit. B_1)	0.05 mg
Riboflavin (vit. B_2)	0.05 mg
Niacin (vit. B_3)	0.3 mg
Pantothenic acid (B_5)	0.398 mg
Vitamin B_6	0.066 mg
Vitamin C	181 mg
Calcium	55 mg
Iron	1.5 mg
Magnesium	24 mg
Phosphorus	59 mg
Potassium	322 mg
Zinc	0.27 mg

Figure 20.1: Flow sheet for preparation of karonda, phalsa and plum juice

Fully Ripe Fruits
↓
Washing
↓
Addition of water (1:0.5)
↓
Crushing
↓
Heating upto 80 °C with continuous crushing
↓
Straining to separate seeds, skin etc.
↓
Juice
↓
Pasteurization
↓
Filling cans/bottles
↓
Sealing
↓
Storage
↓
Marketing

Figure 20.2: Flow sheet for preparation of Karonda pickle

Mature pink Karondas

↓

Washing

↓

Cutting into two pieces

↓

Removal of seed

↓

Mixing with salt

↓

Filling in jar

↓

Keeping in sun for 4 days

↓

Frying spices in oil

↓

Mixed fried spices with Karonda pieces

↓

Labeling

↓

Storage

↓

Marketing

Sl.No.	*Item*	*Quantity*	*Sl.No.*	*Item*	*Quantity*
1.	Karonda	1 kg	6.	Cumin	10 g
2.	Salt	200 g	7.	Cardamom (large)	10 g
3.	Red chilli powder	15 g	8.	Aniseed powder	10 g
4.	Turmeric powder	10 g	9.	Mustard oil	300 ml
5.	Fenugreek	10 g			

Figure 20.3: Flow sheet for preparation of karonda wine

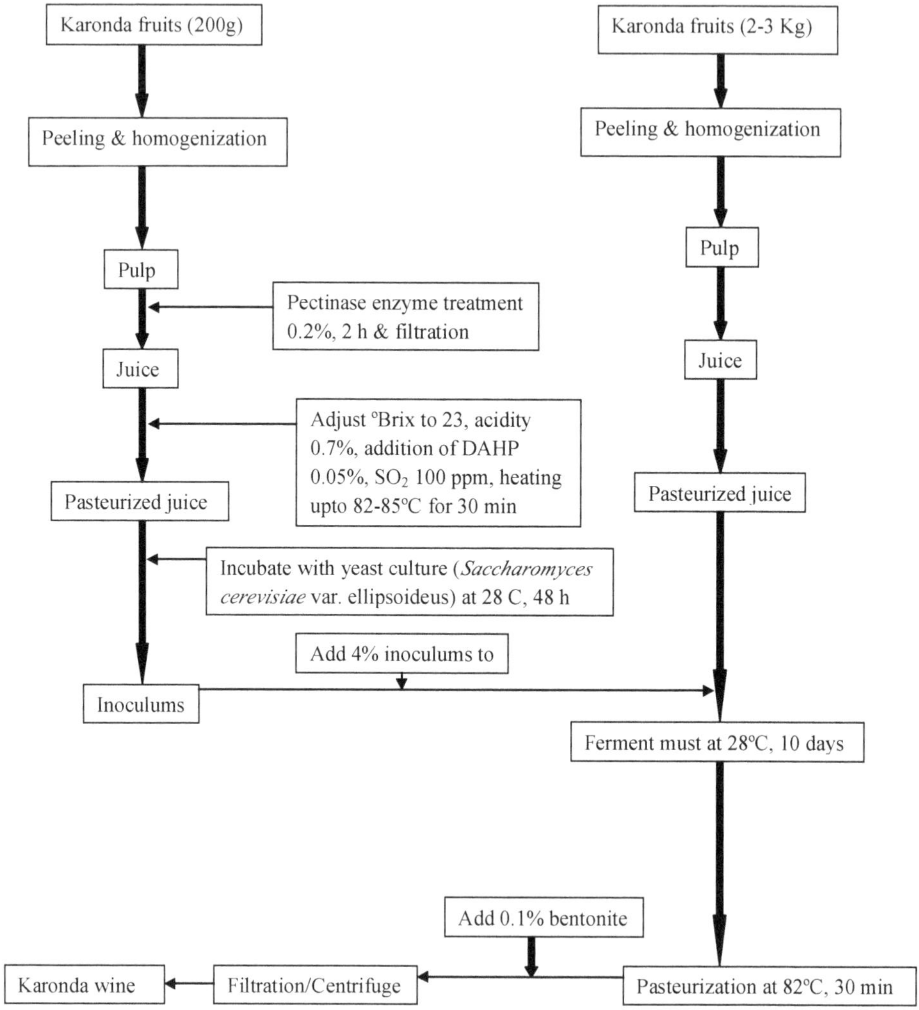

21

Mango

The mango is a fleshy stone fruit belonging to the genus *Mangifera*, consisting of numerous tropical fruiting trees in the flowering plant family Anacardiaceae. The mango is native to India from where it spread all over the world. It is also the most cultivated fruit of the tropical world. While other *Mangifera* species (*e.g.* horse mango, *M. foetida*) are also grown on a more localized basis, *Mangifera indica* – the common mango or Indian mango – is the only mango tree commonly cultivated in many tropical and subtropical regions, and its fruit is distributed essentially worldwide. In several cultures, its fruit and leaves are ritually used as floral decorations at weddings, public celebrations and religious ceremonies.

Mango trees (*Mangifera indica* L.) grow up to 35–40 m (115–130 ft) tall, with a crown radius of 10 m (33 ft). The mango tree is long-lived, as some specimens still fruit after 300 years in deep soil, the taproot descends to a depth of 6 m (20 ft) with profuse, wide-spreading feeder roots; the tree also sends down many anchor roots, which penetrate several feet of soil. The leaves are evergreen, alternate, simple, 15–35 cm (5.9–14 in) long and 6–16 cm (2.4–6.3 in) broad; when the leaves are young they are orange-pink, rapidly changing to a dark glossy red, then dark green as they mature. The flowers are produced in terminal panicles 10–40 cm (3.9–16 in) long; each flower is small and white with five petals 5–10 mm (0.20–0.39 in) long, with a mild sweet odor suggestive of lily of the valley. The fruit takes three to six months to ripen.

The ripe fruit varies in size and color. Cultivars are variously yellow, orange, red or green, and carry a single flat, oblong pit that can be fibrous or hairy on the surface, and which does not separate easily from the pulp. Ripe, unpeeled fruit gives off a distinctive resinous, sweet smell. Inside the pit 1–2 mm (0.039–0.079 in) thick is a thin lining covering a single seed, 4–7 mm (0.16–0.28 in) long. The seed contains the plant embryo.

Food

The mango is generally sweet, although the taste and texture of the flesh varies across cultivars, some having a soft, pulpy texture similar to an overripe plum, while the flesh of others is firmer, like a cantaloupe or avocado, or may have a fibrous texture. For consumption of unripe, pickled or cooked fruit, the mango skin may be consumed comfortably, but has potential to cause contact dermatitis of the lips, gingiva or tongue in susceptible people (see above). Under-ripe mangos can be ripened by placing them in brown paper bags. They will then keep in a plastic bag in the refrigerator for about four or five days. In ripe fruits which are commonly eaten fresh, the skin may thicker and bitter taste, so is typically not eaten.

Cuisine

Mangoes are widely used in cuisine. Sour, unripe mangoes are used in chutneys, *athanu*, pickles, or side dishes, or may be eaten raw with salt, chili, or soy sauce. A cooling summer drink called *panna* or *panha* comes from mangoes. Ripe mangoes are typically eaten fresh; however, they can have many other culinary uses. Mango Lassi, a popular drink made throughout South Asia is created by mixing ripe mangoes or mango pulp with buttermilk and sugar. Ripe mangoes are also used to make curries. *Aamras* is a popular pulp/thick juice made of mangoes with sugar or milk, and is consumed with bread, rice or pooris. The pulp from ripe mangoes is also used to make jam called 'mangada'.

Mangoes are used in preserves like *moramba, amchur* (dried and powdered unripe mango) and pickles, including a spicy mustard-oil pickle. Ripe mangoes are often cut into thin layers, desiccated, folded, and then cut. These bars are similar to dried guava fruit bars available in some countries. The fruit is also added to cereal products like muesli and oat granola. Unripe mango may be eaten with *bagoong* (especially in the Philippines), fish sauce or with dash of salt. Dried strips of sweet, ripe mango (sometimes combined with seedless tamarind to form *mangorind*) are also popular. Mangoes may be used to make juices, mango nectar, and as a flavoring and major ingredient in ice cream and *sorbetes*.

Mango is used to make juices, smoothies, ice cream, fruit bars, *raspados, aguas frescas*, pies and sweet chilli sauce, or mixed with *chamoy*, a sweet and spicy chilli paste. It is popular on a stick dipped in hot chilli powder and salt or also as a main ingredient in fresh fruit combinations. In Central America, mango is either eaten green mixed with salt, vinegar, black pepper and hot sauce, or ripe in various forms. Toasted and ground pumpkin seed (called *pepita*) with lime and salt are the norm when eating green mangoes. Some people also add soy sauce or chilli sauce. Pieces of mango can be mashed and used as a topping on ice cream or blended with milk and ice as milkshakes. Sweet glutinous rice is flavored with coconut, and then served with sliced mango as a dessert. In other parts of Southeast Asia, mangoes are pickled with fish sauce and rice vinegar. Green mangoes can be used in mango salad with fish sauce and dried shrimp. Mango with condensed milk may be used as a topping for shaved ice.

Nutrients and Phytochemicals

Nutritional value per 100 g (3.5 oz) is 272 kJ (65 kcal) and that of the apple mango is slightly higher (79 kcal per 100g). Mango is rich in a variety of phytochemicals and nutrients. The fruit pulp is high in prebiotic dietary fiber, vitamin C, diverse polyphenols and provitamin A carotenoids.

Mango contains essential vitamins and dietary minerals. The antioxidant vitamins A, C and E compose 25 per cent, 76 per cent and 9 per cent of the Dietary Reference Intake (DRI) in a 165-gram (5.8-oz) serving. Vitamin B_6 (pyridoxine, 11 per cent DRI), vitamin K (9 per cent DRI), other B vitamins and essential nutrients, such as potassium, copper and 17 amino acids are at good levels. Mango peel and pulp contain other phytonutrients, such as the pigment antioxidants – carotenoids and polyphenols – and omega-3 and -6 polyunsaturated fatty acids.

Mango peel contains pigments that may have antioxidant properties, including carotenoids, such as the provitamin A compound, β-carotene, lutein and α-carotene, polyphenols such as quercetin, kaempferol, gallic acid, caffeic acid, catechins, tannins, and the unique mango xanthonoid, mangiferin, any of which may counteract free radicals in various disease processes as revealed in preliminary research. Phytochemical and nutrient content appears to vary across mango species. Up to 25 different carotenoids have been isolated from mango pulp, the densest of which was beta-carotene, which accounts for the yellow-orange pigmentation of most mango species. Peel and leaves also have significant polyphenol content, including xanthonoids, mangiferin and gallic acid.

The mango triterpene, lupeol, is an effective inhibitor in laboratory models of prostate and skin cancers. An extract of mango branch bark called Vimang, isolated by Cuban scientists, contains numerous polyphenols with antioxidant properties in vitro and on blood parameters of elderly humans. The pigment euxanthin, known as Indian yellow, is often thought to be produced from the urine of cattle fed mango leaves; the practice is described as having been outlawed in 1908 due to malnutrition of the cows and possible urushiol poisoning. This supposed origin of euxanthin appears to rely on a single, anecdotal source, and Indian legal records do not outlaw such a practice.

Cultural Significance

The mango is the national fruit of India, Pakistan, and the Philippines. The mango tree is the national tree of Bangladesh. In Hinduism, the perfectly ripe mango is often held by Lord Ganesha as a symbol of attainment, regarding the devotee potential perfection. Mango blossoms are also used in the worship of the goddess Saraswati.

Mango leaves are used to decorate archways and doors in Indian houses and during weddings and celebrations like Ganesh Chaturthi. Mango motifs and paisleys are widely used in different Indian embroidery styles, and are found in Kashmiri shawls, Kanchipuram silk sarees, etc. Paisleys are also common to Iranian art, because of its pre-Islamic Zoroastrian past. In Tamil Nadu, the mango is considered, along with banana and jackfruit, as one of the three royal fruits (Mukkani). There are many anecdotes concerning his love for mangoes. In Australia, where mangoes are considered to be a symbol of summer, the first tray of mangoes of the season is traditionally sold at an auction for charity.

Mango, Raw

Nutritional value per 100 g (3.5 oz)	
Energy	272 kJ (65 kcal)
Carbohydrates	17.00 g
Sugars	14.8 g
Dietary fiber	1.8 g
Fat	0.27 g
Protein	0.51 g
Vitamin A equiv.	38 μg
β-carotene	445 μg
Thiamine (vit. B_1)	0.058 mg
Riboflavin (vit. B_2)	0.057 mg
Niacin (vit. B_3)	0.584 mg
Pantothenic acid (B_5)	0.160 mg
Vitamin B_6	0.134 mg
Folate (vit. B_9)	14 μg
Vitamin C	27.7 mg
Calcium	10 mg
Iron	0.13 mg
Magnesium	9 mg
Phosphorus	11 mg
Potassium	156 mg
Zinc	0.04 mg

Figure 21.1: Flow sheet for preparation of mango toffee

Mango fruits fully ripe

↓

Peeling

↓

Homogenization of pulp

↓

Addition of sugar, fat, milk powder
(750g, 100g, 100g/Kg Pulp)

↓

Heating upto 80 °Brix

↓

Addition of salt and citric acid (5g, 2g)

↓

Heating upto 82 °Brix

↓

Spreading in aluminium trays (1 cm thick; already smeared with fat)

↓

Cooling

↓

Cutting into attractive pieces (1.5 × 2.5 cm)

↓

Rapping into metallic paper

↓

Storage

Figure 21.2: Flow sheet for preparation of mango pulp

Fully Ripe Mangos

↓

Washing and Cleaning

↓

Steaming

↓

Pulping

↓

Heating

↓

Preservative addition

↓

Mixing

↓

Can filling and sealing

↓

Pasteurization

↓

Cooling and Labeling

↓

Storage

↓

Marketing

Figure 21.3: Flow sheet for preparation of mango juice

Fully Ripe Mango
↓
Washing
↓
Peeling
↓
Stone removal
↓
Straining of pulp
↓
Addition of water (1 Lit. pulp : 0.5 Lit. water)
↓
Mixing with sugar syrup
↓
Homogenization
↓
Heating at 85 °C, 15 min
↓
Filling hot into cans (A 2 ½ cans)
↓
Sealing of cans
↓
Processing at 100 °C, 20 min
↓
Cooling at room temperature
↓
Storage
↓
Marketing

Syrup is prepared by dissolving 200 g of sugar and 1 g of citric acid in 800 ml of water for 1 kg pulp. Commercially the juice is adjusted to 15 per cent TSS and 0.3 per cent acidity. Prepared juice can also be bottled and crown corked before pasteurization.

Figure 21.4: Flow sheet for preparation of mango pickle with salt

Fully Mature Mangos
↓
Washing
↓
Peeling
↓
Slicing
↓
Putting slices in jar
↓
Sprinkling salt
↓
Putting in sun for one week
↓
Mixing spices
↓
Storage at ambient temperature

Sl.No.	*Item*	*Quantity*	*Sl.No.*	*Item*	*Quantity*
1.	Mango sliced	1 kg	6.	Cumin	10 g
2.	Salt	200 g	7.	Cardamom	10 g
3.	Red chilli powder	10 g	8.	Black pepper powder	10 g
4.	Cinnamon	10 g	9.	Clove headless	6 No
5.	Asafoetida	5 g	10.	Fenugreek	10 g

Figure 21.5: Flow sheet for preparation of mango pickle with oil

Mature Mangoes

↓

Washing

↓

Cutting lengthwise into four/eight pieces

↓

Removal of kernel

↓

Dipping pieces in 2 per cent salt solution (to prevent browning)

↓

Draining off water

↓

Drying in shade for few hrs

↓

Heating oil

↓

Cooling of oil

↓

Mixing spices in a little oil

↓

Mixing with mango pieces

↓

Filling in jar

↓

Keeping in sun for a week

↓

Add remaining oil

↓

Seal the jar

↓

Labeling

↓

Marketing

Sl.No.	*Item*	*Quantity*	*Sl.No.*	*Item*	*Quantity*
1.	Mango pieces	1 kg	8.	Cumin	15 g
2.	Salt	150 g	9.	Cardamom	15 g
3.	Red chilli powder	10 g	10.	Black pepper powder	15 g
4.	Fenugreek powder	25 g	11.	Clove headless	8 No
5.	Asafoetida	2 g	12.	Aniseeds powder	15 g
6.	Turmeric powder	15 g	13.	Mustard oil	350 ml
7.	Nigella seeds	15 g			

Figure 21.6: Flow sheet for preparation of sweet mango chutney

Mature mangoes
↓
Washing
↓
Peeling
↓
Grating or slicing
↓
Cooking with a little water to make highly soft
↓
Mixing with sugar and salt and leaving for an hour
↓
Keeping all ingredients (except vinegar) in cloth bag, tied loosely, putting in mixture and cooking on low flame
↓
During cooking spices bag pressed occasionally
↓
Cooking to consistency of jam (upto 105 °C) with occasionally stirring
↓
Remove of spice bag after squeezing
↓
Addition of vinegar
↓
Cooking for 2-5 min
↓
Filling hot into hot and dry sterilized bottles
↓
Sealing and labeling
↓
Storage and Marketing

Sl.No.	*Item*	*Quantity*	*Sl.No.*	*Item*	*Quantity*
1.	Mango slices/shreds	1 kg	8.	Cardamom (large)	10 g
2.	Salt	45 g	9.	Cinnamon	10 g
3.	Sugar or gul	1 Kg	10.	Aniseed (powder)	10 g
4.	Onion (chopped)	50 g	11.	Ginger (chopped)	15 g
5.	Red chilli	10 g	12.	Clove (headless)	5 No.
6.	Cumin	10 g	13.	Black pepper	10 g
7.	Garlic (chopped)	15 g	14.	Vinegar	170 ml

Figure 21.7: Flow sheet for preparation of mango slices (Amchur)

Mature, green mangoes

↓

Washing and Peeling

↓

Cutting into slices of suitable size

↓

Dipping in 2 per cent salt solution for an hour

↓

Draining out and dipping in 2000 ppm sulphur dioxide solution for 2 hrs

↓

Spreading slices in thin layer on wooden trays or cloth

↓

Sun drying (fresh slice and dried slice ratio 6:1)

↓

Filling into airtight containers

↓

Storage in dry place

↓

Labeling and Marketing

Figure 21.8: Flow sheet for preparation of mango leather

Ripe mango fruits

↓

Washing and Peeling

↓

Extraction of juice

↓

Straining of juice

↓

Adding 0.6 g KMS for every kg of juice

↓

Smearing aluminium or stainless steel trays with mustard oil

↓

Spreading juice in thin layers on trays

↓

Drying at low temperature

↓

After first layer spreading second layer on it and repeating the process until thickness of layer is 1-1.25 cm

↓

Cutting dried sheets into pieces of suitable size

↓

Wrapping in butter or waxed paper

↓

Filling in dry glass jar

↓

Sealing and Labeling

↓

Storage and Marketing

22

Papaya

The papaya (from Carib via Spanish), papaw, or pawpaw is the fruit of the plant *Carica papaya*, the sole species in the genus *Carica* of the plant family Caricaceae. It is native to the tropics of the Americas, and was first cultivated in Mexico several centuries before the emergence of the Mesoamerican classic cultures.

The papaya is a large tree-like plant, with a single stem growing from 5 to 10 meters (16 to 33 ft) tall, with spirally arranged leaves confined to the top of the trunk. The lower trunk is conspicuously scarred where leaves and fruit were borne. The leaves are large, 50–70 centimeters (20–28 in) diameter, deeply palmately lobed with 7 lobes. The tree is usually unbranched, unless lopped. The flowers are similar in shape to the flowers of the *Plumeria*, but are much smaller and wax-like. They appear on the axils of the leaves, maturing into the large 15–45 centimeters (5.9–18 in) long, 10–30 centimeters (3.9–12 in) diameter fruit. The fruit is ripe when it feels soft (like a ripe avocado or a bit softer) and its skin has attained amber to orange hue.

Nutrition

Papaya fruit is a rich source of antioxidants nutrients such as carotenes, vitamin C and flavonoids; the B vitamins, folate and pantothenic acid; and the minerals, potassium and magnesium; and fiber. Together, these nutrients promote the health of the cardiovascular system and also provide protection against colon cancer. Papayas may be very helpful for the prevention of atherosclerosis and diabetic heart disease.

Gastronomy

The ripe fruit of the papaya is usually eaten raw, without skin or seeds. The unripe green fruit can be eaten cooked, usually in curries, salads, and stews. Green papaya is used in Southeast Asian cooking, both raw and cooked. In Thailand, papaya is used to make some tam when still not fully ripe. Papayas have a relatively high amount of pectin, which can be used to make jellies. The smell of ripe, fresh papaya flesh can strike some people as unpleasant.

The black seeds of the papaya are edible and have a sharp, spicy taste. They are sometimes ground and used as a substitute for black pepper. In some parts of Asia, the young leaves of papaya are steamed and eaten like spinach. In some parts of the world, papaya leaves are made into tea as a preventative for malaria, although there is no scientific evidence for the effectiveness of this treatment.

Meat Tenderizing

Green papaya fruit and the tree's latex are both rich in an enzyme called papain, a protease which is useful in tenderizing meat and other proteins. Its ability to break down tough meat fibers was used for thousands of years by indigenous Americans. It is now included as a component in powdered meat tenderizers.

Reported Health Benefits of Papaya

- Papaya contains the digestive enzyme papain and therefore valuable for aiding digestion.
- The unique protein-digesting enzymes; papain and chymopapain have been shown to help lower inflammation and to improve healing from burns in addition to helping in digestion of proteins. The antioxidant nutrients found in papaya, including vitamin C, vitamin E, and β-carotene, are also very good at reducing inflammation.
- The ripe fruit is easily digestible and prevents constipation.
- Case studies indicate that this food taken alone for two or three days has a highly beneficial tonic effect upon the stomach and intestines.
- The juice of the papaya aids in relieving infections of the colon and has a tendency to break down pus and mucus reached by the juice.
- May help prevent cancer in organs and glands with epithelial tissue (ripe papaya). Papaya's fiber is able to bind to cancer-causing toxins in the colon and keep them away from the healthy colon cells. In addition, papaya's folate, vitamin C, β-carotene, and vitamin E have each been associated with a reduced risk of colon cancer.
- Prevents nausea (includes morning sickness and motion sickness)
- The seeds are antihelmintic, for expelling worms and they are given with honey. Chew and swallow two teaspoonfuls of seeds after each principal meal (three times a day).
- Papayas may be very helpful for the prevention of atherosclerosis and diabetic heart disease. Papayas are an excellent source of vitamin C as well as a good source of vitamin E and vitamin A (through their concentration of pro-vitamin A carotenoid phytonutrients), three very powerful antioxidants.
- Papayas are also a good source of fiber, which has been shown to lower high cholesterol levels.

Allergies and Side Effects

Papaya is frequently used as a hair conditioner, but should be used in small amounts. Papaya releases a latex fluid when not quite ripe, which can cause irritation and provoke allergic reaction in some people. The papaya fruit, seeds, latex, and leaves also contains carpaine, an anthelmintic alkaloid (a drug that removes parasitic worms from the body), which can be dangerous in high doses.

It is speculated that the latex concentration of unripe papayas may cause uterine contractions, which may lead to a miscarriage. Papaya seed extracts in large doses have a contraceptive effect on rats and monkeys, but in small doses have no effect on the unborn animals. Excessive consumption of papaya can cause carotenemia, the yellowing of soles and palms, which is otherwise harmless. However, a very large dose would need to be consumed; papaya contains about 6 per cent of the level of β-carotene found in carrots (the most common cause of carotenemia).

Medicinal Potential

- ☆ The juice has an *in vitro* antiproliferative effect on liver cancer cells, probably due to its component of lycopene or immune system stimulation.
- ☆ Papaya seed could be used as an antibacterial agent for *Escherichia coli*, *Staphylococcus aureus* or *Salmonella typhi*, although further research is needed before advocating large-scale therapy·
- ☆ Papaya seed extract may be nephroprotective (protect the kidneys) in toxicity-induced kidney failure.

Papaya, Raw

Nutritional value per 100 g (3.5 oz)	
Energy	163 kJ (39 kcal)
Carbohydrates	9.81 g
Sugars	5.90 g
Dietary fiber	1.8 g
Fat	0.14 g
Protein	0.61 g
Vitamin A equiv.	328 μg
Thiamine (vit. B_1)	0.04 mg
Riboflavin (vit. B_2)	0.05 mg
Niacin (vit. B_3)	0.338 mg
Vitamin B_6	0.1 mg
Folate (vit. B_9)	38 μg
Vitamin C	61.8 mg
Calcium	24 mg
Iron	0.10 mg
Magnesium	10 mg
Phosphorus	5 mg
Potassium	257 mg
Sodium	3 mg
Drying	
Pectin powder, 1.0 per cent (200 jelly grade)	

Figure 22.1: Flow sheet of integrated process for production of papain and pectin from papaya

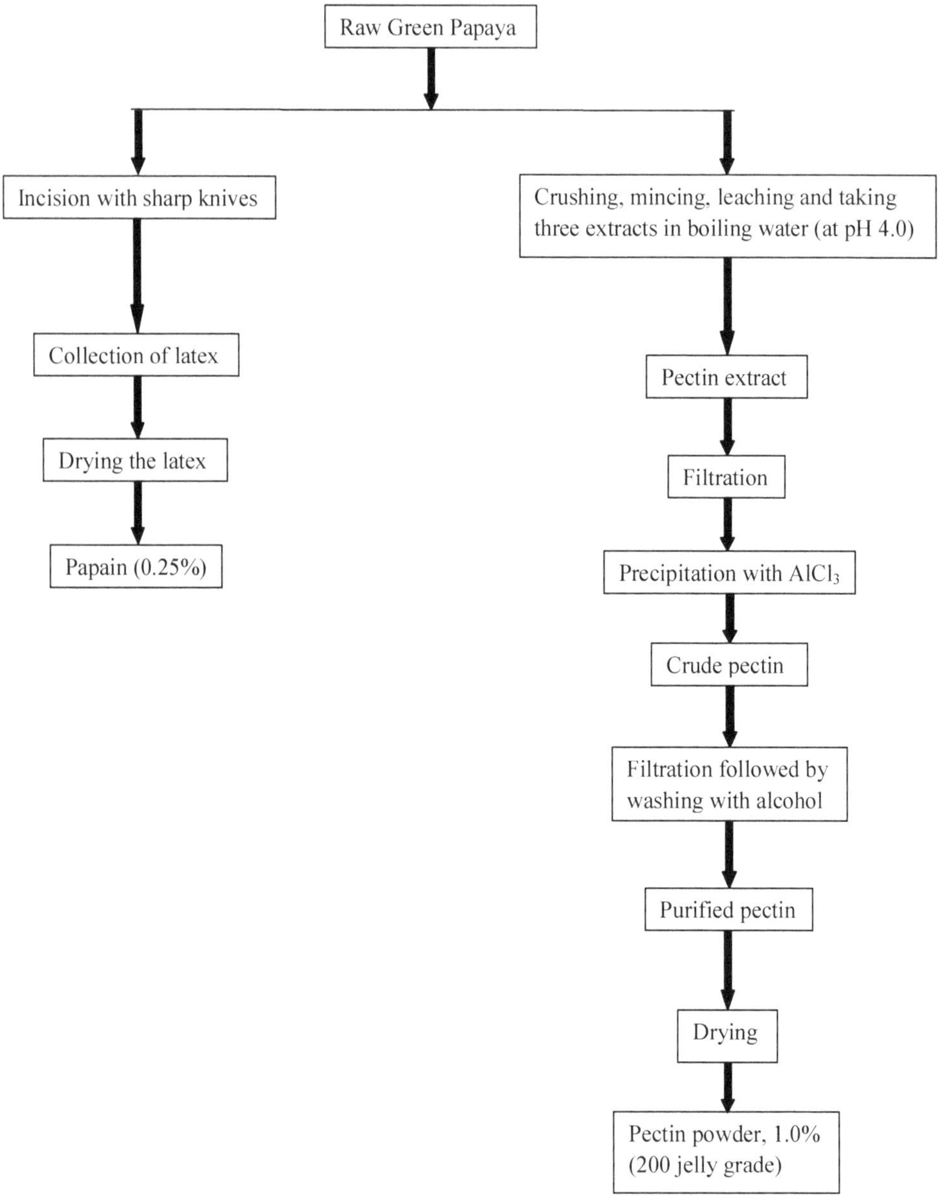

Figure 22.2: Flow sheet for preparation of papaya jam

Fully ripe papaya fruits

↓

Cleaning/Washing

↓

Peeling and removal of seeds

↓

Homogenization of pulp

↓

Mixing of ingredients (pulp 1kg, sugar 750g, citric acid 9g)

↓

Heating upto 68.5 °Brix

↓

Filling into sterilized bottles

↓

Cooling

↓

Capping and sealing

↓

Storage

Figure 22.3: Flow sheet for preparation of papaya toffee

Fully ripe papaya fruits

↓

Cleaning/Washing

↓

Peeling, removal of seeds

↓

Homogenization of pulp

↓

Mixing ingredients; sugar 750g, fat 40g, skim milk powder 50g)

↓

Heating upto 80 °Brix

↓

Addition; salt 5g, citric acid 2g

↓

Heating upto 82 °Brix

↓

Spread in aluminium trays already smeared with fat/oil

↓

Cooling

↓

Cutting (2.5 × 1.5 cm size pieces)

↓

Packaging

↓

Storage

Figure 22.4: Flow sheet for preparation of papaya juice

Ripe papaya fruits

↓

Washing

↓

Peeling

↓

Cutting into slices

↓

Removal of seeds

↓

Homogenization

↓

Filtration (60 mesh)

↓

Pulp

↓

Pectinase enzyme treatment 0.5 per cent, 2 hrs

↓

Filtration/Centrifugation

↓

Pure Juice

Figure 22.5: Flow sheet for preparation of papaya pulp

Fully Ripe Papaya Fruits
↓
Washing
↓
Peeling
↓
Cutting into halves/pieces
↓
Removal of seeds
↓
Passing through pulper
↓
Pulp
↓
Addition of preservative
↓
Filling cans/bottles/pouches
↓
Sealing
↓
Storage
↓
Marketing

Figure 22.6: Flow sheet for preparation of papaya butter

Ripe papaya

↓

Washing

↓

Cutting into thin pieces

↓

Boiling with equal quantity of water (to soften pulp)

↓

Sieving (to remove seeds and skin) and making into fine pulp

↓

Mixing pulp with sugar

↓

Cooking to consistency of jam

↓

Addition spices (fine powdered) and mix well

↓

Remove from fire

↓

Filling into clean sterilized bottles

↓

Sealing

↓

Storage

Figure 22.7: Flow sheet for preparation of papaya petha

Half ripe papaya
↓
Washing
↓
Cutting into thin slice (1" squares)
↓
Pick each piece with fork
↓
In a heavy bottomed pan dissolve sugar in 2 cups of water and boil it
↓
Skim off impurities floating on the surface using a slotted spoon
↓
Simmer till the syrup is of 1 stirring consistency
↓
Remove from the fire
↓
Cool completely in the sugar syrup and drain it
↓
Allow it to harden a little overnight in a cool place
↓
Packing in dry Jar
↓
Sealing
↓
Storage

Figure 22.8: Flow sheet for preparation of papaya nectar

Fully ripe papaya
↓
Peeling, coring and trimming
↓
Cutting into pieces
↓
Pulped in a blender
↓
Pulp
↓
Mixing of pulp (340 g), sugar (140 g), citric acid (3 g), salt (1 g) and water (640 ml)
↓
Heating to boil (82 °C)
↓
Hot filling in glass bottles
↓
Capping and sealing
↓
Sterilization at 90 °C for 15 min
↓
Cooling
↓
Storage

This type of fruit beverage contains at least 20 per cent fruit juice/pulp and 15 per cent total soluble solids and also about 0.3 per cent acidity. It is not diluted before serving. Commercially, nectar can be prepared with 13 per cent TSS and 0.3 per cent acidity.

In homes, nectar can be prepared by employing the following recipes:

Fruit	*Juice (liter)*	*Sugar (Kg)*	*Citric acid (g)*	*Water (liter)*
Mango	1.0	0.65	13	3.3
Papaya	1.0	0.65	13	3.3
Guava	1.0	0.60	13	3.3
Bael	1.0	0.65	15	3.3
Jamun	1.0	0.70	7	3.3
Aonla blend (20 parts aonla + 2 parts lime juice + 1 part ginger)	2.3	1.35	4	7.8

Figure 22.9: Flow sheet for preparation of papaya-carrot cheese

Fresh papaya and carrots

↓

Washing and peeling

↓

Cutting into pieces, boiling separately with equal quantity of water

↓

Mashing and mixing

↓

Adding sugar and butter (10 per cent)

↓

Addition of preservative (Sodium benzoate) and citric acid (5 per cent)

↓

Cooking till mixture become sufficiently thick

↓

Addition of salt (1 per cent)

↓

Removing from fire when mixture starts leaving side of pan

↓

Evenly distribution over butter coated tray and left for 3 h to set

↓

Cutting into pieces with sharp knife

↓

Pre-packing with butter paper

↓

Packing in polythene paper

↓

Store in cool place

Figure 22.10: Flow sheet for preparation of papaya pickle

Fully Mature Papaya
↓
Washing
↓
Peeling
↓
Removal of seeds
↓
Cutting into pieces
↓
Boiling to soften
↓
Removal of pieces from boiling water
↓
Mixing with salt
↓
Drying in shade for few minutes
↓
Mixing spices
↓
Filling in jars
↓
Addition of vinegar (sufficient to cover the pieces)
↓
Keeping for a week
↓
Storage
↓
Marketing

Sl.No.	*Item*	*Quantity*
1.	Peeled papaya pieces	1 kg
2.	Salt	100 g
3.	Red chilli powder	10 g
4.	Cardamom	10 g
5.	Cumin	10 g
6.	Black pepper powder	10 g
7.	Vinegar	750 ml

Figure 22.11: Flow sheet for preparation of papaya chutney

Papaya (Fully Mature)

↓

Washing

↓

Peeling

↓

Cutting into halves

↓

Seed removal

↓

Grating

↓

Rest process similar to that for sweet mango chutney

Sl.No.	*Item*	*Quantity*	*Sl.No.*	*Item*	*Quantity*
1.	Papaya halves	1 Kg	8.	Cinnamon	10 g
2.	Salt	45 g	9.	Cardamom	10 g
3.	Sugar	750 g	10.	Aniseed	10 g
4.	Onion (chopped)	100 g	11.	Black pepper (powder)	10 g
5.	Ginger (chopped)	50 g	12.	Clove (headless)	5 No.
6.	Garlic (chopped)	15 g	13.	Cumin	10 g
7.	Red chilli powder	10 g	14.	Vinegar	200 ml

Figure 22.12: Flow sheet for preparation of papaya sauce

Papaya (ripe)

↓

Washing

↓

Peeling

↓

Cutting into halves

↓

Removal of seeds

↓

Making fine pulp

↓

Straining of pulp

↓

Cooking pulp with one third quantity of sugar

↓

Putting spice bag in pulp and pressing occasionally

↓

Cooking to one third its original volume

↓

Removal of spice bag after squeezing

↓

Adding remaining sugar and salt

↓

Cooking to one third of original volume of pulp

↓

Addition of vinegar and preservative

↓

Filling hot into bottles

↓

Crown corking

↓

Pasteurization at 85-90 °C, 30 min

↓

Cooling

↓

Labeling and Storage

↓

Marketing

Sl.No.	Item	Quantity	Sl.No.	Item	Quantity
1.	Papaya pulp	1Kg	6.	Hot spices	10 g
2.	Salt	14 g	7.	Garlic (chopped)	5 g
3.	Sugar	50 g	8.	Red chilli powder	5 g
4.	Onion (chopped)	50 g	9.	Vinegar/acetic acid	40 ml
5.	Ginger (chopped)	10 g	10.	Sodium benzoate	700mg/kg

Figure 22.13: Flow sheet for preparation of RTS from guava and papaya juice

Guava : Papaya fruit juice

↓

Blending (70 : 30 v/v)

↓

Addition of water, sugar and citric acid

↓

Filtration

↓

Checking °Brix 15, acidity 0.25 per cent

↓

Filling into pasteurized bottles

↓

Pasteruization

↓

Cooling and labelling

↓

Storage

↓

Marketing

23

Pear

The pear is any of several tree species of genus *Pyrus* and also the name of the pomaceous fruit of these trees. Several species of pear are valued by humans for their edible fruit, but the fruit of other species is small, hard, and astringent. The genus *Pyrus* is classified in subtribe Pyrinae within tribe Pyreae. The apple (*Malus domestica*), which it resembles in floral structure, is also a member of this subcategory. The English word "pear" is probably from Common West Germanic *pera*, probably a loanword of Vulgar Latin *pira*, the plural of *pirum*, akin to Greek *apios* (from Mycenaean *ápisos*), which is likely of Semitic origin. The place name *Perry* and *Pharisoulopol* can indicate the historical presence of pear trees. The term "pyriform" is sometimes used to describe something which is "pear-shaped".

The cultivation of the pear in cool temperate climates extends to the remotest antiquity, and there is evidence of its use as a food since prehistoric times. Many traces of it have been found in the Swiss lake-dwellings. The word "pear", or its equivalent, occurs in all the Celtic languages, while in Slavonic and other dialects differ in appellations, but still referring to the same thing, are found–a diversity and multiplicity of nomenclature which led Alphonse de Candolle to infer a very ancient cultivation of the tree from the shores of the Caspian to those of the Atlantic.

Pears are native to coastal and mildly temperate regions of the Old World, from Western Europe and North Africa east right across Asia. They are medium-sized trees, reaching 10–17 m tall, often with a tall, narrow crown; a few species are shrubby. The leaves are alternately arranged, simple, 2–12 cm long, glossy green on some species, densely silvery-hairy in some others; leaf shape varies from broad oval to narrow lanceolate. Most pears are deciduous, but one or two species in Southeast Asia are evergreen. Most are cold-hardy, withstanding temperatures between –25 °C and –40 °C in winter, except for the evergreen species, which only tolerate temperatures down to about –15 °C. The flowers are white, rarely tinted yellow or pink, 2–4 cm diameter, and have five petals. Like that of the related apple, the pear fruit is a pome, in most wild species 1–4 cm diameter, but in some cultivated forms up to 18 cm long and 8 cm broad; the shape varies in most species from oblate or globose, to the classic pyriform 'pear-shape' of the European pear with an elongated basal portion and a bulbous end.

The fruit is composed of the receptacle or upper end of the flower-stalk (the so-called calyx tube) greatly dilated. Enclosed within its cellular flesh is the true fruit: five cartilaginous carpels, known colloquially as the "core". From the upper rim of the receptacle are given off the five sepals the five petals, and the very numerous stamens. The pear is very similar to the apple in cultivation, propagation and pollination. The pear and the apple are also related to the quince. Pears and apples cannot always be distinguished by the form of the fruit; some pears look very much like some apples. One major difference is that the flesh of pear fruit contains stone cells (also called "grit"). Pear trees and apple trees do have several visible differences. According to Pear Bureau Northwest, about 3000 known varieties of pears are grown worldwide. In the United States, only 10 heirloom varieties are widely recognized: Green Bartlett, Red Bartlett, Bosc, Green Anjou, Red Anjou, Comice, Forelle, Seckel, Concorde, and Starkrimson.

Harvest

Summer and autumn cultivars of *Pyrus communis*, being climacteric fruits, are gathered before they are fully ripe, while they are still green, but snap off when lifted. In the case of the 'Passe Crassane', long the favored winter pear in France, the crop should is traditionally gathered at three different times: the first a fortnight or more before it is ripe, the second a week or ten days after that, and the third when fully ripe. The first gathering will come into eating last, and thus the season of the fruit may be considerably prolonged. Nashi pears are allowed to ripen on the tree.

Storage

Pears may be stored at room temperature until ripe. Pears are ripe when flesh around stem gives to gentle pressure. Ripe pears are optimally stored refrigerated, uncovered in a single layer, where they have a shelf life of 2 to 3 days.

Uses

Pears are consumed fresh, canned, as juice, and dried. The juice can also be used in jellies and jams, usually in combination with other fruits or berries. Fermented pear juice is called perry or pear cider. Pears ripen at room temperature. They will ripen faster if placed next to bananas in a fruit bowl. Refrigeration will slow further ripening. Pear Bureau Northwest offers tips on ripening and judging ripeness: Although the skin on Bartlett pears changes from green to yellow as they ripen, most varieties show little color change as they ripen. Because pears ripen from the inside out, the best way to judge ripeness is to "Check the Neck." To Check the Neck for ripeness, apply gentle thumb pressure to the neck, or stem end of the pear. If it yields to gentle pressure, then the pear is ripe, sweet, and juicy. If it is firm, leave pear at room temperature and Check the Neck daily for ripeness. Source: Pear Bureau Northwest. The culinary or cooking pear is green but dry and hard and only edible after several hours of cooking. Two Dutch cultivars are "Gieser Wildeman" and "Saint Remy". They are traditionally stewed in wine with spices and served both warm and cold. Pear wood is one of the preferred materials in the manufacture of high-quality woodwind instruments and furniture. It is also used for wood carving, and as a fire wood to produce aromatic smoke for smoking meat or tobacco. Pear wood is valued for kitchen spoons, scoops and stirrers, as it does not contaminate food with color, flavor or smell and resists warping and splintering despite repeated soaking and drying cycles. Lincoln describes it as "a fairly tough, very stable wood... (used for) carving... brush backs, umbrella handles, measuring instruments such as set squares and T-squares... recorders... violin and guitar fingerboards and piano keys... decorative veneering." Pear wood is the favored wood for architect's rulers because it does not warp. It is similar to the wood of its relative, the apple tree, (*Pyrus malus* [also called *Malus domestica*]) and used for many of the same purposes. Pear leaves were smoked in Europe before tobacco was introduced.

Health Benefits

Pears are a good source of dietary fiber and a good source of vitamin C. Most of the vitamin C, as well as the dietary fiber, are contained within the skin of the fruit. Pears are less allergenic than many other fruits, and pear juice is therefore sometimes used as the first juice introduced to infants. However, caution is recommended for all fruit juice consumption by infants, as studies have suggested a link between excessive fruit juice consumption and reduced nutrient intake, as well as a tendency towards obesity. Pears are low in salicylates and benzoates, so are recommended in exclusion diets for allergy sufferers. Along with lamb and rice, pears may form part of the strictest exclusion diet for allergy sufferers. Most of the fiber is insoluble, making pears a good laxative. The gritty fiber content may reduce the number of cancerous colon polyps.

Pear, Raw

Nutritional value per 100 g (3.5 oz)	
Energy	242 kJ (58 kcal)
Carbohydrates	15.46 g
Sugars	9.80 g
Dietary fiber	3.1 g
Fat	0 g
Protein	0.38 g
Thiamine (vit. B_1)	0.012 mg
Riboflavin (vit. B_2)	0.025 mg
Niacin (vit. B_3)	0.157 mg
Pantothenic acid (B_5)	0.048 mg
Vitamin B_6	0.028 mg
Folate (vit. B_9)	7 μg
Vitamin C	4.2 mg
Calcium	9 mg
Iron	0.17 mg
Magnesium	7 mg
Phosphorus	11 mg
Potassium	119 mg
Zinc	0.10 mg

Figure 23.1: Flow sheet for preparation of pear toffee

Pear fruits fully ripe

↓

Cleaning/Peeling

↓

Homogenization of pulp

↓

Addition of sugar, fat, milk powder
(750g, 100g, 100g/Kg Pulp)

↓

Heating upto 80 °Brix

↓

Addition of salt and citric acid (5g, 2g)

↓

Heating upto 82 °Brix

↓

Spreading in aluminium trays
(1 cm thick; already smeared with fat)

↓

Cooling

↓

Cutting into attractive pieces (1.5 × 2.5 cm)

↓

Rapping into metallic paper

↓

Storage

Figure 23.2: Flow sheet for preparation of pear pickle

Ripe Firm Pears
↓
Washing
↓
Peeling
↓
Removal of cone
↓
Cutting lengthwise into four pieces
↓
Placing immediately in salt water (to prevent discoloration)
↓
Cooking with sugar, vinegar and spices 5-10 min
↓
Mixing with pieces
↓
Heating till pieces tender but still firm
↓
Cooling
↓
Filling in jars
↓
Storage at ambient temperature
↓
Marketing

Sl.No.	*Item*	*Quantity*	*Sl.No.*	*Item*	*Quantity*
1.	Pear	1 kg	5.	Cumin	10 g
2.	Cinnamon	10 g	6.	Black pepper powder	10 g
3.	Sugar	250 g	7.	Vinegar	250 ml
4.	Cardamom	10 g	8.	Clove headless	5 No

24

Pineapple

Pineapple (*Ananas comosus*) is the common name for a tropical plant and its edible fruit, which is actually a multiple fruit consisting of coalesced berries. It was given the name *pineapple* due to its resemblance to a pine cone. The pineapple is by far the most economically important plant in the *Bromeliaceae*. Besides being produced for consumption, it can be grown as an ornamental or houseplant, obtained from the crown of a supermarket fruit. Some sources say that the plant will flower after about 24 months and produce a fruit during the following six months, while others indicate a 20-month timetable.

Pineapple can be consumed fresh, canned or juiced and can be used in a variety of ways. It is popularly used in desserts, fruit salads (usually tropical fruit salads, but it can vary), jams, yogurts, ice creams, various candies, as a complement to meat dishes and in fruit cocktail. The popularity of the pineapple is due to its sweet-sour taste. The core of the pineapple is continuous with the stem supporting the fruit and with the crown, a feature unique among cultivated fruits.

Pineapple is used for many other things aside from consumption. In the Philippines, pineapple leaves are used as the source of a textile fiber called piña. This fiber can use in a variety of ways such as a wall paper and a component of furnishings.

The pineapple is a herbaceous perennial which grows to 1.0 to 1.5 meters (3.3 to 4.9 ft) tall, although sometimes it can be taller. In appearance, the plant itself has a short, stocky stem with tough, waxy leaves. When creating its fruit, it usually produces up to 200 flowers, although some large-fruited cultivars can exceed this. Once it flowers, the individual fruits of the flowers join together to create what are commonly referred to as a pineapple. After the first fruit is produced, side shoots (called 'suckers' by commercial growers) are produced in the leaf axils of the main stem. These may be removed for propagation, or left to produce additional fruits on the original plant. Commercially, suckers that appear around the base are cultivated. It has 30 or more long, narrow, fleshy, trough-shaped leaves with sharp spines along the margins that are 30 to 100 centimeters (1.0 to 3.3 ft) long, surrounding a thick stem. In the first year of growth the axis lengthens and thickens, bearing numerous leaves in close spirals. After 12 to 20 months the stem grows into a spike-like inflorescence up to 15 cm long with over 100 spirally arranged, trimerous flowers, each subtended by a bract. Flower colors vary, depending on variety, from lavender, through light purple to red.

The ovaries develop into berries which coalesce into a large, compact, multiple accessory fruit. The fruit of a pineapple is arranged in two interlocking helices, eight in one direction, thirteen in the other, each being a Fibonacci number. Pineapple carries out CAM photosynthesis, fixing carbon dioxide at night and storing it as the acid malate and then releasing it during the day, aiding photosynthesis.

Nutrition

Raw pineapple is an excellent source of manganese (45 per cent DV in a 100 g serving) and vitamin C (80 per cent DV per 100 g). Mainly from its stem, pineapple contains a proteolytic enzyme, bromelain, which breaks down protein. If having sufficient bromelain content, pineapple juice can thus be used as a marinade and tenderizer for meat. Pineapple enzymes can interfere with the preparation of some foods, such as jelly or other gelatin-based desserts, but would be destroyed during cooking and the canning process. The quantity of bromelain in the fruit is probably not significant, being mostly in the inedible stalk. Furthermore, an ingested enzyme like bromelain is unlikely to survive intact the proteolytic processes of digestion.

Cultivars

There are many cultivars. The leaves of the commonly grown 'Smooth Cayenne' are smooth and is the most commonly grown worldwide. Many cultivars have become distributed from its origins in Paraguay and the southern part of Brazil and later improved stocks were introduced into the Americas, the Azores, Africa, India, Malaysia and Australia. Varieties include:

- 'Hilo': A compact 1–1.5 kg (2–3 lb) Hawaiian variant of 'Smooth Cayenne'. The fruit is more cylindrical and produces many suckers but no slips.
- 'Kona Sugarloaf': 2.5–3 kg (5–6 lb), white flesh with no woodiness in the center. Cylindrical in shape, it has high sugar content but no acid. An unusually sweet fruit.
- 'Natal Queen': 1–1.5 kg (2–3 lb), golden yellow flesh, crisp texture and delicate mild flavor. Well adapted to fresh consumption. Keeps well after ripening. Leaves spiny. Grown in Australia, Malaysia, and South Africa.
- 'Pernambuco' ('Eleuthera'): 1–2 kg (2–4 lb) with pale yellow to white flesh. Sweet, melting and excellent for eating fresh. Poorly adapted for shipping. Leaves spiny. Grown in Latin America.
- 'Red Spanish': 1–2 kg (2–4 lb), pale yellow flesh with pleasant aroma; squarish in shape. Well adapted for shipping as fresh fruit to distant markets. Leaves spiny. Grown in Latin America.
- 'Smooth Cayenne': 2.5–3 kg (5–6 lb), pale yellow to yellow flesh. Cylindrical in shape and with high sugar and acid content. Well adapted to canning

and processing. Leaves without spines. This is the variety from Hawaii, and the most easily obtainable in U.S. grocery stores. Both 73–114 and 73-50 are of this cultivar.

Traditional Medicine and Preliminary Research

Both the root and fruit may be eaten or applied topically as an anti-inflammatory or as a proteolytic agent. In some practices, it may be used to induce abortion or menstruation or as an antihelminthic agent. Bromelain purified from pineapple stem or fresh juice, then provided in the diet over 6 months, decreased the severity of colonic inflammation in mice with experimental colitis.

Storage and Transport

Some buyers prefer green fruit, others ripened or off-green. A plant growth regulator Ethephon is typically sprayed onto the fruit one week before harvest, developing ethylene, which turns the fruit golden yellow. After cleaning and slicing they are typically canned in sugar syrup with added preservative.

A pineapple will never become any riper than it was when harvested, though a fully ripe pineapple can bruise and rot quickly. The fruit itself is quite perishable and storage of it should be taken seriously. If it is stored at room temperature, it should be used within two days; however, if it is refrigerated, the time span is extended to five to seven days.

Pineapple, Raw

Nutritional value per 100 g (3.5 oz)	
Energy	202 kJ (48 kcal)
Carbohydrates	12.63 g
Sugars	9.26 g
Dietary fiber	1.4 g
Fat	0.12 g
Protein	0.54 g
Thiamine (vit. B_1)	0.079 mg
Riboflavin (vit. B_2)	0.031 mg
Niacin (vit. B_3)	0.489 mg
Pantothenic acid (B_5)	0.205 mg
Vitamin B_6	0.110 mg
Folate (vit. B_9)	15 μg
Vitamin C	36.2 mg
Calcium	13 mg
Iron	0.28 mg
Magnesium	12 mg
Manganese	0.9 mg
Phosphorus	8 mg
Potassium	115 mg
Zinc	0.10 mg

Figure 24.1: Flow sheet for preparation of pineapple-sapota toffee

Pineapple : Sapota (50 : 50) fruits fully ripe

↓

Cleaning and Peeling

↓

Homogenization of pulp

↓

Addition of sugar, fat, milk powder
(750g, 100g, 100g/Kg Pulp)

↓

Heating upto 80 °Brix

↓

Addition of salt and citric acid (5g, 2g)

↓

Heating upto 82 °Brix

↓

Spreading in aluminium trays
(1 cm thick; already smeared with fat)

↓

Cooling

↓

Cutting into attractive pieces (1.5 × 2.5 cm)

↓

Rapping into metallic paper

↓

Storage

Figure 24.2: Flow sheet for preparation of pineapple powder

Firm and ripped pineapple fruit

↓

Trimming and peeling by knife

↓

Cutting into 3 mm thick slices by knife

↓

Coring by knife

↓

Washing

↓

Osmotic dehydration of slices (60 °Brix sucrose solution for 6 h)

↓

Drying in fluidized bed dryer at 50 °C

↓

Grinding

↓

Sieving by 100 micron sieve

↓

Fine powder

Figure 24.3: Flow sheet for preparation of pineapple juice

Ripe pineapple fruit

↓

Trimming and peeling by knife

↓

Cutting into small pieces

↓

Passing through pulper

↓

Filtration

↓

Juice

25

Pomegranate

A pomegranate (*Punica granatum*) is a fruit-bearing deciduous shrub or small tree growing between five and eight meters tall. The pomegranate is native to Iran and from there it spread to Asian areas such as the Caucasus, the Himalayas in Northern India and in Kerala.

It has been cultivated in the Caucasus since ancient times, and today, is widely cultivated throughout Iran, Azerbaijan, Armenia, Afghanistan, India, Pakistan, Bangladesh, Iraq, Lebanon, Egypt, China, Burma, Saudi Arabia, Israel, Jordan, the drier parts of southeast Asia, the Mediterranean region of Southern Europe, and tropical Africa. Introduced into Latin America and California by Spanish settlers in 1769, pomegranate is now cultivated in parts of California and Arizona for juice production.

In the northern hemisphere, the fruit is typically in season from September to February. In the southern hemisphere, it is in season from March to May. The pomegranate has been mentioned in many ancient texts, notably the Homeric Hymns and the Book of Exodus, and is valued by many cultures for its beauty. It has, in recent years, reached mainstream prominence in the commercial markets of North America and the Western Hemisphere.

Culinary Uses

After opening the pomegranate by scoring it with a knife and breaking it open, the arils (seed casings) are separated from the peel and internal white pulp membranes. Separating the red arils is easier in a bowl of water, because the arils sink and the inedible pulp floats. Freezing the entire fruit also makes it easier to separate. Another very effective way of quickly harvesting the arils is to cut the pomegranate in half, score each half of the exterior rind four to six times, hold the pomegranate half over a bowl and smack the rind with a large spoon. The arils should eject from the pomegranate directly into the bowl, leaving only a dozen or more deeply embedded arils to remove.

The entire seed is consumed raw, though the watery, tasty aril is the desired part. The taste differs depending on the subspecies of pomegranate and its ripeness. The pomegranate juice can be very sweet or sour, but most fruits are moderate in taste, with sour notes from the acidic tannins contained in the aril juice. Pomegranate juice has long been a popular drink in Persian and Indian cuisine, and began to be widely distributed in the United States and Canada in 2002.

Grenadine syrup is thickened and sweetened pomegranate juice used in cocktail mixing. Before tomatoes (a New World fruit) arrived in the Middle East, grenadine was widely used in many Iranian foods, and is still found in traditional recipes such as *fesenjân*, a thick sauce made from pomegranate juice and ground walnuts, usually spooned over duck or other poultry and rice, and in *ash-e anar* (pomegranate soup).

Wild pomegranate seeds are used as a spice known as *anardana* (from Persian: *anar* + *dana*, *pomegranate seed*), most notably in Indian and Pakistani cuisine, but also as a substitute for pomegranate syrup in Persian cuisine. Dried whole arils can often be obtained in ethnic Indian subcontinent markets. These seeds are separated from the flesh, dried for 10–15 days and used as an acidic agent for chutney and curry preparation. Ground *anardana* is also used, which results in a deeper flavoring in dishes and prevents the seeds from getting stuck in teeth. Seeds of the wild pomegranate variety known as *daru* from the Himalayas are regarded as quality sources for this spice. Dried pomegranate arils, found in some natural specialty food markets, still contain the seed and residual aril water, maintaining a natural sweet and tart flavor. Dried arils can be used in several culinary applications, such as trail mix, granola bars, or as a topping for salad, yogurt, or ice cream. Chocolate covered arils, also available in gourmet food stores like Trader Joes, may be added to desserts and baked items.

In the Caucasus, pomegranate is used mainly as juice. In Azerbaijan a sauce from pomegranate juice (narsharab) is usually served with fish or tika kabab. In Turkey, pomegranate sauce, (Turkish: *nar ekþisi*) is used as a salad dressing, to marinate meat, or simply to drink straight. Pomegranate seeds are also used in salads and sometimes as garnish for desserts such as güllaç. Pomegranate syrup or molasses is used in *muhammara*, a roasted red pepper, walnut, and garlic spread popular in Syria and Turkey.

In Greece, pomegranate (Greek: ñüäé, rodi) is used in many recipes, including *kollivozoumi*, a creamy broth made from boiled wheat, pomegranates and raisins, legume salad with wheat and pomegranate, traditional Middle Eastern lamb kebabs with pomegranate glaze, pomegranate eggplant relish, and avocado-pomegranate dip. Pomegranate is also made into a liqueur and popular fruit confectionery used as ice cream topping or mixed with yogurt or spread as jam on toast. In Cyprus as well as in Greece and among the Greek Orthodox Diaspora, ñüäé is used to make kolliva, a mixture of wheat, pomegranate seeds, sugar, almonds and other seeds served at memorial services.

Prominence in Ayurvedic medicine

In the Indian subcontinent's ancient Ayurveda system of medicine, the pomegranate has extensively been used as a source of traditional remedies for thousands of years. The rind of the fruit and the bark of the pomegranate tree are used as a traditional remedy against diarrhea, dysentery and intestinal parasites. The seeds and juice are considered a tonic for the heart and throat, and classified as a bitter-astringent (*pitta* or fire) component under the Ayurvedic system, and considered a healthful counterbalance to a diet high in sweet-fatty (*kapha* or earth) components. The astringent qualities of the flower juice, rind and tree bark are considered valuable for a variety of purposes, such as stopping nose bleeds and gum bleeds, toning skin, (after blending with mustard oil) firming-up sagging breasts and treating hemorrhoids. Pomegranate juice (of specific fruit strains) is also used as eyedrops as it is believed to slow the development of cataracts.

Nutrients and Phytochemicals

Pomegranate aril juice provides about 16 per cent of an adult's daily vitamin C requirement per 100 ml serving, and is a good source of vitamin B_5 (pantothenic acid), potassium and polyphenols, such as tannins and flavonoids.

Pomegranates are listed as high-fiber in some charts of nutritional value. That fiber, however, is entirely contained in the edible seeds which also supply unsaturated oils. People who choose to discard the seeds forfeit nutritional benefits conveyed by the seed fiber, oils and micronutrients. The most abundant polyphenols in pomegranate juice are the hydrolyzable tannins called ellagitannins formed when ellagic acid binds with a carbohydrate. Punicalagins are tannins with free-radical scavenging properties in laboratory experiments and with potential human effects. Punicalagins are absorbed into the human body and may have dietary value as antioxidants, but conclusive proof of efficacy in humans has not yet been shown. During intestinal metabolism by bacteria, ellagitannins and punicalagins are converted to urolithins which have unknown biological activity *in vivo*.

Other phytochemicals include polyphenolic catechins, gallocatechins, and anthocyanins, such as prodelphinidins, delphinidin, cyanidin, and pelargonidin. The ORAC (antioxidant capacity) of pomegranate juice was measured at 2,860 units per 100 grams. Many food and dietary supplement makers use pomegranate phenolic extracts as ingredients in their products instead of the juice. One of these extracts is ellagic acid, which may become bioavailable only after parent molecule punicalagins are metabolized. However, ingested ellagic acid from pomegranate juice does not accumulate in the blood in significant quantities and is rapidly excreted. Accordingly, ellagic acid from pomegranate juice does not appear to be biologically important *in vivo*.

Potential Health Benefits

In preliminary laboratory research and clinical trials, juice of the pomegranate may be effective in reducing heart disease risk factors, including LDL oxidation, macrophage oxidative status, and foam cell formation. In an article published in the *American Journal of Clinical Nutrition* in 2000, researchers detailed an experiment in which healthy adult men and unhealthy mice consumed pomegranate juice daily. After two weeks, the healthy men experienced increased antioxidant levels, which resulted in a ninety percent drop in LDL cholesterol oxidation. In the mice, "oxidation of LDL by peritoneal macrophages was reduced by up to 90 per cent after pomegranate juice consumption". In a limited study of hypertensive patients, consumption of pomegranate juice for two weeks was shown to reduce systolic blood pressure by inhibiting serum angiotensin-converting enzyme. Juice consumption may also inhibit viral infections while pomegranate extracts have antibacterial effects against dental plaque.

Despite some positive research data, manufacturers and marketers of pomegranate juice have liberally used evolving research results for product promotion, especially for putative antioxidant health benefits. In February 2010, the FDA issued a Warning Letter to one such manufacturer, POM Wonderful, for using published literature to make illegal claims of unproven antioxidant and anti-disease benefits.

Clinical Trial Rationale and Activity

Metabolites of pomegranate juice ellagitannins localize specifically in the prostate gland, colon, and intestinal tissues of mice, leading to clinical studies of pomegranate juice or fruit extracts for efficacy against several diseases. In 2011, 32 clinical trials were registered with the National Institutes of Health to examine effects of pomegranate extracts or juice consumption on a list of diseases (prostate cancer; prostatic hyperplasia; diabetes; lymphoma; rhinovirus infection; common cold ; oxidative stress in diabetic hemodialysis; atherosclerosis; coronary artery disease; infant brain injury; hemodialysis for kidney disease); consumption of pomegranates and pomegranate juice appear to correlate with preventing such diseases.

Pomegranate, Arils Only

Nutritional value per 100 g (3.5 oz)	
Energy	346 kJ (83 kcal)
Carbohydrates	18.7 g
Sugars	13.7 g
Dietary fiber	4.0 g
Fat	1.2 g
Protein	1.7 g
Thiamine (vit. B_1)	0.07 mg
Riboflavin (vit. B_2)	0.05 mg
Niacin (vit. B_3)	0.29 mg
Pantothenic acid (B_5)	0.38 mg
Vitamin B_6	0.08 mg
Folate (vit. B_9)	38 μg
Vitamin C	10 mg
Calcium	10 mg
Iron	0.30 mg
Magnesium	12 mg
Phosphorus	36 mg
Potassium	236 mg
Zinc	0.35 mg

Health and Nutrition Benefits of Pomegranates/Pomegranate Juice

- Being rich in antioxidants, pomegranate can keep LDL (bad cholesterol) from oxidizing and thus, prevents atherosclerosis.
- Pomegranate has been found to help keep blood platelets from clumping together and forming unnecessary clots.
- Regular consumption of pomegranate juice is said to be beneficial for those suffering from heart diseases.
- Researchers have indicated that pomegranate compounds might prevent prostate cancer and even slowdown its growth.
- Pomegranate juice has also been associated with reducing the risk of breast cancer.
- Pomegranate juice is pretty good for those suffering from diarrhea, but it over-consumption can lead to constipation.
- Pomegranate is said to be good for reducing plaque in the arteries and raising levels of HDL (good cholesterol).
- It has been seen that pomegranate juice helps prevent heart attacks and even stroke.
- Studies have shown that maternal consumption of pomegranate juice might protect the neonatal brain from damage after injury.
- Pomegranate is believed to help prevent cartilage deterioration and thus, keep osteoarthritis at bay.
- Consumption of pomegranate might prevent and even slow down Alzheimer's disease.
- Pomegranate juice is associated with lowering of blood pressure as well as better oral health.

Figure 25.1: Flow sheet for preparation of pomegranate juice

Mature Pomegranate fruits
↓
Selection of fruits
↓
Cleaning/Washing
↓
Separation of arils
↓
Squeezing the arils through muslin cloth
↓
Centrifugation/Filtration
↓
Clarification of juice
↓
Juice

Figure 25.2: Flow sheet for preparation of pomegranate juice

Pomegranate fruits

↓

Washing

↓

Cutting into 4 equal pieces

↓

Separation of seeds/areal

↓

Pressing of seeds/areal (basket press)

↓

Straining

↓

Pasteurization of juice

↓

Bottling

↓

Crown corking

↓

Bottles pasteurization

↓

Cooling

↓

Storage

↓

Marketing

Figure 25.3: Flow sheet for preparation of pomegranate carbonated beverage

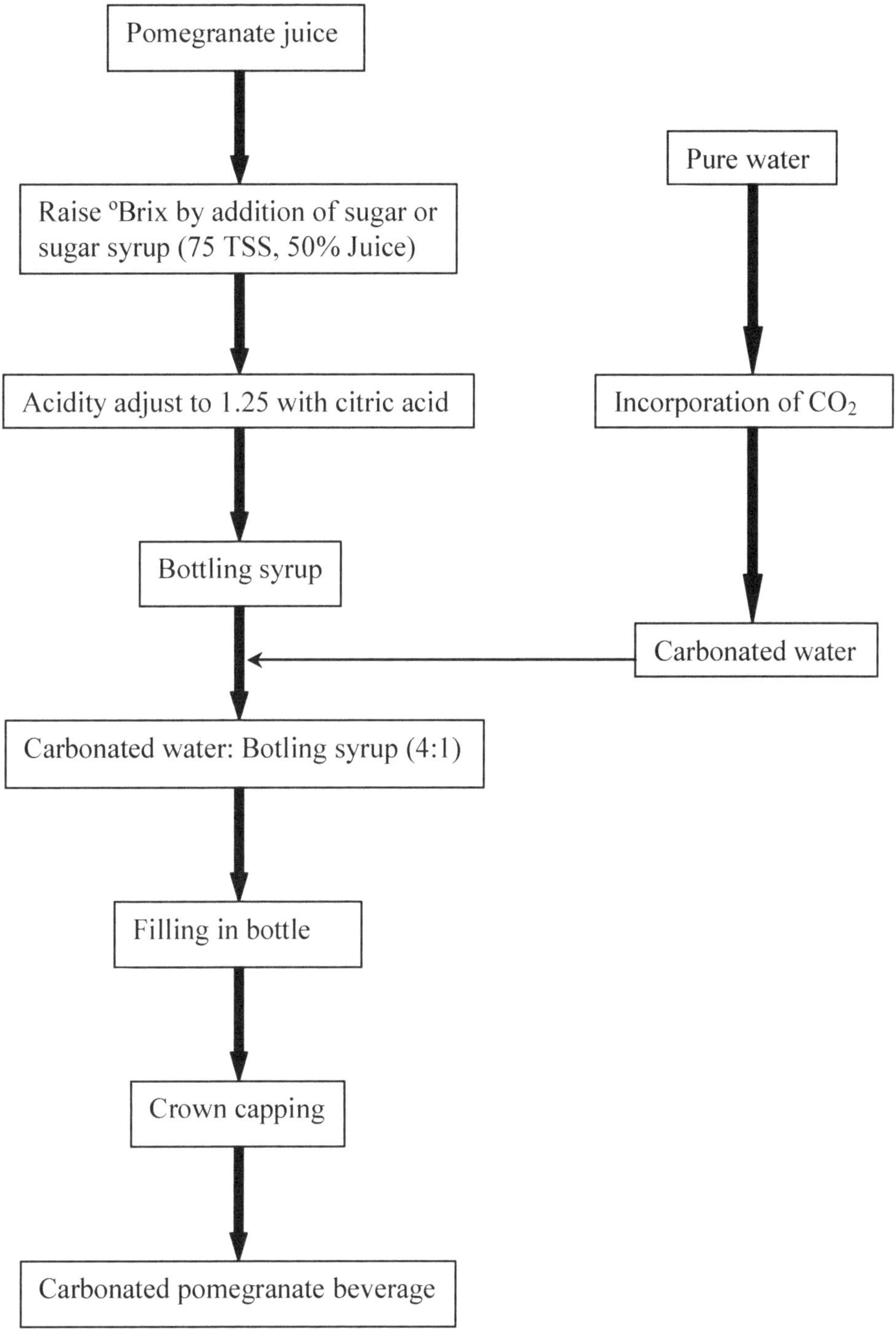

Figure 25.4: Flow sheet for preparation of pomegranate anardana

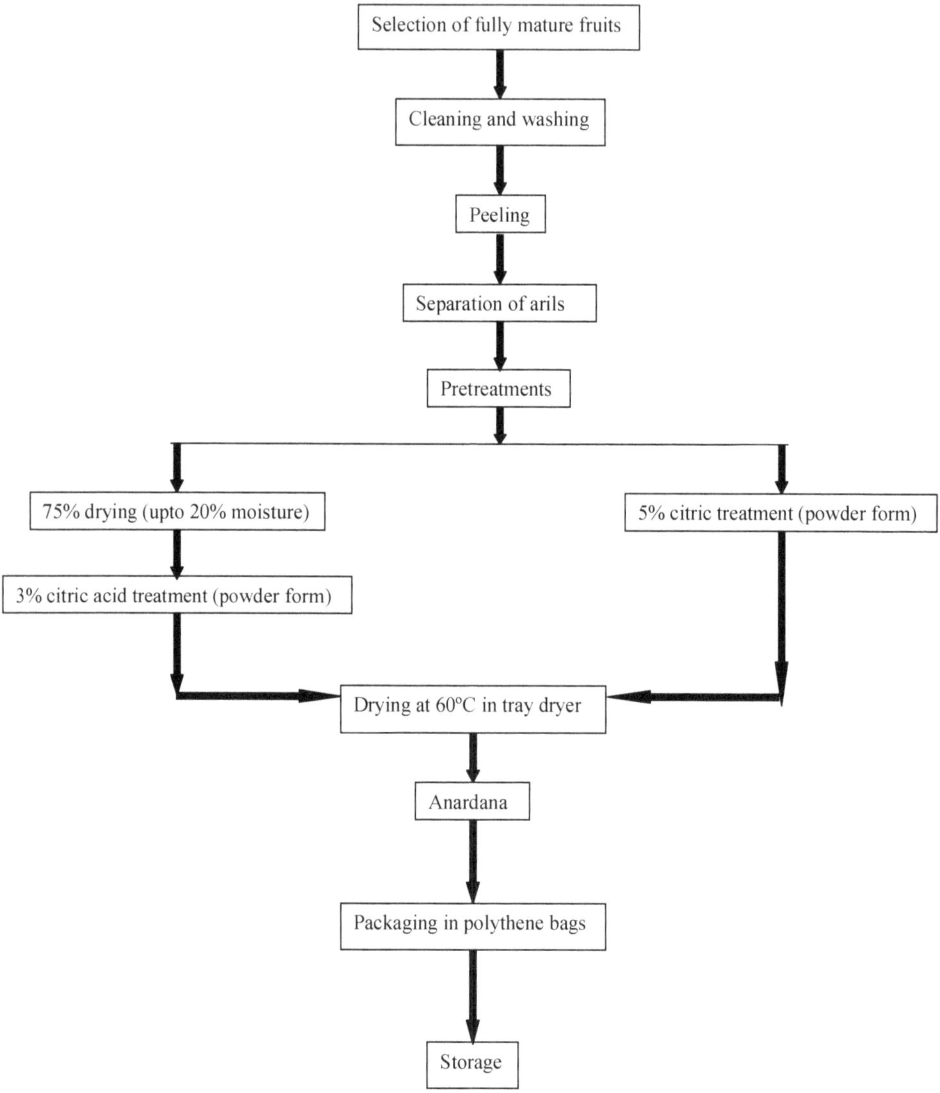

Figure 25.5: Flow sheet for preparation of pomegranate rind powder

Selection of mature healthy fruits

⬇

Washing

⬇

Separation of seeds and skin

⬇

Cutting of skin into small pieces

⬇

Drying in tray dryer/Sun

⬇

Powdering with the help of grinder

⬇

Sieving (60 mesh)

⬇

Pomegranate powder

⬇

Packaging

⬇

Storage

Figure 25.6: Flow sheet for preparation of pomegranate wine

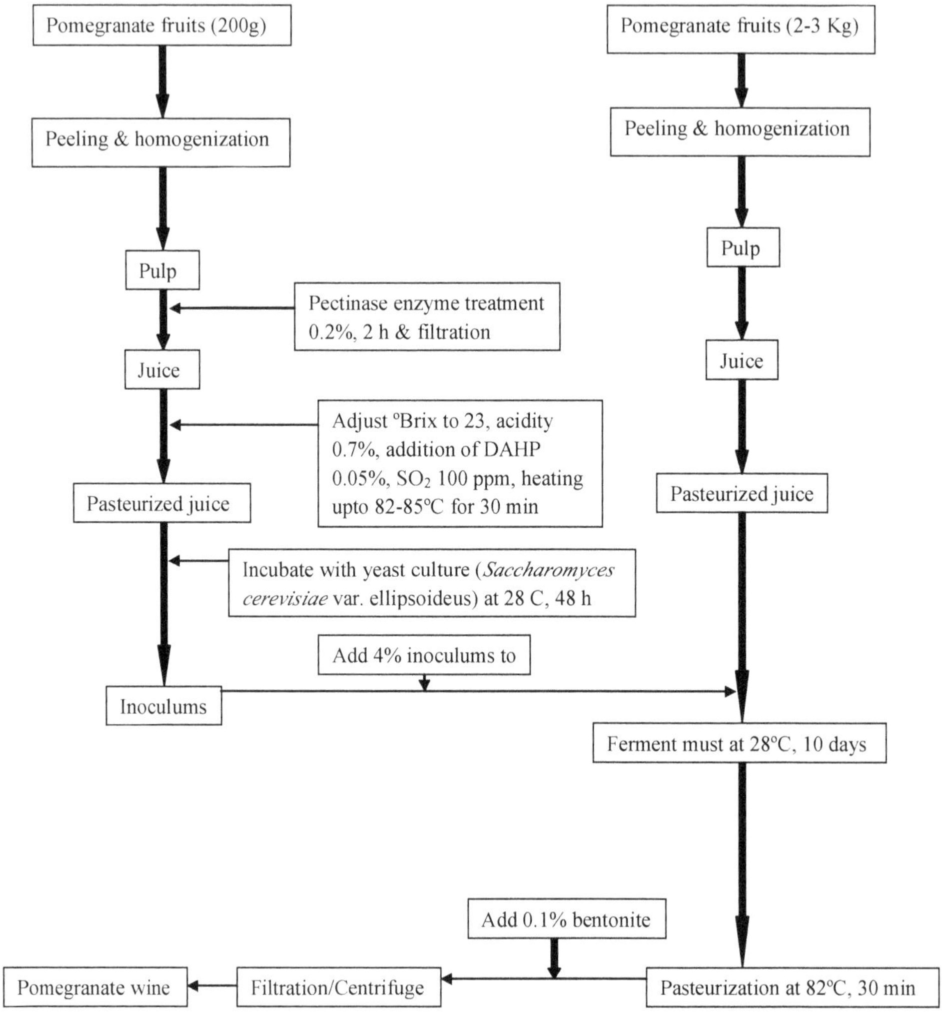

26

Sapota

Manilkara zapota, commonly known as the sapodilla, is a long-lived, evergreen tree native to southern Mexico, Central America and the Caribbean. An example natural occurrence is in coastal Yucatan in the Petenes mangroves ecoregion, where it is a subdominant plant species. It is grown in huge quantities in India, Pakistan and Mexico, and was introduced to the Philippines during Spanish colonisation.

The mamey sapote (*Pouteria sapota*) is a species of tree native to southern Mexico. Today, the tree is cultivated not only in Mexico, but also in Central America, the Caribbean, and South Florida for its fruit, which is commonly eaten in many Latin American countries. Mamey sapote is a large and highly ornamental evergreen tree that can reach a height of 15 to 45 meters (60 to 140 feet) at maturity. Like most fruit trees, it is mainly propagated by grafting, which ensures the new plant has the same characteristics as the parent, especially its fruit. It is also considerably faster than growing trees by seed. The fruit is about 10 to 25 cm (4 to 10 inches) long and 8 to 12 cm (3 to 5 inches) wide and has orange flesh.

The fruit is eaten raw or made into milkshakes, smoothies, ice cream and fruit bars. Some consider the fruit to be an aphrodisiac. Some beauty products use oil pressed from the seed, otherwise known as sapayul oil. The brown skin has a texture somewhat between sandpaper and the fuzz on a peach. The fruit's texture is creamy and soft. A mamey sapote is ripe when the flesh is pink when a fleck of the skin is removed. The flesh should give slightly, as with a ripe kiwifruit.

The mamey sapote is related to other sapotes such as *sapodilla* (*Manilkara zapota*), abiu (*P. caimito*) and canistel (*P. campechiana*), but unrelated to the black sapote (*Diospyros digyna*) and white sapote (*Casimiroa edulis*). It should not be confused with the mammee apple (*Mammea americana*). It has different names depending on the country: mamey, *Zapote colorado* (Costa Rica), and *Zapote rojo* (South America), among others. Prominent varieties of sapota are Pala, Kirtibarthi, Cricket Ball, Dwarapudi, Guthi, Jonavalasa, Kalipatti, Bhuripatti, Pilipatti, Dhola Diwani, Jhaumakhia, Cricket

Ball, Murraba, Kirtibarthi, DHS1, DHS2, Guthi, Kirtibarthi, Pala, Co1, Co2, PKM1, Calcutta Special, Round, Oval and Baramasi.

Health Benefits of Sapodilla/Sapota

- ✰ Sapodilla is rich in dietary fiber (5.6 g/100g), which makes it a good bulk laxative. The fiber content helps relieve constipation episodes and also helps protect the mucous membrane of the colon from cancer causing toxins by firmly binding to them.
- ✰ The fruit is rich in antioxidant poly-phenolic compound tannin. Tannins are a complex family of naturally occurring polyphenols that neutralize acids by precipitating proteins. Research studies found that tannins have shown to have potential anti-inflammatory, antiviral, anti-bacterial and anti-parasitic effects. Hence, these compounds have many useful medicinal applications such as anti-diarrheal, hemostatic (stops bleeding) and as a remedy for hemorrhoids.
- ✰ Furthermore, the anti-inflammatory effect of tannins help limit conditions like erosive gastritis, reflux-esophagitis, enteritis and irritating bowel disorders. Some other fruits that are rich in tannins include pomegranate, persimmon, grapes etc.
- ✰ Sapote contains good amount of antioxidant vitamins like vitamin C (24.5 per cent of recommended daily intake per 100 g of fruit) and vitamin A. Vitamin A is essential for vision. It is also required for maintaining healthy mucus membranes and skin. Consumption of natural fruits rich in vitamin A known to protect from lung and oral cavity cancers. So also, consumption of foods rich in vitamin C helps body develop resistance against infectious agents and scavenge harmful free radicals.
- ✰ Fresh ripen sapodilla are good source of minerals like potassium, copper, iron and vitamins like folate, niacin and pantothenic acid. These compounds are essential for optimal health as they involve in various metabolic processes in the body as cofactors for the enzymes.

Harvesting and Postharvest Management

Sapota takes about 7-101/2 months from anthesis to maturity of fruits depending on variety and climate. Fruits follow double sigmoid pattern of growth. Properly developed fruits have high TSS and sugar, and reduced acidity, astringency, latex and vitamin C. Maturity is decided on the basis of ease with which brown scruff gets off the fruit surface and development of yellowish tinge intermixed wit corky brown color on the surface of the fruit. At this stage, practically no green tissue and milky latex are seen on fruits when scratched with nails. The fruits are handpicked or harvested with special harvester which has a round ring with a net bag fixed onto a long bamboo. Depending on management level, 15-20 tons fruits are harvested from a hectare.

Since sapota is a climacteric fruit, it has to be ripening artificially. Fruits are highly perishable and they undergo rapid ripening changes within 5-7days during which the fruits become soft, sweet and develop excellent aroma with decline in tannins, latex sapotin, aldehydes and acidity. These changes are associated with increase in production of ethylene, rate of respiration, catalase, peroxidase and PME activities. These changes can be regulated through chemicals, temperature and storage gas composition.

Harvested fruits should be cleaned of latex and scurf by washing in clean water to make them look attractive. Such fruits should be graded into big, medium and small sizes. Fruits should be tightly packed in cardboard boxes of 10 kg capacity with rice straw as padding material and with ethylene absorbents and transported quickly to wholesale markets. For extending shelf life and to avoid storage rots, fruits can be dipped in GA300 ppm + bavistin 1,000ppm solution at pre-packing stage. For uniform and rapid ripening ethephon (1,000ppm) can be utilized at 20º-25ºC. Modified storage with 5-10 per cent (c/c) CO_2 can be employed for long storage (21-25days). Refrigerated vans (12-23'C) should be utilized for long distance and export markets.

Sapota Fresh Nutritive Value per 100 g

Principle	*Nutrient Value*	*Percentage of RDA*
Energy	83 Kcal	4 per cent
Carbohydrates	19.9 g	15 per cent
Protein	0.44 g	<1 per cent
Total Fat	1.10 g	3.5 per cent
Cholesterol	0 mg	0 per cent
Dietary Fiber	5.3 g	14 per cent
Vitamins		
Folates	14 mcg	3.5 per cent
Niacin	0.200 mg	1 per cent
Pantothenic acid	0.252 mg	5 per cent
Pyridoxine	0.037 mg	3 per cent
Riboflavin	0.020 mg	1.5 per cent
Thiamin	0.058 mg	5 per cent
Vitamin A	60 IU	2 per cent
Vitamin C	14.7 mg	24.5 per cent
Electrolytes		
Sodium	12 mg	1 per cent
Potassium	193 mg	4 per cent
Minerals		
Calcium	21 mg	2 per cent
Copper	0.086 mg	9 per cent
Iron	0.80mg	10 per cent
Magnesium	12mg	3 per cent
Phosphorous	12 mg	2 per cent
Zinc	0.10 mg	1 per cent

Figure 26.1: Flow sheet for preparation of sapota preserves (candy)

Selection of fully mature sapota fruits

↓

Washing

↓

Peeling

↓

Pricking

↓

Blanching, 3 min

↓

Syruping in 40 °Brix, 24 h

↓

Syruping in 50 °Brix, 24 h

↓

Syruping in 60 °Brix, 24 h

↓

Syruping in 70 °Brix, 7 days

↓

Draining

↓

Packaging

↓

Storage

Figure 26.2: Flow sheet for preparation of sapota powder

Selection of fully mature sapota fruits
↓
Washing
↓
Peeling and slicing
↓
Blanching, 3 min
↓
Sugar syruping 40 per cent for 20 min,
1 per cent KMS dipping for 2 min
↓
Drying in cabinet dryer at 50-55 °C
↓
Grinding
↓
Sieving, 80 mesh
↓
Packaging
↓
Storage

Figure 26.3: Flow sheet for preparation of Sapota toffee

Sapota fruits

↓

Washing

↓

Peeling and deseeding

↓

Mixing of ingredients
(Sugar 1000g, fat 100g, milk powder 75g)

↓

Heating upto 80 °Brix

↓

Adding salt 5g, citric acid 1g

↓

Heating upto 82 °Brix

↓

Spreading in aluminium trays (already smeared with fat/oil)

↓

Cooling

↓

Cutting into attractive pieces

↓

Packaging into metallic paper

↓

Storage

Figure 26.4: Flow sheet for preparation of custard apple: Sapota jams

Custard apple : Sapota fruits (50:50)

↓

Cleaning/Washing

↓

Peeling

↓

Extraction of pulp

↓

Addition of sugar, 750g/kg

↓

Heating upto 65 °Brix

↓

Addition of citric acid, 3g

↓

Heating upto 68.5 °Brix

↓

Filling in sterilized glass jars

↓

Cooling

↓

Sealing

↓

Storage

27

Strawberry

Strawberry is one of the fruits that almost everybody loves. First grown in Europe in the early 18th century, the fruit is today cultivated in almost all the cold countries of the world. Strawberries are used as one of the major ingredients in a number of food items, like ice creams, milkshakes, smoothies, pies, yogurts, and so on. The fruit not only tastes extremely delicious, it also packs a lot of nutritional value within itself, according the consumers a large number of health benefits. The only problem with strawberries is that they perish pretty quickly and easily. In the following lines, we tell you about the nutrition benefits of eating strawberries.

Nutritional Value of Strawberry

1 cup (around 145-150 gm) of strawberry comprises of the following nutrients:

Nutrient	*Amount*
Protein	1 g
Carbohydrates	10.5 g
Fat	0.6 g
Fiber	1.6 g
Vitamin A	3 RE
Vitamin C	84.5 mg
Thiamin	0.1 mg
Riboflavin	0.1 mg
Niacin	0.4 mg
Vitamin B_6	0.1 mcg
Vitamin B_{12}	0.1 mg
Folic Acid	26.4 mcg

Nutrient	*Amount*
Sodium	2 mg
Calcium	21 mg
Iron	0.6 mg
Potassium	247 mg
Zinc	0.2 mg
Magnesium	16 mg
Calories	45 Kcal

Health and Nutrition Benefits of Eating Strawberries

- ☆ Rubbing a cut strawberry on teeth and gums helps remove tartar, apart from strengthening and healing the gums.
- ☆ Since strawberries are rich in fiber, they help the body in absorbing nutrients.
- ☆ Strawberries are good for diabetics, as they help stabilize the level of blood glucose.
- ☆ Strawberries are said to help in regulating blood pressure and thus, diminishing the risk of heart disease.
- ☆ Strawberries have been associated with inhibiting the production of cholesterol in the liver.
- ☆ Strawberries have been known to help the body in getting rid of harmful toxins.
- ☆ The high antioxidant levels in strawberries can help the body neutralize the destructive effects of free radicals.
- ☆ The potassium in strawberries helps regulate the electrolytes in the body, in turn lowering the risk of stroke.

Figure 27.1: Flow sheet for preparation of strawberry pulp

Fully Ripe Strawberry

↓

Washing and Cleaning

↓

Sorting for good fruits

↓

Pulping

↓

Heating

↓

Preservative addition

↓

Mixing

↓

Can filling and sealing

↓

Pasteurization

↓

Cooling and Labeling

↓

Storage

↓

Marketing

Figure 27.2: Flow sheet for preparation of strawberry toffee

Strawberry fruits fully ripe

↓

Cleaning and Pulping

↓

Homogenization of pulp

↓

Addition of sugar, fat, milk powder
(750g, 100g, 100g/Kg Pulp)

↓

Heating upto 80 °Brix

↓

Addition of salt and citric acid (5g, 2g)

↓

Heating upto 82 °Brix

↓

Spreading in aluminium trays
(1 cm thick; already smeared with fat)

↓

Cooling

↓

Cutting into attractive pieces (1.5 × 2.5 cm)

↓

Rapping into metallic paper

↓

Storage

Figure 27.3: Flow sheet for preparation of strawberry jam

Strawberry ripe fruits

↓

Washing

↓

Extraction of pulp

↓

Addition of sugar 1000g, pectin 8g/kg

↓

Heating upto 68 °Brix

↓

Filling into sterilized bottles

↓

Cooling

↓

Waxing

↓

Sealing

↓

Storage

Figure 27.4: Flow sheet for preparation of strawberry crush

Selection of fully ripe fruits

⬇

Washing

⬇

Extraction of juice by juicer

⬇

Filtration

⬇

Addition of sugar (500g/kg)

⬇

Mixing thoroughly

⬇

Addition of 500 ppm sodium benzoate

⬇

Mixing thoroughly

⬇

Filling into sterilized bottles

⬇

Sealing

⬇

Storage

28

Tamarind

Botanically, the tree is among the large tropical trees belonging to the family of *fabaceae*; of the genus: *Tamarindus*. Scientific name: *Tamarindus indica*. Sweet and tangy tamarind is one of the widely used condiment spice found in every South Asian kitchens. The tree is very large with long heavy drooping branches and dense foliage. Fully grown up tree reaches up to 80 feet in height. During each season, the tree bears irregularly curved pods in abundance all along its branches. Each pod has thick outer shell encasing deep brown color sticky pulp enveloping 2-10 hard dark brown colored seeds.

Health Benefits of Tamarind

- ☆ Tamarind contains many health benefiting essential volatile chemical compounds, minerals, vitamins and dietary fiber.
- ☆ Its sticky pulp is rich source of non-starch polysaccharides (NSP) or dietary fiber such as *gums, hemicelluloses, mucilage, pectin* and *tannins*. 100 g of fruit pulp provides 5.1 or over 13 per cent of dietary fiber. NSP or dietary fiber in the food increases its bulk and augments bowel movements thereby help prevent constipation. The fiber also binds to toxins in the food thereby help protect the colon mucus membrane from cancer causing chemicals.
- ☆ In addition, dietary fibers in the pulp bind to bile salts (produced from cholesterol) and decrease their re-absorption in the colon; thereby help excretion of "bad" or LDL cholesterol levels from the body.
- ☆ While lemon contains *citric acid,* tamarind is rich in tartaric acid. Tartaric acid gives a sour taste to food but is also a very powerful antioxidant. (Anti-oxidant E-number is E334). It helps body protect from harmful free radicals.

- Tamarind fruit contains many volatile phytochemicals such as *limonene, geraniol, safrole, cinnamic acid, methyl salicylate, pyrazine* and *alkyl-thiazoles*. Together these compounds account for the medicinal properties of tamarind.
- This prized spice is a good source of minerals like copper, potassium, calcium, iron, selenium, zinc and magnesium. Potassium in an important component of cell and body fluids that helps control heart rate and blood pressure. Iron is essential for red blood cell production and as a co-factor for *cytochrome oxidases* enzymes.
- It is also rich in many vital vitamins including thiamin, vitamin A, folic acid, riboflavin, niacin, vitamin-C that are essential for optimum health.

Medicinal Uses of Tamarind

- Its pulp has been used in many traditional medicines as laxative, digestive, and as a remedy for biliousness and bile disorders.
- The spice is also used as emulsifying agent in syrups, decoctions, etc in different pharmaceutical products.

Culinary Uses

- Delicately sweet and sour, tamarind is one of the most sought after ingredient in Indian, Middle Eastern and south-East Asian cooking. In some Indian households, the pods are cut open and fresh pulp is used as and when required. The seeds are then removed by beating the pulp with "wooden stick" kept at home especially for this purpose. One may also carefully use paring knife to separate seeds.
- In general, a small slice of the pulp is soaked in half a cup of warm water for about 10 minutes. Swirl the pulp with your fingers so that the pulp is dissolved evenly in water to make thin sauce. Strain the juice through a thin cloth sieve and use for cooking.

Selection and Storage

Fresh tamarind pods are available in late spring and early summer seasons. However several different forms of processed tamarind such as compressed tamarind blocks, ready-to-use slice, paste, concentrates, balls etc. are made available in the markets. Choose fresh unbroken pods packed in boxes. If you are purchasing processed form, buy the product from a well reputed authentic brand. Avoid old, desiccated pulp, or off-smelling products. Once at home store the pods or pulp in the refrigerator where it will stay fresh for several months.

Tamarind (*Tamarindus indica*), Nutritional Value per 100 g

Principle	*Nutrient Value*	*Percentage of RDA*
Energy	239 Kcal	12 per cent
Carbohydrates	62.50 g	40 per cent
Protein	2.80 g	5 per cent
Total Fat	0.60 g	3 per cent
Cholesterol	0 mg	0 per cent
Dietary Fiber	5.1 g	13 per cent
Vitamins		
Folates	14 µg	3.5 per cent
Niacin	1.938 mg	12 per cent
Pantothenic acid	0.143 mg	3 per cent
Pyridoxine	0.066 mg	5 per cent
Thiamin	0.428 mg	36 per cent
Vitamin A	30 IU	1 per cent
Vitamin C	3.5 mg	6 per cent
Vitamin E	0.10 mg	<1 per cent
Vitamin K	2.8 µg	2 per cent
Electrolytes		
Sodium	28 mg	2 per cent
Potassium	628 mg	13 per cent
Minerals		
Calcium	74 mg	7 per cent
Copper	0.86 mg	9.5 per cent
Iron	2.80 mg	35 per cent
Magnesium	92 mg	23 per cent
Phosphorus	113 mg	16 per cent
Selenium	1.3 µg	2 per cent
Zinc	0.10 mg	1 per cent
Phyto-nutrients		
Carotene-β	18 µg	–
Crypto-xanthin-β	0 µg	–
Lutein-zeaxanthin	0 µg	–

Figure 28.1: Flow sheet for preparation of tamarind toffee

Tamarind fruits fully ripe

↓

Removing seeds and sheath, Cleaning and Pulping
(1:0.5 pulp; water)

↓

Homogenization of pulp and adjust acidity

↓

Addition of sugar, fat, milk powder
(750g, 100g, 100g/Kg Pulp)

↓

Heating upto 80 °Brix

↓

Addition of salt (5g)

↓

Heating upto 82 °Brix

↓

Spreading in aluminium trays
(1 cm thick; already smeared with fat)

↓

Cooling

↓

Cutting into attractive pieces (1.5 × 2.5 cm)

↓

Rapping into metallic paper

↓

Storage

Figure 28.2: Flow sheet for the preparation of tamarind powder

Mature tamarind fruits
↓
Removal of rags and seeds
↓
Flesh
↓
Drying in cabinet drier for 4-6 h at 50-55 °C
↓
Grinding in electric grinder by mixing 0.5 per cent calcium sterate
↓
Sieving through 60 mesh sieve
↓
Packaging in 30 gauge polythene bags or glass bottles
↓
Storage

Figure 28.3: Flow sheet for preparation of tamarind pulp by cold method

Selection of mature tamarind pods
↓
Deshelling, removal of seeds and rags
↓
Flesh
↓
Addition of water (1:2)
↓
Soaking
↓
Homogenization
↓
Filtration
↓
Pulp

Figure 28.4: Flow sheet for preparation of tamarind pulp by hot enzymatic process

Selection of mature tamarind pods
↓
Deshelling, removal of seeds and rags
↓
Flesh
↓
Addition of water (1:2)
↓
Heating (70°C for 10 min) and cooling to room temperature
↓
Addition of *Biotropicase* L. (pectinase) enzyme (0.5 per cent)
↓
Soaking (9 h)
↓
Homogenization
↓
Filtration
↓
Pulp

Figure 28.5: Flow sheet for preservation of tamarind pulp

Tamarind pulp
↓
Addition of sodium benzoate (650 ppm)
↓
Homogenization
↓
Heating upto 80 °C for 30 min
↓
Filling sterilized bottles
↓
Cooling
↓
Capping and sealing
↓
Storage

Figure 28.6: Flow sheet for preparation of tamarind RTS

Tamarind juice

↓

Addition of sugar, water

↓

Adjust °brix (21) and acidity (0.35 per cent)

↓

Filtration

↓

RTS beverage

↓

Filling in sterilized glass bottles

↓

Capping and sealing

↓

Storage

Parameter	*RTS*	*Syrup*	*Concentrate*
TSS	21	55	65
Acidity, per cent	0.35	1.0	12.0
Ascorbic acid, mg/100g	0.27	1.16	10.6
Reducing sugars, per cent	13.6	46.2	55.6
Non-reducing sugars, per cent	6.6	4.9	2.0
Total sugars, per cent	20.2	51.1	57.8

Figure 28.7: Flow sheet for preparation of tamarind syrup from pulp

Tamarind juice
↓
Addition of sugar, water
↓
Heating
↓
Checking °Brix, acidity
↓
Adjust °Brix and acidity
↓
Filtration
↓
Syrup
↓
Filling in sterilized glass bottles
↓
Capping and sealing
↓
Storage

Figure 28.8: Flow sheet for preparation of tamarind juice concentrate

Tamarind juice
↓
Concentrate in vacuum flash evaporator for 2-3 h
↓
Concentrate
↓
Bottling
↓
Capping and sealing
↓
Pasteurization (80 °C, 20 min)
↓
Cooling
↓
Storage

Figure 28.9: Flow sheet for preparation of tamarind jam

Tamarind pulp
↓
Adjust acidity
↓
Addition of sugar (1:0.75)
↓
Heating (30 min)
↓
Adding pectin 0.5 per cent
↓
Heating (68 °Brix)
↓
Jam
↓
Filling in sterilized glass bottles
↓
Cooling
↓
Waxing
↓
Capping and sealing
↓
Storage

Figure 28.10: Flow sheet for preparation of jelly from tamarind pulp

Tamarind pulp
↓
Addition of sugar (pulp:water, 1:1)
↓
Heating 30 min
↓
Filtration
↓
Extract
↓
Adjust acidity
↓
Adding sugar, 1:1
↓
Heating
↓
Add pectin, 1 per cent
↓
Heating, 68-69 °Brix
↓
Jelly
↓
Filling in sterilized glass bottles
↓
Cooling
↓
Waxing
↓
Capping and sealing
↓
Storage

Figure 28.11: Flow sheet for preparation of tamarind sauce

Tamarind pulp

↓

Adjust acidity

↓

Add 1/3 sugar

↓

Heating to 1/3 volume with spices and onion

↓

Adding other ingredients, heating (32 °Brix)

↓

Sauce

↓

Filling in sterilized glass bottles

↓

Capping and sealing

↓

Storage

29

Umber

Ficus racemosa (syn. *Ficus glomerata* Roxb.) is a species of plant in the Moraceae family. Popularly known as the Cluster Fig Tree or Goolar (Gular) Fig, this is native to Australasia, South-East Asia and the Indian Subcontinent. It is unusual in that its figs grow on or close to the tree trunk. In India the tree and its fruit are called *gular* in the north and *atti* in the south. The fruits are a favorite staple of the common Indian macaque. In Vietnam, it is called sung. It serves as a food plant for the caterpillars of the butterfly the Two-brand Crow (*Euploea sylvester*) of northern Australia.

In Hinduism

In the Atharva Veda, this fig tree (Sanskrit: *uumbara* or *udumbara*) is given prominence as a means for acquiring prosperity and vanquishing foes. For instance, regarding an amulet of the *udumbara* tree, a hymn (AV xix, 31) extols:

The Lord of amulets art thou, most mighty: in thee wealth's ruler hath engendered riches; these gains are lodged in thee, and all great treasures. Amulet, conquer thou: far from we banish malignity and indigence, and hunger. Vigour art thou, in me do thou plant vigour: riches art thou, so Do thou grant me riches? Plenty art thou, so prosper me with plenty: House-holder, hear a householder's petition.

It has been described in the story of Raja Harischandra of the Ikshvaku dynasty, that the crown was a branch of this Udumbura tree, set in a circlet of gold. Additionally, the Throne (simhasana) was constructed out of this wood and the royal personage would ascend it on his knee, chanting to the gods to ascend it with him, which they did so, albeit unseen.

Uses

In historical times both Hindu and Buddhist ascetics on their way to Taxila, (Original name is Taksha Sila) travelling through vast areas of Indian forests used to consume the fruit during their travels. One challenge to vegetarians were the many fig wasps that one finds when opening a gular fig. One way to get rid of them was to break the figs into halves or quarters, discard most of the seeds and then place the figs into the midday sun for an hour. Gular fruit are almost never sold commercially because of this problem

Nutritive Value

Principle	*Nutrient Value*
Moisture	13.6 per cent
Albuminoids	7.4 per cent
Fat	5.6 per cent
Carbohydrates	49.0 per cent
Coloring matter	8.5 per cent
Fiber	17.9 per cent
Ash	6.5 per cent
Silica (SiO_2)	0.25 per cent
Phosphorus	0.91 per cent

Figure 29.1: Flow sheet for preparation of umber powder

Fully Ripe Umber fruits
↓
Washing and Cleaning
↓
Sorting
↓
Pulping
↓
Heating
↓
Preservative addition
↓
Mixing
↓
Drying at low temperature
↓
Grinding, sieving with 60 mesh
↓
Pouch/bag filling and Labeling
↓
Storage
↓
Marketing

Figure 29.2: Flow sheet for preparation of umber (*Ficus glomerata*) jam

Selection of fully ripe umber fruits

↓

Cut into pieces and keep in sunlight for half an hour

↓

Wash in clean water for removal of external impurities

↓

Heat with sugar and citric acid upto 68–69 °Brix
(During heating crush all fruit pieces)

↓

Fill into clean sterilized glass bottles

↓

Cool to room temperature

↓

Waxing

↓

Close with sterilized lid

↓

Storage

30

Woodapple

The woodapple, *Feronia limonia* Swingle (syns. *F. elephantum* Correa; *Limonia acidissima* L.; *Schinus limonia* L.) is the only species of its genus, in the family Rutaceae. Besides wood-apple, it may be called elephant apple, monkey fruit, curd fruit, *kath bel* and other dialectal names in India. In Malaya it is *gelinggai* or *belinggai;* in Thailand, *ma-khwit;* in Cambodia, *kramsang;* in Laos, *ma-fit.* In French, it is *pomme d' elephant, pomme de bois,* or *citron des mois.*

The slow-growing tree is erect, with a few upward-reaching branches bending outward near the summit where they are subdivided into slender branch lets drooping at the tips. The bark is ridged, fissured and scaly and there are sharp spines: 3/4 to 2 in (2-5 cm) long on some of the zigzag twigs. The deciduous, alternate leaves, 3 to 5 in (7.5-12.5 cm) long, dark-green, leathery, often minutely toothed, blunt or notched at the apex, are dotted with oil glands and slightly lemon-scented when crushed. Dull-red or greenish flowers to 1/2 in (1.25 cm) wide are borne in small, loose, terminal or lateral panicles. They are usually bisexual. The fruit is round to oval, 2 to 5 in (5-12.5 cm) wide, with a hard, woody, grayish-white, scurfy rind about 1/4 in (6 mm) thick. The pulp is brown, mealy, odorous, resinous, astringent, acid or sweetish, with numerous small, white seeds scattered through it.

Harvesting

The fruit is tested for maturity by dropping onto a hard surface from a height of 1 ft (30 cm). Immature fruits bounce, while mature fruits do not. After harvest, the fruit is kept in the sun for 2 weeks to fully ripen.

Food Uses

The rind must be cracked with a hammer. The scooped-out pulp, though sticky, is eaten raw with or without sugar, or is blended with coconut milk and palm-sugar syrup and drunk as a beverage, or frozen as an ice cream. It is also used in chutneys and for making jelly and jam. The jelly is purple and much like that made from black currants.

Bottled nectar is made by diluting the pulp with water, passing through a pulper to remove seeds and fiber, further diluting, straining, and pasteurizing. A clear juice for blending with other fruit juices has been obtained by clarifying the nectar with Pectinol R-10. Pulp sweetened with syrup of cane or palm sugar, has been canned and sterilized. The pulp can be freeze-dried for future use but it has not been satisfactorily dried by other methods.

The pulp represents 36 per cent of the whole fruit. The pectin content of the pulp is 3 to 5 per cent (16 per cent yields on dry-weight basis). The seeds contain bland, non-bitter, oil high in unsaturated fatty acids.

Other Uses

Pectin

The pectin has potential for multiple uses in pectin-short India, but it is reddish and requires purification.

Rind

The fruit shell is fashioned into snuffboxes and other small containers.

Gum

The trunk and branches exude a white, transparent gum especially following the rainy season. It is utilized as a substitute for, or adulterant of, gum arabic, and is also used in making artists' watercolors, ink, dyes and varnish. It consists of 35.5 per cent arabinose and xylose, 42.7 per cent *d*-galactose, and traces of rhamnose and glucuronic acid.

Wood

The wood is yellow-gray or whitish, hard, heavy, durable, and valued for construction, pattern-making, agricultural implements, rollers for mills, carving, rulers, and other products. It also serves as fuel.

The heartwood contains ursolic acid and a flavanone glycoside, 7-methylporiol-b-D-xylopyranosyl-D-glucopyranoside.

Medicinal Uses

The fruit is much used in India as a liver and cardiac tonic, and, when unripe, as an astringent means of halting diarrhea and dysentery and effective treatment for hiccough, sore throat and diseases of the gums. The pulp is poultice onto bites and stings of venomous insects, as is the powdered rind.

Juice of young leaves is mixed with milk and sugar candy and given as a remedy for biliousness and intestinal troubles of children. The powdered gum, mixed with honey, is given to overcome dysentery and diarrhea in children.

Oil derived from the crushed leaves is applied on itch and the leaf decoction is given to children as an aid to digestion. Leaves, bark, roots and fruit pulp are all used against snakebite. The spines are crushed with those of other trees and an infusion taken as a remedy for menorrhagia. The bark is chewed with that of *Barringtonia* and applied on venomous wounds. The unripe fruits contain 0.015 per cent stigmasterol. Leaves contain stigmasterol (0.012 per cent) and bergapten (0.01 per cent). The bark contains 0.016 per cent marmesin. Root bark contains aurapten, bergapten, isopimpinellin and other coumarins.

Food Value Per 100 g of Edible Pulp*

	Pulp (ripe)	*Seeds*
Moisture	74.0 per cent	4.0 per cent
Protein	8.00 per cent	26.18 per cent
Fat	1.45 per cent	27 per cent
Carbohydrates	7.45 per cent	35.49 per cent
Ash	5.0 per cent	5.03 per cent
Calcium	0.17 per cent	1.58 per cent
Phosphorus	0.08 per cent	1.43 per cent
Iron	0.07 per cent	0.03 per cent
Tannins	1.03 per cent	0.08 per cent

Nutritive Values (per 100 g edible portion)

Energy	134.0 kcal
Protein	7.1 g
Fat	3.7 g
Carbohydrates	18.1 g
Calcium	130.0 mg
Phosphorus	110.0 mg
Iron	61.0 ug
Carotene	40.0 ug
Thiamine	170.0 ug
Vitamin C	3.0 ug

Figure 30.1: Flow sheet for preparation of woodapple jelly

Fully mature woodapple fruits

↓

Washing and Cleaning

↓

Separation of pulp and shell

↓

Addition of water (1:1) and 2 g citric acid/kg pulp

↓

Heating for 30 min

↓

Straining the pulp

↓

Again heating same pulp with ½ quantity of water

↓

Second time straining the pulp

↓

Pulp extract keep for sedimentation 12 hrs

↓

Take supernant

↓

Pectin test (add pectin/sugar as per requirement)

↓

Boiling to get jelly test

↓

Filling pasteurized hot bottles

↓

Waxing and Sealing

↓

Pasteurization

↓

Labeling and Storage

↓

Marketing

Figure 30.2: Flow sheet for preparation of woodapple fruit pulp

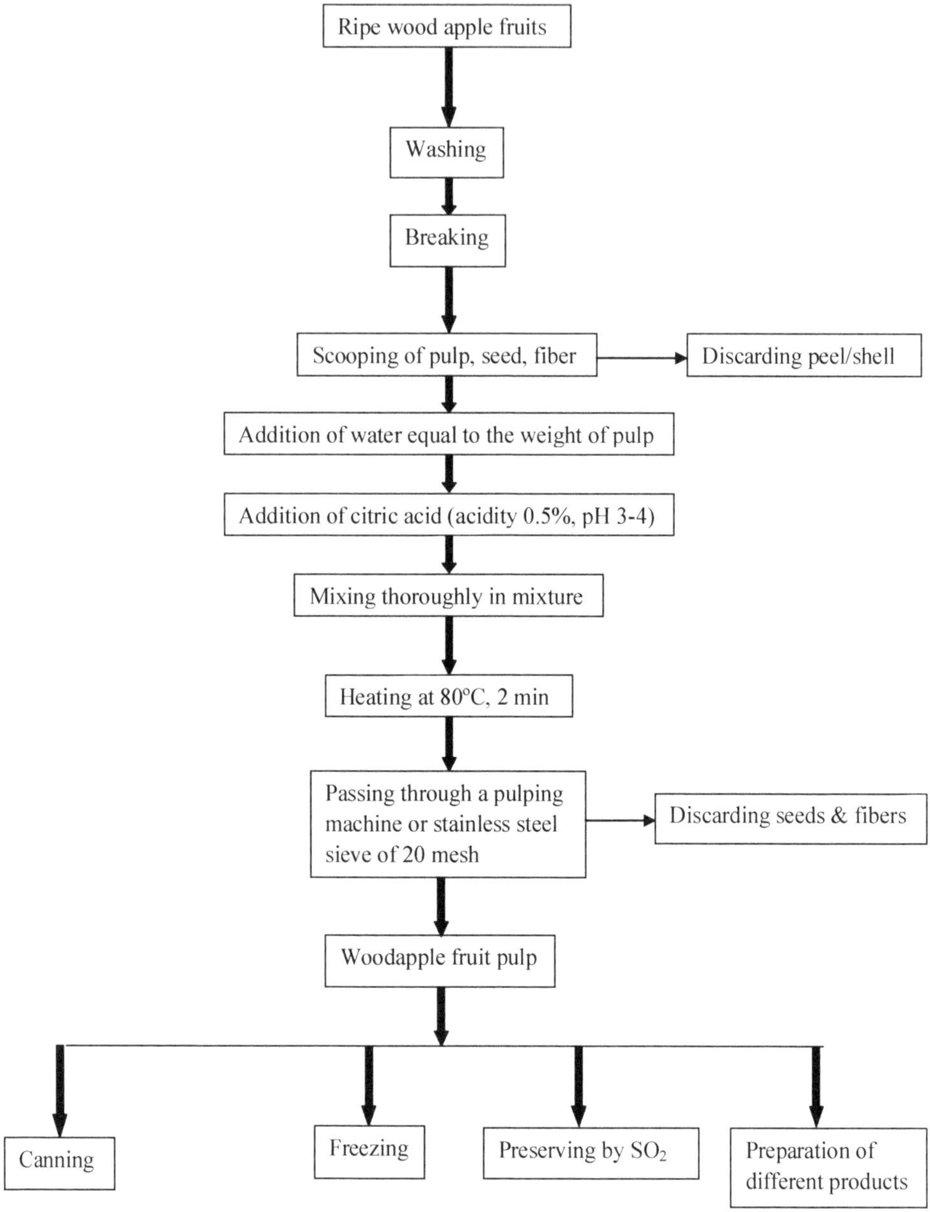

Figure 30.3: Flow sheet for preparation of woodapple chutney

Woodapples (Ripe)

↓

Washing

↓

Breaking of shell

↓

Removal of pulp

↓

Mixing with sugar, salt and leaving for hour

↓

Rest of process similar to that for sweet mango chutney

Sl.No.	*Item*	*Quantity*	*Sl.No.*	*Item*	*Quantity*
1.	Wood apple pulp	1 Kg	8.	Cinnamon	10 g
2.	Salt	45 g	9.	Cardamom	10 g
3.	Sugar	1 Kg	10.	Aniseed	10 g
4.	Onion (chopped)	50 g	11.	Black pepper (powder)	10 g
5.	Ginger (chopped)	50 g	12.	Clove (headless)	5 No.
6.	Garlic (chopped)	15 g	13.	Vinegar	100 ml
7.	Red chilli powder	10 g			

31

Bitter Gourd

Momordica charantia, called bitter melon or bitter gourd in English, is a tropical and subtropical vine of the family Cucurbitaceae, widely grown in Asia, Africa, and the Caribbean for its edible fruit, which is among the most bitter of all fruits. There are many varieties that differ substantially in the shape and bitterness of the fruit.

This herbaceous, tendril-bearing vine grows to 5 meters. It bears simple, alternate leaves 4–12 cm across, with 3–7 deeply separated lobes. Each plant bears separate yellow male and female flowers. In the Northern Hemisphere, flowering occurs during June to July and fruiting during September to November. The fruit has a distinct warty exterior and an oblong shape. It is hollow in cross-section, with a relatively thin layer of flesh surrounding a central seed cavity filled with large flat seeds and pith. The fruit is most often eaten green, or as it is beginning to turn yellow. At this stage, the fruit's flesh is crunchy and watery in texture, similar to cucumber, chayote or green bell pepper, but bitter. The skin is tender and edible. Seeds and pith appear white in unripe fruits; they are not intensely bitter and can be removed before cooking. As the fruit ripens, the flesh becomes tougher, bitterer, and too distasteful to eat. On the other hand, the pith becomes sweet and intensely red; it can be eaten uncooked in this state, and is a popular ingredient in some Southeast Asian salads. When the fruit is fully ripe it turns orange and mushy, and splits into segments which curl back dramatically to expose seeds covered in bright red pulp.

Varieties

Bitter melon comes in a variety of shapes and sizes. The China phenotype is 20–30 cm long, oblong with bluntly tapering ends and pale green in color, with a gently undulating, warty surface. The bitter melon more typical of India has a narrower shape with pointed ends, and a surface covered with jagged, triangular "teeth" and ridges. It is green to white in color. Between these two extremes are any numbers of intermediate forms. Some bear miniature fruit of only 6–10 cm in length, which may be served individually as stuffed vegetables. These miniature fruit are popular in India and elsewhere in Southeast Asia.

Culinary Uses

Bitter melon is generally consumed cooked in the green or early yellowing stage. The young shoots and leaves of the bitter melon may also be eaten as greens.

Bitter melon is often used in Chinese cooking for its bitter flavor, typically in stir-fries (often with pork and *douchi*), soups, and also as tea. It has also been used in place of hops as the bittering ingredient in some Chinese beers. It is very popular throughout South Asia. In Northern India, it is often prepared with potatoes and served with yogurt on the side to offset the bitterness, or used in *sabji*. In North Indian cuisine it is stuffed with spices and then cooked in oil. In Southern India it is used in the dishes *thoran*/ *thuvaran* (mixed with grated coconut), *theeyal* (cooked with roasted coconut) and *pachadi* (which is considered a medicinal food for diabetics). Other popular recipes include preparations with curry, deep fried with peanuts or other ground nuts, and *pachi pulusu* a soup with fried onions and other spices.

Medicinal Uses

Bitter melon has been used in various Asian and African traditional medicine systems for a long time. In Turkey it has been used as a folk remedy for a variety of ailments, particularly stomach complaints. The fruit is broken up and soaked in either olive oil or honey.

Active Substances

The plant contains several biologically active compounds, chiefly momordicin I and II, and cucurbitacin B. The plants contains also several bioactive glycosides (including momordin, charantin, charantosides, goyaglycosides, momordicosides) and other terpenoid compounds (including momordicin-28, momordicinin, momordicilin, momordenol, and momordol). It also contains cytotoxic (ribosome-inactivating) proteins such as momorcharin and momordin.

Antihelmintic

Bitter melon is used as a folk medicine in Togo to treat gastrointestinal diseases, and extracts have shown activity in vitro against the nematode worm *Caenorhabditis elegans*.

Antimalarial

Bitter melon is traditionally regarded in Asia as useful for preventing and treating malaria. Tea from its leaves is used for this purpose also in Panama and Colombia. In Guyana, bitter melons are boiled and stir-fried with garlic and onions. This popular side dish known as corolla is served to prevent malaria. Laboratory studies have confirmed that species related to bitter melon have anti-malarial activity, though human studies have not yet been published.

Antiviral

In Togo the plant is traditionally used against viral diseases such as chickenpox and measles. Tests with leaf extracts have shown *in vitro* activity against the herpes simplex type 1 virus, apparently due to unidentified compounds other than the momordicins.

Laboratory tests suggest that compounds in bitter melon might be effective for treating HIV infection. As most compounds isolated from bitter melon that impact HIV have either been proteins or lectins, neither of which are well-absorbed, it is unlikely that oral intake of bitter melon will slow HIV in infected people. It is possible oral ingestion of bitter melon could offset negative effects of anti-HIV drugs, if a test tube study can be shown to be applicable to people.

Cardioprotective

Studies in mice indicate that bitter melon seed may have a cardioprotective effect by down-regulating the NF-B inflammatory pathway.

Diabetes

In 1962, Lolitkar and Rao extracted from the plant a substance, which they called charantin, which had hypoglycaemic effect on normal and diabetic rabbits. Another principle, active only on diabetic rabbits, was isolated by Visarata and Ungsurungsie in 1981. Bitter melon has been found to increase insulin sensitivity. In 2007, a study by the Philippine Department of Health determined that a daily dose of 100 mg per kilogram of body weight is comparable to 2.5 mg/kg of the anti-diabetes drug glibenclamide taken twice per day. Tablets of bitter melon extract are sold in the Philippines as a food supplement and exported to many countries. Other compounds in bitter melon have been found to activate the AMPK, the protein that regulates glucose uptake (a process which is impaired in diabetics).

Bitter melon also contains a lectin that has insulin-like activity due to its non-protein-specific linking together to insulin receptors. This lectin lowers blood glucose concentrations by acting on peripheral tissues and, similar to insulin's effects in the brain, suppressing appetite. This lectin is likely a major contributor to the hypoglycemic effect that develops after eating bitter melon.

Anticancer

Two compounds extracted from bitter melon, α-eleostearic acid (from seeds) and 15,16-dihydroxy-α-eleostearic acid (from the fruit) have been found to induce apoptosis of leukemia cells *in vitro*. Diets containing 0.01 per cent bitter melon oil (0.006 per cent as α-eleostearic acid) were found to prevent azoxymethane-induced colon carcinogenesis in rats. Researchers at Saint Louis University claims that an extract from bitter melon, commonly eaten and known as karela in India, causes a chain of events which helps to kill breast cancer cells and prevents them from multiplying.

Other Uses

Bitter melon has been used in traditional medicine for several other ailments, including dysentery, colic, fevers, burns, painful menstruation, scabies and other skin problems. It has also been used as abortifacient, for birth control, and to help childbirth.

Health Benefits of Bitter Gourd

- ☆ Very low in calories provides only 17 cal per 100g. The pods are rich in phytonutrients like dietary fiber, minerals, vitamins and anti-oxidants.
- ☆ Bitter melon notably contains phyto-nutrient, polypeptide-P; a plant insulin known to lower blood sugar levels. In addition it also contain hypoglycemic agent called charantin. *Charantin* increases glucose uptake and glycogen synthesis in the cells of liver, muscle and adipose tissue. Together, these compounds are thought to be responsible for reduction of blood sugar levels in the treatment of type-2 diabetes.
- ☆ Fresh pods are an excellent source of foliates; contains about 72 mcg/100g (Provides 18 per cent of RDA). Foliate helps reduce incidence of neural tube defects in pregnant mothers when taken during early pregnancy.
- ☆ Fresh bitter melon is an excellent source of vitamin-C (100g of raw pod provides about 140 per cent of RDI). Vitamin-C, one of the powerful natural antioxidant, helps body scavenge deleterious free radicals one of the reasons for cancers development.
- ☆ The vegetable also an excellent source of health benefiting flavonoids such as β-carotene, α- carotene, lutein, zeaxanthins. It also contains good amount of vitamin A. These compounds help act as protective scavengers against oxygen-derived free radicals and reactive oxygen species (ROS) that play a role in aging, cancers and various disease processes.
- ☆ Bitter melon stimulates digestion and peristalsis which can be helpful in relieving indigestion and constipation problems.
- ☆ The vegetable is also good source of Niacin (vitamin B_3), Pantothenic acid (vitamin B_5), Pyridoxine (vitamin B_6) and minerals such as iron, zinc, potassium, manganese and magnesium.
- ☆ Early laboratory tests suggest that compounds in bitter melon might be effective for treating HIV infection.

Cautions

The seeds of bitter melon contain vicine and therefore can trigger symptoms of favism in susceptible individuals. In addition, the red arils of the seeds are reported to be toxic to children, and the fruit is contraindicated during pregnancy.

Bitter Gourd (Boiled, Drained, No Salt)

Nutritional value per 100 g (3.5 oz)	
Energy	79 kJ (19 kcal)
Carbohydrates	4.32 g
Sugars	1.95 g
Dietary fiber	2.0 g
Fat	0.18 g
Saturated	0.014 g
Monounsaturated	0.033 g
Polyunsaturated	0.078 g
Protein	0.84 g
Water	93.95 g
Vitamin A equiv.	6 μg
Thiamine (vit. B_1)	0.051 mg
Riboflavin (vit. B_2)	0.053 mg
Niacin (vit. B_3)	0.280 mg
Vitamin B_6	0.041 mg
Folate (vit. B_9)	51 μg
Vitamin B_{12}	0 μg
Vitamin C	33.0 mg
Vitamin E	0.14 mg
Vitamin K	4.8 μg
Calcium	9 mg
Iron	0.38 mg
Magnesium	16 mg
Phosphorus	36 mg
Potassium	319 mg
Sodium	6 mg
Zinc	0.77 mg

Figure 31.1: Flow sheet for preparation of canned bitter gourd

Mature bitter gourd fruits
↓
Washing
↓
Grading
↓
Cutting similar size pieces or use whole fruits
↓
Blanching for 2-3 min
↓
Cooling in 1.5 per cent salt solution
↓
Filling cans with bitter gourd in 2 per cent salt solution
↓
Pasteurization at 79-82 °C for 30-50 min
↓
Sealing cans and labeling
↓
Storage and marketing

Figure 31.2: Flow sheet for preparation of dried bitter gourd

Mature bitter gourd fruits
↓
Cleaning and washing
↓
Grading
↓
Whole fruits or similar size cut pieces
↓
Blanching for 5 min
↓
Drying at 55 °C 14 hrs
↓
Packing in plastic bags
↓
Sealing and labeling
↓
Storage and marketing

Figure 31.3: Flow sheet for preparation of bitter gourd pickle

Mature Green Bitter Gourds
↓
Washing
↓
Making incisions
↓
Removal of pulp and seed
↓
Mixing a little salt
↓
Keeping in sun for 3 hrs
↓
Draining off water
↓
Shade drying, 2 hrs
↓
Frying spices in oil
↓
Cooling
↓
Mixing with salt and lime juice
↓
Filling into bitter gourds
↓
Packaging in jar
↓
Addition of oil after heating and cooling it
↓
Sealing
↓
Labeling
↓
Marketing

Sl.No.	*Item*	*Quantity*	*Sl.No.*	*Item*	*Quantity*
1.	Bitter gourds	1 kg	6.	Cumin	10 g
2.	Salt	150 g	7.	Cardamom (large)	10 g
3.	Red chilli powder	10 g	8.	Tamarind pulp or amchur	250 g
4.	Turmeric powder	10 g	9.	Mustard ground	100 g
5.	Aniseed powder	10 g	10.	Mustard oil	500 ml

32

Bottle Gourd

The bottle gourd (*Lagenaria siceraria*; Family, Cucurbitaceae) is a common vegetable in India. Indian Names are Lauki, Doodhi Bhopala, Ghia and Kaddu, cucuzzi squash. It is yellowish green, having the shape of a bottle. It has white pulp, with white seeds embedded in spongy flesh.

Origin and Distribution

Not very familiar to the western world, bottle gourd is one of the favorite vegetables in Indian cuisine. Known by various names, the bottle gourd is yellowish-green in color and usually in the shape of a bottle. The vegetable has a white pulp, with white seeds embedded in the spongy flesh. It is also known as bottle squash, white gourd, trumpet gourd, lauki, doodhi, ghia and calabash gourd. The bottle gourds have been cultivated since time immemorial and they are probably one of the earliest vegetables cultivated by man. It appears to have originated in Africa where it occurs spontaneously as it also does in India. It is now widely cultivated throughout the tropics, especially India, Sri Lanka, Indonesia, Malaysia, the Philippines, China, tropical Africa and South America. The bottle gourd is a warm season crop and grows best in a warm humid climate.

The bottle gourd is a hairy, rapid growing trailing or climbing annual herb, extending 30 feet or more in length. It has stout 5-angled stems, long-stalked, 5-lobed leaves and large yellow flowers. The fruits which may exceed 5 feet in length vary in shape. In India, their woody shells serve as utensils and as sounding boxes for the sitar and ektara used by the bauls of Bengal. The pulp of the fruit is eaten as a cooked vegetable. Raw juice of this fruit is used.

Benefits of Bottle Gourd and Bottle Gourd Juice

- ✰ The cooked vegetable is cooling, diuretic, sedative and anti-bilious. It gives a feeling of relaxation after eating it.

- ☆ Bottle gourd is very valuable in urinary disorders. A glassful of fresh juice prepared by grating the whole fruit should be mixed with teaspoonful of lime juice. It should be given once daily in the treatment of burning sensation in urinary passage due to high acidity of urine. It serves as an alkaline mixture.
- ☆ The juice of bottle gourd is a valuable medicine for excessive thirst due to severe diarrhea, diabetes and excessive use of fatty or fried foods. A glassful of plain juice with a pinch of salt should be taken every day in this condition. Its use during summer prevents excessive loss of sodium, quenches thirst and helps in preventing fatigue.
- ☆ Mixture of bottle gourd juice and sesame oil acts as an effective medicine for insomnia. It should be massaged over scalp every night.
- ☆ The Gourd fruit juice is used in the treatment of insanity, epilepsy and other nervous diseases.
- ☆ The juice of the fruit is used in the treatment of stomach acidity, indigestion and ulcers.
- ☆ Bottle Gourd helps in constipation. Bottle gourd is not only rich in essential minerals, iron, protein and trace elements; it is also rich in fiber. Fiber is known to be missing in the modern diet, the absence of which is the cause of not only constipation but other digestive disorders like flatulence and even piles. Hence diet using different preparations made from this vegetable is useful.

Nutritive Value per 100 gm edible portion

Food Value		*Minerals and Vitamins*	
Moisture	96.1 per cent	Calcium	20 mg
Protein	0.2 per cent	Phosphorus	10 mg
Fat	0.1 per cent	Iron	0.7 mg
Minerals	0.5 per cent	Small amounts of Vitamin B Complex	
Fiber	0.6 per cent	Calorific Value	12
Carbohydrates	2.5 per cent		

Nutrition and Health Benefits of Eating Bottle Gourds

- ☆ High in sodium and potassium, bottle gourd is an excellent vegetable for hypertension patients.
- ☆ It is a suitable vegetable for light, low-cal diets as well as for children, people with digestive problems, diabetics and convalescents.
- ☆ High in water content, calabash gourd is cooling, calming, diuretic and easy to digest when cooked.
- ☆ The vegetable is very effective in treating constipation and other digestive disorders.

- It helps in treating urinary disorders, since it serves as an alkaline mixture for treating burning sensation in urinary passage, caused by high acidity of urine.
- Bottle gourd is recommended by ayurvedic physicians for balancing the liver function, when the liver is inflamed and cannot efficiently process food for maximum nutrition and assimilation.
- Drinking a glass of bottle gourd juice once in the early morning helps in treating grey hair.
- Mixing the juice with sesame oil provides an effective medicine for insomnia. Massage the scalp with this preparation every night.
- Bottle gourd juice is very helpful in treating insanity, epilepsy, stomach acidity, indigestion, ulcers and other nervous diseases.
- The bitter variety of calabash gourd is considered as a cardiac tonic, an antidote for poisoning and a tonic for alleviating bronchitis, cough, asthma, and biligenic affections.
- A glass of bottle gourd juice, mixed with a little salt, helps in preventing excessive loss of sodium from the body. It quenches thirst, prevents fatigue and keeps one refreshed during summers.
- The juice extracted from the leaves of bottle gourd is beneficial in jaundice.
- Bottle gourd juice is also very useful for shedding extra calories and maintaining optimum health.

Nutrients	*Amount*
Basic Components	
Protein	0.4 g
Carbohydrate	2 g
Water	55.4 g
Ash	0.2 g
Total Omega-6 Fatty Acids	5.2 mg
Calories	
Total Calories	8.1
Calories from Carbohydrate	7.1
Calories from Fat	0.1
Calories from Protein	0.9
Vitamins	
Vitamin A	9.3 IU
Vitamin C	5.9 mg
Niacin	0.2 mg
Folate	3.5 mcg
Pantothenic Acid	0.1 mg

Nutrients	*Amount*
Minerals	
Calcium	15.1 mg
Iron	0.1 mg
Magnesium	6.4 mg
Phosphorus	7.5 mg
Potassium	87 mg
Sodium	1.2 mg
Zinc	0.4 mg
Selenium	0.1 µg

Figure 32.1: Flow sheet for preparation of bottle gourd juice

Tender bottle gourd fruits

↓

Washing

↓

Sorting and trimming (removing green portion and stem)

↓

Cutting and chopping into small pieces

↓

Heating at 70-90 °C, 3-5 min to soften

↓

Pulping or extraction of juice/pulp (mechanically or sieving)

↓

Addition of salt or citric acid

↓

Homogenization

↓

Heating at 82-88 °C, 3 min

↓

Filling hot into bottles or cans

↓

Pasteurization in boiling water for 30 min

↓

Cooling and Labeling

↓

Storage and Marketing

Figure 32.2: Flow sheet for preparation of bottle gourd powder

Tender bottle gourd fruits

↓

Washing

↓

Sorting and trimming
(removing green portion and stem)

↓

Cutting and chopping into small pieces

↓

Heating at 70–90 °C, 3–5 min to soften

↓

Pulping or extraction of juice/pulp
(mechanically or sieving)

↓

Homogenization

↓

Drying at 40–50 °C

↓

Grinding and sieving (60 mesh)

↓

Filling into plastic pouch

↓

Sealing and Labeling

↓

Storage and Marketing

Figure 32.3: Flow sheet for preparation of tutti-frutti from bottle gourd

Mature bottle gourd fruits
↓
Washing
↓
Peeling
↓
Cutting into long slices
↓
Removal of seeds
↓
Cutting into cubes by machine
↓
Blanching 3 min
↓
Soaking in 1 per cent $CaCl_2$ solution 3 h
↓
Syruping in 70 per cent + 1 per cent citric acid and boiling 2 h
↓
Cooling
↓
Washing under running water
↓
Drying
↓
Packaging
↓
Storage

33

Brinjal

The eggplant, aubergine, melongene, brinjal or guinea squash (*Solanum melongena*) is a plant of the family Solanaceae (also known as the nightshades) and genus *Solanum*. It bears a fruit of the same name, commonly used in cooking. As a nightshade, it is closely related to the tomato and potato and is native to India.

It is a delicate perennial often cultivated as an annual. It grows 40 to 150 cm (16 to 57 in) tall, with large coarsely lobed leaves that are 10 to 20 cm (4–8 in) long and 5 to 10 cm (2–4 inch) broad. Semi wild types can grow much larger, to 225 cm (7 ft) with large leaves over 30 cm (12 in) long and 15 cm (6 in) broad. The stem is often spiny. The flowers are white to purple, with a five-lobed corolla and yellow stamens. The fruit is fleshy, has a meaty texture, and is less than 3 cm (1.2 in) in diameter on wild plants, but much larger in cultivated forms. The fruit is botanically classified as a berry, and contains numerous small, soft seeds, which are edible, but have a bitter taste because they contain nicotinoid alkaloids; this is unsurprising as it is a close relative of tobacco.

The plant is native to the Indian subcontinent. It has been cultivated in southern and eastern Asia since prehistory but appears to have become known to the Western world no earlier than ca. 1500. The first known written record of the plant is found in Qí mín yào shù, an ancient Chinese agricultural treatise completed in 544. The numerous Arabic and North African names for it, along with the lack of the ancient Greek and Roman names, indicate it was introduced throughout the Mediterranean area by the Arabs in the early Middle Ages. The scientific name *Solanum melongena* is derived from a 16th century Arabic term for one variety.

The name *eggplant*, rather than *aubergine*, is used in the United States, Australia, New Zealand, and Canada, and refers to the fruits of some 18th century European cultivars which were yellow or white and resembled goose or hen's eggs. In Indian, South African, Malaysian and Singaporean English, the fruit is known as a "vengan", "baingan" or *brinjal*, with the latter being derived directly from the Portuguese *beringela*. A less common British English word is *melongene*, which is also from French (derived) from Italian "melanzana" from Greek "μåλéôæÜíα". In the Caribbean Trinidad, it also goes by "meloongen" from melongene.

Cooking

The raw fruit can have a somewhat bitter taste, but becomes tender when cooked and develops a rich, complex flavor. Traditionally, recipes would advise the salting, rinsing and draining of the sliced fruit (known as "degorging") to soften it and to reduce the amount of fat absorbed during cooking, but mainly to remove the bitterness of the earlier cultivars. Some modern varieties - including those large, purple varieties commonly imported into Western Europe - do not need this treatment. The fruit is capable of absorbing large amounts of cooking fats and sauces, allowing for very rich dishes, but the salting process will reduce the amount of oil absorbed. The fruit flesh is smooth; as in the related tomato, the numerous seeds are soft and edible along with the rest of the fruit. The thin skin is also edible, so peeling is not required.

The plant is used in cuisines from Japan to Spain. It is often stewed, as in the French ratatouille, the Italian *parmigiana di melanzane*, the Turkish *musakka*, and Middle-Eastern and South Asian dishes. Eggplants can also be battered before deep-frying and served with a sauce made of tahini and tamarind. In Iranian cuisine, it can be blended with whey as *kashk e-bademjan*, tomatoes as *mirza ghasemi* or made into stew as *khoresh-e-bademjan*. It can be sliced and deep-fried, then served with plain yoghurt, (optionally) topped with a tomato and garlic sauce, such as in the Turkish dish *patlýcan kýzartmasý* or without yoghurt as in *patlýcan pakþuka*. However, arguably the most famous Turkish eggplant dish is *Imam baytldi.*

It may also be roasted in its skin until charred, so the pulp can be removed and blended with other ingredients, such as lemon, tahini, and garlic, as in the Middle Eastern dish *baba ghanoush* and the similar Greek dish *melitzanosalata*. Grilled, mashed and mixed with onions, tomatoes and spices make the Indian dish *baingan ka Bhartha* or *gojju*, similar to salatã de vinete in Romania, while a mix of roasted eggplant, roasted red peppers, chopped onions, tomatoes, mushrooms, carrots, celery and spices is called *zacuscã* in Romania or *ajvar* in Serbia and the Balkans. A simpler version of the dish, Begun-Pora (eggplant-charred or burnt), is popular in the east Indian state of Bengal and Bangladesh where the charred pulp is mixed with raw chopped onions, green chillies, salt and mustard oil. As a native plant, it is widely used in Indian cuisine, for example in *sambhar, dalma* (a dal preparation with vegetables, native to Orissa), chutney, curry, and *achaar*. Owing to its versatile nature and wide use in both everyday and festive Indian food, it is often described (under the name brinjal) as the 'King of Vegetables'. In one dish, brinjal is stuffed with ground coconut, peanuts, and masala, and then cooked in oil.

Health Properties

Studies of the Institute of Biology of São Paulo State University, Brazil, have shown eggplant is effective in the treatment of high blood cholesterol. Another study from Heart Institute of the University of São Paulo found no effects at all and does not recommend eggplant as a replacement to statins. It helps to block the formation of free radicals and is also a source of folic acid and potassium. Eggplant is richer in nicotine than any other edible plant, with a concentration of 100 ng/g (or 0.01 mg/100g). However, the amount of nicotine from eggplant or any other food is negligible compared to passive smoking. On average, 20 lbs (9 kg) of eggplant contains about the same amount of nicotine as a cigarette.

Allergies

Case reports of itchy skin and/or mouth after handling and/or eating eggplant have been reported anecdotally and published in medical journals (see also oral allergy syndrome). A recent (2012) study of a sample of 741 people in India (where eggplant is commonly consumed) found nearly 10 per cent reported some allergic symptoms after consuming eggplant, while 1.4 per cent showed symptoms in less than 2 hours. Contact dermatitis from eggplant leaves and allergy to eggplant flower pollen have also been reported. Individuals who are atopic (genetically predisposed to hypersensitivity, such as hay fever) are more likely to have a reaction to eggplant, which may be because eggplant is high in histamines. A few proteins and at least one secondary metabolite have been identified as potential allergens. Cooking eggplant thoroughly seems to preclude reactions in some individuals, but at least one of the allergenic proteins survives the cooking process.

Eggplant, Raw Fruits

Nutritional value per 100 g (3.5 oz)	
Energy	102 kJ (24 kcal)
Carbohydrates	5.7 g
Sugars	2.35 g
Dietary fiber	3.4 g
Fat	0.19 g
Protein	1.01 g
Thiamine (vit. B_1)	0.039 mg
Riboflavin (vit. B_2)	0.037 mg
Niacin (vit. B_3)	0.649 mg
Pantothenic acid (B_5)	0.281 mg
Vitamin B_6	0.084 mg
Folate (vit. B_9)	22 μg
Vitamin C	2.2 mg
Calcium	9 mg
Iron	0.24 mg
Magnesium	14 mg
Manganese	0.25 mg
Phosphorus	25 mg
Potassium	230 mg
Zinc	0.16 mg

Figure 33.1: Flow sheet for preparation of brinjal pickle

Mature Brinjal Fruits
↓
Washing
↓
Cutting into suitable size pieces
↓
Boiling in 2 per cent salt solution for 5 min
↓
Draining off water
↓
Shade drying
↓
Mixing pieces with spices
↓
Filling in jar
↓
Addition of oil after heating and cooling
↓
Keeping in sun for a week
↓
Sealing
↓
Labeling
↓
Storage
↓
Marketing

Sl.No.	*Item*	*Quantity*	*Sl.No.*	*Item*	*Quantity*
1.	Brinjal	1 kg	6.	Cumin	10 g
2.	Salt	100 g	7.	Cardamom (large)	10 g
3.	Red chilli powder	10 g	8.	Black pepper	10 g
4.	Turmeric powder	10 g	9.	Mustard ground	50 g
5.	Aniseed powder	10 g	10.	Mustard oil	350 ml

34

Cabbage

Cabbage is a popular cultivar of the species *Brassica oleracea* Linn. (Capitata Group) of the Family Brassicaceae (or Cruciferae) and is a leafy green vegetable. It is a herbaceous, biennial, dicotyledonous flowering plant distinguished by a short stem upon which is crowded a mass of leaves, usually green but in some varieties red or purplish, which while immature form a characteristic compact, globular cluster (*cabbagehead*). The plant is also called *head cabbage* or *heading cabbage*, and in Scotland a *bowkail*, from its rounded shape. The Scots call its stalk a *castock*, and the British occasionally call its head a *loaf*. It is in the same genus as the turnip – *Brassica rapa*. Cabbage leaves often have a delicate, powdery, waxy coating called *bloom*. The occasionally sharp or bitter taste of cabbage is due to glucosinolate(s). Cabbages are also a good source of riboflavin.

The cultivated cabbage is derived from a leafy plant called the wild mustard plant, native to the Mediterranean region, where it is common along the seacoast. Also called *sea cabbage* and *wild cabbage*, it was known to the ancient Greeks and Romans; Cato the Elder praised this vegetable for its medicinal properties, declaring that "It is the cabbage that surpasses all other vegetables".

Uses

The only part of the plant that is normally eaten is the leafy head; more precisely, the spherical cluster of immature leaves, excluding the partially unfolded outer leaves. Cabbage is used in a variety of dishes for its naturally spicy flavor. The so-called "cabbage head" is widely consumed raw, cooked, or preserved in a great variety of dishes. It is the principal ingredient in coleslaw.

Cabbage is often added to soups or stews. Cabbage soup is popular in Central and Eastern Europe, and cabbage is an ingredient in some kinds of borscht. *Garbure* (from Provençal *garburo*) is a thick soup of cabbage or other vegetables with bacon. Cabbage may be an ingredient in *kugel*, a baked pudding served as a side dish or dessert. Cabbage is also used in many popular dishes in India. Boiling tenderizes the leaves and releases sugars, which leads to the characteristic "cabbage" aroma. Boiled cabbage has become stigmatized because of its strong cooking odor and the fact that it causes flatulence. Moreover, boiling reduces the cabbage's anticancer properties. It is often prepared and served with boiled meat and other vegetables as part of a *boiled dinner*. Harold McGee has studied the development of unpleasant smells when cooking brassicas and reports that they develop with prolonged cooking. According to Corriher's Compendium, smell doubles when prolonging cooking from five to seven minutes; for best results cabbage should be sliced thinly and cooked for four minutes.

Fermented and Preserved

Cabbage is the basis for German *Sauerkraut*. Chinese *suan cai* and Korean *kimchi* are produced using the related Chinese cabbage. To pickle cabbage, it is covered with brine made of its own juice with salt, and left in a warm place for several weeks to ferment. Sauerkraut (colloquially referred to as "kraut") was historically prepared at home in large batches, as a way of storing food for the winter. The word comes from German *sauer* (sour) and *kraut* (plant or cabbage) (Old High German *sûr* and *krût*). Cabbage can also be pickled in vinegar with various spices, alone or in combination with other vegetables (turnips can be cured in the same way). Korean *baechu kimchi* is usually sliced thicker than its European counterpart, and the addition of onions, chilies, minced garlic and ginger is common.

Medicinal Properties

Cabbage is an excellent source of vitamin C. It also contains significant amounts of glutamine, an amino acid that has anti-inflammatory properties. Cabbage can also be included in dieting programs, as it is a low calorie food. Along with broccoli and other *Brassica* vegetables, cabbage is a source of indole-3-carbinol, a chemical which boosts DNA repair in cells and appears to block the growth of cancer cells. The compound is also used as an adjuvant therapy for recurrent respiratory papillomatosis, a disease of the head and neck caused by human papillomavirus (usually types 6 and 11) that causes growths in the airway that can lead to death. Boiling reduces anticancer properties. In European folk medicine, cabbage leaves are used to treat acute inflammation. A paste of raw cabbage may be placed in a cabbage leaf and wrapped around the affected area to reduce discomfort. Some claim it is effective in relieving painfully engorged breasts in breastfeeding women. Fresh cabbage juice has been shown to promote rapid healing of peptic ulcers.

Effect on the Thyroid Gland

Cabbage may also act as a goitrogen. It blocks organification in thyroid cells, thus inhibiting the production of the thyroid hormones (thyroxine and triiodothyronine). The result is an increased secretion of thyroid-stimulating hormone (TSH) due to low thyroid hormone levels. This increase in TSH results in an enlargement of the thyroid gland (goiter).

Varieties

There are many varieties of cabbage based on shape and time of maturity. Cabbages grown late in autumn and in the beginning of winter are called "coleworts"; their leaves do not form a compact head. "Colewort" may also refer to a young cabbage. The word comes from Latin *caulis* (stalk of a plant, cabbage) and Old English *wyrt* (herb, plant, and root). A "drumhead" cabbage has a rounded, flattened head. An "oxheart" cabbage has an oval or conical head. A "pickling" cabbage, such as the red-leafed cabbage, is especially suitable for pickling; 'Krautman' is the most common variety for commercial production of sauerkraut. Red cabbage is a small, round-headed type with dark red leaves. Savoy cabbage has a round, compact head with crinkled and curled leaves. Winter cabbage will survive the winter in the open in mild regions such as the southern United States; the name is also used for Savoy cabbage. Other traditional varieties include white cabbage, 'Late Flat Dutch', 'Early Jersey Wakefield' (a conical variety), 'Danish Ballhead' (late, round-headed), 'Cuor di Bue Grosso' (conical heads, from Italy), 'Copenhagen Market Cabbage' (large round heads, from Denmark), and 'Mammoth Red Rock' (large round heads with deep red leaves).

Cabbage, Raw

Nutritional value per 100 g (3.5 oz)	
Energy	103 kJ (25 kcal)
Carbohydrates	5.8 g
Sugars	3.2 g
Dietary fiber	2.5 g
Fat	0.1 g
Protein	1.28 g
Thiamine (vit. B_1)	0.061 mg
Riboflavin (vit. B_2)	0.040 mg
Niacin (vit. B_3)	0.234 mg
Pantothenic acid (B_5)	0.212 mg
Vitamin B_6	0.124 mg
Folate (vit. B_9)	53 μg
Vitamin C	36.6 mg
Vitamin K	76 μg
Calcium	40 mg
Iron	0.47 mg
Magnesium	12 mg
Phosphorus	26 mg
Potassium	170 mg
Zinc	0.18 mg

Figure 34.1: Flow sheet for preparation of sauerkraut (acid cabbage)

Cabbage (mature, white, sound headed)
↓
Removal of core
↓
Trimming
↓
Washing and slicing
↓
Putting shreds in jars or vats
↓
Addition of 2.25 per cent salt
↓
Covering mouth of jar with plastic cover
↓
Allowing fermenting for a week at 16-22 °C
↓
Removal of scum
↓
Draining out juice
↓
Pasteurizing shreds (sauerkraut at 75-80 °C for 10 min)
↓
Filling hot into sterilized jars or wide mouthed bottles
↓
Pasteurizing previously drained out juice at 90 °C
↓
Pouring hot on to sauerkraut
↓
Sealing and Labeling
↓
Storage and Marketing

35

Cardamom

Cardamom is a seed pod, known since antiquity for its culinary and medicinal properties. The spice is native to evergreen rain forest of southern India and now grown in only few tropical countries. Botanicallay, it belongs to the family of *"zingiberaceae"* and consists of two genera; *elettaria* and *amomum.*

Genarally, the plant grows up to 4 meters in length and starts bearing fruits known as seed pods after about two years of implantation. Each pod measuring about 1-2 cm in length.

Both varieties of cardamom feature three sided pods with a thin papery outer cover and small black seeds that are arranged in vertical rows. Elettaria pods are small and light green in color, while Amomum pods are larger and dark brown. The pods are being used as flavoring base in both food and drink, in cooking recipes and as well as in medicine.

Black cardamom (*Amomum subulatum*) also known as *Nepal cardamom* is relatively big sized pod of same zingiberaceae family. The pod has dark brown rough outer coat, measure about 2-4 cm in length and 1-2 cm in diameter. The pods have camphor like intense flavor commonly used in spicy stews in sub Himalayan plains of India, Pakistan, Nepal as well as in China.

Health Benefits of cardamom

- ✰ This exotic spice contains many plant derived chemical compounds that are known to have anti-oxidant, disease preventing and health promoting properties.
- ✰ The spicy pods contain many essential volatile oils that include *pinene, sabinene, myrcene, phellandrene, limonene, 1, 8-cineole, terpinene, p-cymene, terpinolene, linalool, linalyl acetate, terpinen-4-oil, α-terpineol, α-terpineol acetate, citronellol, nerol, geraniol, methyl eugenol,* and *trans-nerolidol.*

- The therapeutic properties of cardamom-oil have found application in many traditional medicines as antiseptic, antispasmodic, carminative, digestive, diuretic, expectorant, stimulant, stomachic and tonic.
- Cardamom is a good source of minerals like potassium, calcium, and magnesium. Potassium in an important component of cell and body fluids that helps control heart rate and blood pressure. Copper is required in the production of red blood cells.
- It is also an excellent source of manganese and iron. Manganese is a co-factor for the enzyme superoxide dismutase, which is a very powerful free radical scavenger. Iron is required for red blood cell formation.
- The pods are rich in many vital vitamins including riboflavin, niacin, vitamin-C that is essential for optimum health.

Selection and Storage

Fresh cardamom pods as well as its powder are available in the markets year around. Both varieties of pods are available in the markets. "Elettaria" pods are small and light green in color, while 'Amomum" pods are larger and dark brown. Fresh pods feature heavy and give sweet aroma when scratched with finger tip. Avoid pods that appear light and that are stain or spots which may feature mold. Store the pods in airtight containers in cool, dark place. They remain fresh for longer period when placed in the refrigerator. Powdered form should be kept in airtight containers and used as early as possible since it may lose its flavor rather quickly.

Medicinal Use

The therapeutic properties of cardamom oil have found application in many traditional medicines as antiseptic and local anesthetic, antioxidant and; health promoting and disease preventing roles.

Culinary Uses

Cardamom pods are usually split and seeds are crushed using hand mill just before their use.

- This delicate spice is being used as flavoring base in both food and drink, as well as in cooking spices.
- The pods have been in use in the preparation of sweet dishes in many Asian countries.
- It is used as a flavoring base in the preparation of tea, coffee, and liquors.

Cardamom, Nutritional value per 100 g (*Source*: USDA National Nutrient data base)

Principle	*Nutrient Value*	*Percentage of RDA*
Energy	311 Kcal	15.5 per cent
Carbohydrates	68.47 g	52.5 per cent
Protein	10.76 g	19 per cent
Total Fat	6.7 g	23 per cent
Cholesterol	0 mg	0 per cent
Dietary Fiber	28 g	70 per cent

Principle	*Nutrient Value*	*Percentage of RDA*
Vitamins		
Niacin	1.102 mg	7 per cent
Pyridoxine	0.230 mg	18 per cent
Riboflavin	0.182 mg	14 per cent
Thiamin	0.198 mg	16.5 per cent
Vitamin A	0 IU	0 per cent
Vitamin C	21 mg	35 per cent
Electrolytes		
Sodium	18 mg	1 per cent
Potassium	1119 mg	24 per cent
Minerals		
Calcium	383 mg	38 per cent
Copper	0.383 mg	42.5 per cent
Iron	13.97 mg	175 per cent
Magnesium	229 mg	57 per cent
Manganese	28 mg	1217 per cent
Phosphorus	178 mg	25 per cent
Zinc	7.47 mg	68 per cent

Figure 35.1: Flow sheet for extraction of cardamom oil

Cardamom seeds

↓

Cleaning

↓

Blending with water

↓

Steam distillation

↓

115-120°C oil distilled and condenses at 30°C

↓

Collection of oil

↓

Packing in aluminium container/glass container

↓

Check quality of oil

↓

Store at cool place

↓

Marketing

36

Carrot

The carrot (*Daucus carota* subsp. *sativus*, Etymology: Middle French *carotte*, from Late Latin *carôta*, from Greek êαñüôïí *karôton*, originally from the Indo-European root *ker-* (horn), due to its horn-like shape) is a root vegetable, usually orange in color, though purple, red, white, and yellow varieties exist. It has a crisp texture when fresh. The most commonly eaten part of a carrot is a taproot, although the greens are edible as well. It is a domesticated form of the wild carrot *Daucus carota*, native to Europe and southwestern Asia. The domestic carrot has been selectively bred for its greatly enlarged and more palatable, less woody-textured edible taproot.

Methods of Consumption and Uses

Carrots can be eaten in a variety of ways. Only 3 per cent of the β-carotene in raw carrots is released during digestion: this can be improved to 39 per cent by pulping, cooking and adding cooking oil. Alternatively they may be chopped and boiled, fried or steamed, and cooked in soups and stews, as well as baby and pet foods. A well known dish is *carrots julienne*. Grated carrots are used in carrot cakes, as well as carrot puddings, an old English dish thought to have originated in the early 19th century. The greens are edible as a leaf vegetable, but are rarely eaten by humans. Together with onion and celery, carrots are one of the primary vegetables used in a *mirepoix* to make various broths.

In India carrots are used in a variety of ways, as salads or as vegetables added to spicy rice or daal dishes. The most popular variation in north India is the Gaajar Kaa Halwaa carrot dessert, which has carrots grated and cooked in milk until the whole mixture is solid, after which nuts and butter are added. Carrot salads are usually made with grated carrots in western parts with a seasoning of mustard seeds and green chillies popped in hot oil, while adding carrots to rice usually is in julienne shape.

The variety of carrot found in north India is rare everywhere except in Central Asia and other contiguous regions, and is now growing in popularity in larger cosmopolitan cities in South India. The north Indian carrot is pink-red comparable to plum or raspberry or deep red apple in color (without a touch of yellow or blue) while most other carrot varieties in world are from orange to yellow in color, comparable to Halloween pumpkins. Ever since the late 1980s, baby carrots or mini-carrots (carrots that have been peeled and cut into uniform cylinders) have been a popular ready-to-eat snack food available in many supermarkets. Carrot juice is also widely marketed, especially as a health drink, either stand-alone or blended with fruits and other vegetables.

Nutrition

The carrot gets its characteristic and bright orange color from β-carotene, which is metabolized into vitamin A in humans when bile salts are present in the intestines. Massive overconsumption of carrots can cause carotenosis, a benign condition in which the skin turns orange. Carrots are also rich in dietary fiber, antioxidants, and minerals.

Lack of vitamin A can cause poor vision, including night vision, and vision can be restored by adding it back into the diet. An urban legend says eating large amounts of carrots will allow one to see in the dark. The legend developed from stories of British gunners in World War II, who were able to shoot down German planes in the darkness of night. The legend arose during the Battle of Britain when the RAF circulated a story about their pilots' carrot consumption as an attempt to cover up the discovery and effective use of radar technologies in engaging enemy planes, as well as the use of red light (which does not destroy night vision) in aircraft instruments. It reinforced existing German folklore and helped to encourage Britons–looking to improve their night vision during the blackouts–to grow and eat the vegetable. Ethnomedically, the roots are used to treat digestive problems, intestinal parasites, and tonsillitis or constipation.

Medicinal Uses

Carrots are credited with many medicinal properties; they are said to cleanse the intestines and to be diuretic, remineralizing, antidiarrheal, an overall tonic and antianemic. Carrot is rich in alkaline elements which purify and revitalize the blood. They nourish the entire system and help in the maintenance of acid-alkaline balance in the body. The carrot also has a reputation as a vegetable that helps to maintain good eyesight. Raw grated carrot can be applied as a compress to burns for a soothing effect. Its highly energizing juice has a particularly beneficial effect on the liver. Consumed in excessive quantities, carrots can cause the skin to turn yellow; this phenomenon, which is called Carotenemia and caused by the carotene, contained in carrots, is frequently seen in young children but is not at all dangerous. An infusion of carrot seeds (1 teaspoon per cup of boiling water) is believed to be diuretic, to stimulate the appetite, reduce colic, aid fluid retention and help alleviate menstrual cramps. The dried flowers are also used as a tea as a remedy for dropsy. Taken in wine, or boiled in wine and taken, the seeds help conception. Strangely enough the seeds made into a tea have been used for centuries as a contraceptive. Applied with honey, the leaves cleanse running sores or ulcers. Carrots are also supposed to help break wind and remove stitches in the side. Chewing a carrot immediately after food kills all the harmful germs in the mouth. It cleans the teeth, removes the food particles lodged in the crevices and prevents bleeding of the gums and tooth decay. Carrot soup is supposed to relieve diarrhea and help with tonsilitis. In days gone by the grated raw carrot and gave it to children to expel worms. Pulped carrot is used as a cataplasm for application to ulcers and sores. They were also supposed to improve your memory abilities and relieve nervous tension. An Old English superstition is that the small purple flower in the centre of the Wild Carrot (Queen Annes Lace) was of benefit in curing epilepsy. Queen Annes Lace (the Wild Carrot) was also considered toxic. The leaves contain furocoumarins that may cause allergic contact dermatitis from the leaves, especially when wet. Later exposure to the sun may cause mild photodermatitis. Wild Carrot seed is also an early abortifacient, historically, sometimes used as a natural "morning after" contraceptive tea. Queen Annes Lace has long been used because of its contraceptive properties.

Medicinally the Carrot was used as a diuretic, stimulant, in the treatment of dropsy, flatulence, chronic coughs, dysentery, windy colic, chronic renal diseases and a host of other uses.

Eating carrots is also good for allergies, anemia, rheumatism, tonic for the nervous system. Everyone knows they improve vision; But it does not stop there the delicious carrot is good for diarrhea, constipation (very high in fiber), intestinal inflammation, cleansing the blood (a liver tonic), an immune system tonic. Carrot is traditionally recommended to weak, sickly or rickety children, to convalescents or pregnant women, its anti-anemic properties having been famous for a long time. Tea made the seeds can promote the onset of menstruation. It is effective on skin problems including broken veins/capillaries, burns, creeping impetigo, wrinkles and sun damage. Carrots also help in stimulating milk flow during lactation. Believe it or not the carrot is also effective against roundworms and dandruff. Pureed carrots are good for babies with diarrhea, providing essential nutrients and natural sugars.

Scientists have given us another reason to eat carrots - Falcarinol a compound found in the popular root vegetable has been found to have an effect on the development of cancer.

Alternative Medicinal Uses

The alternative medicine believers consider the carrot (the whole plant or its seeds) to have the following properties:

- ☆ Anihelmintic *(destroying or expelling worms).*
- ☆ Carminative *(expelling flatulence).*
- ☆ Contraceptive.
- ☆ Deobstruent.
- ☆ Diuretic *(promoting the discharge of urine).*

- ☆ Emmenagogue *(producing oils which stimulate the flow of menstrual blood).*
- ☆ Galactogogue *(promoting the secretion of milk).*
- ☆ Ophthalmic *(pertaining to the eye).*
- ☆ Stimulant.
- ☆ Oedema *(water retention).*

Cultivars

Carrot cultivars can be grouped into two broad classes, eastern carrots and western carrots. More recently, a number of novelty cultivars have been bred for particular characteristics. The city of Holtville, California, promotes itself as "Carrot Capital of the World", and holds an annual festival devoted entirely to the carrot.

Eastern Carrots

Eastern carrots were domesticated in Central Asia, probably in modern-day Iran and Afghanistan in the 10th century, or possibly earlier. Specimens of the eastern carrot that survive to the present day are commonly purple or yellow, and often have branched roots. The purple color common in these carrots comes from anthocyanin pigments.

Western Carrots

The western carrot emerged in the Netherlands in the 17th century, from Iran with violet color, its orange color making it popular in those countries as an emblem of the House of Orange and the struggle for Dutch independence. The orange color results from abundant carotenes in these cultivars. While orange carrots are the norm in the West, other colors do exist, including white, yellow, red, and purple. These other colors of carrot are raised primarily as novelty crops.

Western carrot cultivars are commonly classified by their root shape:

- ☆ Chantenay carrots are shorter than other cultivars, but have greater girth, sometimes growing up to 8 centimeters (3 inch) in diameter. They have broad shoulders and taper towards a blunt, rounded tip. They are most commonly diced for use in canned or prepared foods.
- ☆ Danvers carrots have a conical shape, having well-defined shoulders and tapering to a point at the tip. They are somewhat shorter than Imperator cultivars, but more tolerant of heavy soil. Danvers cultivars are often puréed as baby food. They were developed in 1871 in Danvers, Ma.
- ☆ Imperator carrots are the carrots most commonly sold whole in United States supermarkets; their roots are longer than other cultivars of carrot, and taper to a point at the tip.
- ☆ Nantes carrots are nearly cylindrical in shape, and are blunt and rounded at both the top and tip. Nantes cultivars are often sweeter than other carrots.

While any carrot can be harvested before reaching its full size as a tenderer "baby" carrot, some fast-maturing cultivars have been bred to produce smaller roots.

The most extreme examples produce round roots about 2.5 centimeters (1 inch) in diameter. These small cultivars are also more tolerant of heavy or stony soil than long-rooted cultivars such as 'Nantes' or 'Imperator'. The "baby carrots" sold ready-to-eat in supermarkets are, however, often not from a smaller cultivar of carrot, but are simply full-sized carrots that have been sliced and peeled to make carrot sticks of a uniform shape and size.

Carrot flowers are pollinated primarily by bees. Seed growers use honeybees or mason bees for their pollination needs. Carrots are used as food plants by the larvae of some Lepidoptera species, including Common Swift, Garden Dart, Ghost Moth, Large Yellow Underwing and Setaceous Hebrew Character. One particular variety lacks the usual orange pigment from carotenes, owing its white color to a recessive gene for tocopherol (Vitamin E). Derived from *Daucus carota* L. and patented (US patent #6,437,222) at the University of Wisconsin–Madison, the variety is intended to supplement the dietary intake of Vitamin E.

Carrot, Raw

Nutritional value per 100 g (3.5 oz)	
Energy	173 kJ (41 kcal)
Carbohydrates	9 g
Sugars	5 g
Dietary fiber	3 g
Fat	0.2 g
Protein	1 g
Vitamin A equiv.	835 μg
β-carotene	8285 μg
Lutein and zeaxanthin	256 μg
Thiamine (vit. B_1)	0.04 mg
Riboflavin (vit. B_2)	0.05 mg
Niacin (vit. B_3)	1.2 mg
Vitamin B_6	0.1 mg
Folate (vit. B_9)	19 μg
Vitamin C	7 mg
Calcium	33 mg
Iron	0.66 mg
Magnesium	18 mg
Phosphorus	35 mg
Potassium	240 mg
Sodium	2.4 mg

Figure 36.1: Flow sheet for preparation of carrot *halwa*

Carrots fully mature
↓
Cleaning and shredding
↓
Removing big shreds
↓
Addition of sugar, fat, milk powder (1000g, 100g, 100g/Kg Pulp)
↓
Heating for getting good consistency
↓
Addition of salt and citric acid (5g, 2g)
↓
Heating few minutes
↓
Hot serving
↓
Cooling
↓
Storage

Figure 36.2: Flow sheet for preparation of carrot toffee

Whole carrot shreds/Carrot pomace
↓
Add sugar and fat
↓
Heating upto 80 °Brix
↓
Add citric acid and salt
↓
Heating upto 82 °Brix
↓
Spread on aluminium/Stainless steel trays
(Already smeared with oil/fat)
↓
Cool to room temperature
↓
Cut into small pieces
↓
Wrap into metallic paper
↓
Storage

Figure 36.3: Flow sheet for preparation of carrot papaya cheese

Fresh carrots and papaya

↓

Washing and peeling

↓

Cutting into pieces, boiling separately with equal quantity of water

↓

Mashing and mixing

↓

Adding sugar and butter (10 per cent)

↓

Addition of preservative (Sodium benzoate) and citric acid (5 per cent)

↓

Cooking till mixture become sufficiently thick

↓

Addition of salt (1 per cent)

↓

Removing from fire when mixture starts leaving side of pan

↓

Evenly distribution over butter coated tray and left for 3 h to set

↓

Cutting into pieces with sharp knife

↓

Pre-packing with butter paper

↓

Packing in polythene paper

↓

Store in cool place

Figure 36.4: Flow sheet for preparation of carrot pickle

Mature Carrots
↓
Washing
↓
Keeping in sun for 2 hrs
↓
Cutting into 2-2.5 cm pieces
↓
Frying all spices in a little oil
↓
Mixing pieces with fried spices
↓
Frying mixture for 5 min
↓
Cooling
↓
Making paste of vinegar and tamarind pulp
↓
Adding to carrot pieces
↓
Filling in jar
↓
Keeping in sun for a week
↓
Addition of oil after heating and cooling
↓
Sealing
↓
Labeling
↓
Storage
↓
Marketing

Sl.No.	*Item*	*Quantity*	*Sl.No.*	*Item*	*Quantity*
1.	Carrot pieces	1 kg	10.	Cumin	15 g
2.	Salt	150 g	11.	Cardamom (large)	15 g
3.	Ginger chopped	25 g	12.	Aniseed powder	15 g
4.	Onion chopped	50 g	13.	Clove headless	6 Nos
5.	Garlic chopped	10 g	14.	Tamarind pulp	50 g
6.	Red chilli powder	15 g	15.	Vinegar	150 ml
7.	Turmeric powder	15 g	16.	Mustard ground	50 g
8.	Cinnamon	15 g	17.	Mustard oil	400 ml
9.	Black pepper	15 g			

37

Cauliflower

Cauliflower is one of several vegetables in the species *Brassica oleracea*, in the family Brassicaceae. It is an annual plant that reproduces by seed. Typically, only the head (the *white curd*) of aborted floral meristems is eaten, while the stalk and surrounding thick, green leaves are used in vegetable broth or discarded.

Its name is from Latin *caulis* (cabbage) and flower, an acknowledgment of its unusual place among a family of food plants which normally produces only leafy greens for eating. *Brassica oleracea* also includes cabbage, brussels sprouts, kale, broccoli, and collard greens, though they are of different cultivar groups.

Asian Varieties

A tropical cauliflower used in China and India, it was developed in India during the 19th century from the now-abandoned Cornish type, and includes old varieties Early Patna and Early Benaras. A comprehensive list of varieties is maintained at North Carolina State University. Traditional varieties include 'Snowball', 'Hybrid White', 'Super Snowball', 'Snow Crown', 'Mayflower', Candid Charm', 'Mormon', 'Agrahani', 'poushi', 'maghi', 'Snow White' and 'Snow Grace'. Self-blanching varieties are 'Self Blanche', 'Early Tuscan' and 'Late Tuscan'. Heirloom varieties include 'All the Year Round', 'Early Pearl', 'Early Snowball', 'Igloo', 'Violetta Italia' and 'Walcheren Winter'. Commercial varieties include 'Fremont', 'Igloo' and 'Snow Crown'

Colors

White

White cauliflower is the most common color of cauliflower.

Orange

Orange cauliflower (*B. oleracea* L. var. *botrytis*) contains 25 times the level of vitamin A of white varieties. This trait came from a natural mutant found in a cauliflower field in Canada. Cultivars include 'Cheddar' and 'Orange Bouquet'.

Green

Green cauliflower, of the *B. oleracea* botrytis group, is sometimes called broccoflower. It is available both with the normal curd shape and a variant spiky curd called Romanesco broccoli. Both types have been commercially available in the U.S. and Europe since the early 1990s. Green-curded varieties include 'Alverda', 'Green Goddess' and 'Vorda'. Romanesco varieties include 'Minaret' and 'Veronica'.

Purple

Purple color in cauliflower is caused by the presence of the antioxidant group anthocyanins, which can also be found in red cabbage and red wine. Varieties include 'Graffiti' and 'Purple Cape'. In Great Britain and southern Italy, a broccoli with tiny flower buds is sold as a vegetable under the name "purple cauliflower". It is not the same as standard cauliflower with a purple curd.

Nutrition

Cauliflower is low in fat, low in carbs but high in dietary fiber, foliate, water, and vitamin C, possessing a high nutritional density. Cauliflower contains several phytochemicals, common in the cabbage family that may be beneficial to human health.

- ☆ Sulforaphane, a compound released when cauliflower is chopped or chewed, may protect against cancer.
- ☆ Other glucosinolates
- ☆ Carotenoids
- ☆ Indole-3-carbinol, a chemical that enhances DNA repair, and acts as an estrogen antagonist, slowing the growth of cancer cells.
- ☆ Boiling reduces the levels of these compounds, with losses of 20–30 per cent after five minutes, 40–50 per cent after ten minutes, and 75 per cent after thirty minutes.
- ☆ However, other preparation methods, such as steaming, microwaving, and stir frying, had no significant effect on the compounds. A high intake of cauliflower has been associated with reduced risk of aggressive prostate cancer.

Cauliflower, Raw (Edible Parts)

Nutritional value per 100 g (3.5 oz)	
Energy	103 kJ (25 kcal)
Carbohydrates	5 g
Sugars	1.9 g
Dietary fiber	2 g
Fat	0.3 g
Protein	2 g
Water	92 g

Thiamine (vit. B_1)	0.05 mg
Riboflavin (vit. B_2)	0.06 mg
Niacin (vit. B_3)	0.507 mg
Pantothenic acid (B_5)	0.667 mg
Vitamin B_6	0.184 mg
Folate (vit. B_9)	57 μg
Vitamin C	48 mg
Vitamin K	15.5 μg
Calcium	22 mg
Iron	0.42 mg
Magnesium	15 mg
Phosphorus	44 mg
Potassium	299 mg
Sodium	30 mg
Zinc	0.27 mg

Figure 37.1: Flow sheet for preparation of dried cauliflower

Fresh fully matured Cauliflower

↓

Removal of leaves and hard stem

↓

Washing

↓

Cutting (2-3 cm long and 1-2 cm wide)

↓

Blanching (98 °C, 5 min)

↓

Cooling

↓

Pretreatment (5 per cent salt + 5 per cent sugar + 0.1 per cent KMS, 1 h)

↓

Draining

↓

Drying at 60°C in tray dryer

↓

Dehydrated product (9.0 per cent moisture content)

↓

Packaging

↓

Storage

Figure 37.2: Flow sheet for preparation of cauliflower pickle

Mature Cauliflower
↓
Washing
↓
Keeping in sun for 2 hrs
↓
Cutting into 2-2.5 cm pieces
↓
Frying all spices in a little oil
↓
Mixing pieces with fried spices
↓
Frying mixture for 5 min
↓
Cooling
↓
Making paste of vinegar and tamarind pulp
↓
Adding to cauliflower pieces
↓
Filling in jar
↓
Keeping in sun for a week
↓
Addition of oil after heating and cooling
↓
Sealing
↓
Labeling
↓
Storage
↓
Marketing

Sl.No.	Item	Quantity	Sl.No.	Item	Quantity
1.	Cauliflower pieces	1 kg	10.	Cumin	15 g
2.	Salt	150 g	11.	Cardamom (large)	15 g
3.	Ginger chopped	25 g	12.	Aniseed powder	15 g
4.	Onion chopped	50 g	13.	Clove headless	6 No
5.	Garlic chopped	10 g	14.	Tamarind pulp	50 g
6.	Red chilli powder	15 g	15.	Vinegar	150 ml
7.	Turmeric powder	15 g	16.	Mustard ground	50 g
8.	Cinnamon	15 g	17.	Mustard oil	400 ml
9.	Black pepper	15 g			

38

Chilli

Chilli pepper (also chile pepper or chilli pepper, from Nahuatl *chilli*) is the fruit of plants from the genus *Capsicum*, members of the nightshade family, *Solanaceae*. The term in British English and in Australia, New Zealand, India, Malaysia and other Asian countries is just chilli without pepper. Chili peppers originated in the Americas. After the Columbian Exchange, many cultivars of chili pepper spread across the world, used in both food and medicine.

History

In 2007, Guinness World Records certified the *bhut jolokia*, also known as the ghost pepper/chili pepper, as being the world's hottest chili pepper at 401.5 times hotter than Tabasco sauce. On December 3, 2010, the Bhut Jolokia was replaced as the hottest known chili pepper by the Naga Viper pepper, which has an average peak Scoville rating more than 300,000 points higher than an average Bhut Jolokia - but still not higher than the hottest ever recorded Dorset Naga. In February 2011, *Guinness World Records* awarded the title of "World's Hottest Chilli" to the Infinity chilli grown in Grantham, England. This chilli rates at 1,067,286 units on the Scoville scale. Later the same month, on February 25, 2011, the title returned to the Naga Viper pepper with a rating of 1,382,118 Scoville Heat Units (SHU). The current "world's hottest" is the Trinidad Scorpion Butch T, officially tested at 1,463,700 SHU. These figures are highly controversial among the pepper growing community and tests with more rigorous scientific standards are yet to be conducted on the many various peppers vying for "world's hottest" status. Currently, Scoville ratings are highly controversial among the pepper growing community and tests with more rigorous scientific standards are yet to be conducted on the many peppers vying for the title of the "world's hottest".

Culinary Uses

Chili pepper pods, which are berries, are used fresh or dried. Chillies are often dried to preserve them for long periods of time. Preserving may also be done by pickling fresh chilies.

Dried chilies are often ground to powders, although some Mexican dishes including variations on chilies rellenos may use whole reconstituted chilies, and others may reconstitute dried chilies before grinding to a paste. Chilies may be dried using smoke, such as the chipotle, which is the smoked, dried form of the jalapeño.

Many fresh chilies such as poblano have a tough outer skin which does not break down on cooking. For recipes where chilies are used whole or in large slices, roasting, or other means of blistering or charring the skin are usually performed so as not to entirely cook the flesh beneath. When cooled, the skins will usually slip off easily.

Chili pepper plant leaves, mildly bitter but nowhere near as hot as the fruits that come from the same plant, are cooked as greens in Filipino cuisine, where they are called *dahon ng sili* (literally "chili leaves"). They are used in the chicken soup, *tinola*. In Korean cuisine, the leaves may be used in kimchi. In Japanese cuisine, the leaves are cooked as greens, and also cooked in *tsukudani* style for preservation.

Chili is by far the most important fruit in Bhutan. Local markets are never without chili, always teemed with different colors and sizes, in fresh and dried form. Bhutanese call this crop ema (in Dzongkha) or solo (in Sharchop). Chili is a staple fruit in Bhutan; the *ema* datsi recipe is entirely made of chili mixed with local cheese. Chili is also an important ingredient in almost all curries and food recipes in the country.

Chilies are present in many cuisines. Some notable dishes other than the ones mentioned elsewhere in this article include:

- Paprikash from Hungary: uses significant amounts of mild, ground, dried chilies, aka paprika, in a braised chicken dish
- Paprykarz szczeciñski is a Polish fish paste with rice, onion, tomato concentrate, vegetable oil, chilli pepper powder and other spices.
- Chiles en nogada from the Puebla region of Mexico: uses fresh mild chilies stuffed with meat and covered with a creamy nut-thickened sauce.
- Mole poblano from the city of Puebla in Mexico: uses several varieties of dried chilies, nuts, spices, and fruits to produce a thick, dark sauce for poultry or other meats.
- Puttanesca sauce from Italy: a tomato-based sauce for pasta including dried hot chilies.
- Kung Pao Chicken (also spelled Gong Bao) from the Sichuan region of China: small hot dried chilies are briefly fried in oil to add spice to the oil then used for frying.
- Nigerian dishes and those in many parts of Africa.

Fresh or dried chilies are often used to make hot sauce, a bottled condiment to add spice to other dishes. Hot sauces are found in many cuisines including harissa from the Middle East, chili oil from China (known as râyu in Japan), and sriracha from Thailand.

Nutritive Value

Red chilies contain high amounts of vitamin C and carotene (provitamin A). Yellow and especially green chilies (which are essentially unripe fruit) contain a considerably lower amount of both substances. In addition, peppers are a good source of most B vitamins, and vitamin B_6 in particular. They are very high in potassium, magnesium, and iron. Their high vitamin C content can also substantially increase the uptake of non-heme iron from other ingredients in a meal, such as beans and grains.

Peppers, Hot Chili, Red, Raw

Nutritional value per 100 g (3.5 oz)	
Energy	166 kJ (40 kcal)
Carbohydrates	8.8 g
Sugars	5.3 g
Dietary fiber	1.5 g
Fat	0.4 g
Protein	1.9 g
Water	88 g
Vitamin A equiv.	48 μg
β-carotene	534 μg
Vitamin B_6	0.51 mg
Vitamin C	144 mg
Iron	1 mg
Magnesium	23 mg
Potassium	322 mg

Figure 38.1: Flow sheet for preparation of chilli drying

Selection of mature chillies

↓

Cleaning and grading

↓

Removing stems

↓

Blanching (96 °C), 5 min

↓

Drying in tray dryer at 50 °C

↓

Dried product (9.0 per cent M. C.)

↓

Sensory evaluation

↓

Packaging

↓

Storage

Figure 38.2: Flow sheet for preparation of red chillies pickle

Red chillies
↓
Washing
↓
Drying for an hr
↓
Removal of stem
↓
Making incisions
↓
Removal of seeds
↓
Mixing all spices and seeds
↓
Frying in oil
↓
Cooling
↓
Filling into jar
↓
Addition of oil after heating and cooling
↓
Keeping in sun for a week
↓
Sealing
↓
Labeling
↓
Marketing

Sl.No.	*Item*	*Quantity*	*Sl.No.*	*Item*	*Quantity*
1.	Red chillies	1 kg	6.	Cumin	15 g
2.	Salt	100 g	7.	Cardamom (large)	15 g
3.	Amchur	250 g	8.	Black pepper	15 g
4.	Turmeric powder	10 g	9.	Mustard ground	100 g
5.	Clove (headless)	6 No	10.	Mustard oil	700 ml

Figure 38.3: Flow sheet for preparation of green chillies pickle

Green chillies
↓
Washing
↓
Drying
↓
Making incisions
↓
Mixing all spices in a little lime juice
↓
Mixing with chillies
↓
Filling into jar
↓
Adding lime juice and oil
↓
Keeping in sun for a week
↓
Sealing
↓
Labeling
↓
Marketing
↓
Storage

Sl.No.	*Item*	*Quantity*	*Sl.No.*	*Item*	*Quantity*
1.	Green chillies	1 kg	6.	Cumin	15 g
2.	Salt	150 g	7.	Cardamom (large)	15 g
3.	Amchur/lime juice	200 g/200 ml	8.	Aniseed	15 g
4.	Turmeric powder	15 g	9.	Mustard ground	100 g
5.	Fenugreek	15 g	10.	Mustard oil	400 ml

39

Cucumber

The cucumber (*Cucumis sativus*) is a widely cultivated plant in the gourd family Cucurbitaceae, which includes squash, and in the same genus as the muskmelon. The plant is a creeping vine which bears cylindrical edible fruit when ripe. There are three main varieties of cucumber: "slicing", "pickling", and "burpless". Within these varieties, several different cultivars have emerged. The cucumber is originally from India but is now grown on most continents.

The cucumber is a creeping vine that roots in the ground and grows up trellises or other supporting frames, wrapping around supports with thin, spiraling tendrils. The plant has large leaves that form a canopy over the fruit. The fruit of the cucumber is roughly cylindrical, elongated with tapered ends, and may be as large as 60 centimeters (24 in) long and 10 centimeters (3.9 in) in diameter. Having an enclosed seed and developing from a flower, botanically speaking, and cucumbers are classified as fruits. However, much likes tomatoes and squash they are often perceived, prepared and eaten as vegetables. Cucumbers are usually more than 90 per cent water.

History

Cucumbers originated in India. Large genetic variety of cucumber has been observed in different parts of India. It has been cultivated for at least 3,000 years in Western Asia, and was probably introduced to other parts of Europe by the Romans. Records of cucumber cultivation appear in France in the 9th century, England in the 14th century, and in North America by the mid-16th century.

Pickling

Cucumbers can be pickled for flavor and longer shelf-life. Although any cucumber can be pickled, commercial pickles are made from cucumbers specially bred for uniformity of length-to-diameter ratio and lack of voids in the flesh. Those cucumbers intended for pickling, called picklers, grow to about 7 centimeters (2.8 in) to 10 centimeters (3.9 in) long and 2.5 centimeters (0.98 in) wide. As compared to slicers, picklers tend to be shorter, thicker, less regularly shaped, and have bumpy skin with tiny white or black-dotted spines. They are never waxed. Color can vary from creamy yellow to pale or dark green. Pickling cucumbers are sometimes sold fresh as "Kirby" or "Liberty" cucumbers. The pickling process removes or degrades much of the nutrient content, especially that of vitamin C. Pickled cucumbers are soaked in brine or a combination of vinegar and brine, although not vinegar alone, often along with various spices. Pickled cucumbers are called "pickles" in the US or "gherkins" or "wallies" in the UK, the latter name being more common in the north of England, where it refers to the large vinegar-pickled cucumbers commonly sold in fish and chip shops.

Several varietals exist and are sold commercially:

- Lebanese cucumbers are small, smooth-skinned and mild. Like the English cucumber, Lebanese cucumbers are nearly seedless.
- East Asian cucumbers are mild, slender, deep green, and have a bumpy, ridged skin. They can be used for slicing, salads, pickling, etc., and are available year-round. They are usually burpless as well.
- Armenian cucumbers (also known as yard long) have very long, ribbed fruit with a thin skin that does not require peeling, but are actually an immature melon. This is the variety sold in Middle Eastern markets as "pickled wild cucumber". In North America, the term "wild cucumber" refers to manroot.
- Persian cucumber, better known as mini seedless cucumbers, is available from Canada during the summer, and all year-round from the Dominican. Its popularity is increasing 30 to 40 per cent a year. Easy to cut, it is on average 5-8 in. long. Vines are parthenocarpic, requiring no pollinators for fruit set.
- Beit Alpha cucumbers are small, sweet parthenocarpic cucumbers adapted to the dry climate of the Middle East.
- Apple cucumbers are short, round cucumbers grown in New Zealand and parts of Europe, known for their light yellow-green color and mildly sweet flavor. When mature, the fruit may grow tiny spines, and contains numerous edible green seeds. The fruit is usually eaten raw, with skin.
- *Dosakai* is a yellow cucumber available in parts of India. These fruits are generally spherical in shape. It is commonly cooked as curry, added in *sambar* or soup, *daal* and also in making *dosa-aavakaaya* (Indian pickle) and chutney; it is also grown and available through farms in Central California.
- *Kekiri* is a smooth skinned cucumber, relatively hard, and not used for salads. It is cooked as spicy curry. It is found in dry zone of Sri Lanka. It becomes orange colored when the fruit is matured.
- In May 2008, British supermarket chain Sainsbury's unveiled the 'c-thru-cumber', a thin-skinned variety which reportedly does not require peeling.

Taste

There appears to be variability in the human olfactory response to cucumbers, with the majority of people reporting a mild, almost watery flavor or a light melon taste, while a small but vocal minority report a highly repugnant taste, some say almost perfume-like. Various practices have arisen with regard to how bitterness may be removed from cucumbers. Among these a very common practice popular in India includes slicing off the ends of a cucumber, sprinkling some salt, and rubbing the now-exposed ends of said cucumber with the sliced-off ends until it appears to froth. Another such urban legend states that one ought to peel a cucumber away from the end that was once attached to a vine; otherwise one risked spreading the bitterness throughout the cucumber.

Storage

Cucumbers can best be stored sealed in shrink foil. Sealing a cucumber increases its shelf life considerably.

Cucumber, with Peel, Raw

Nutritional value per 100 g (3.5 oz)	
Energy	65 kJ (16 kcal)
Carbohydrates	3.63 g
Sugars	1.67 g
Dietary fiber	0.5 g
Fat	0.11 g
Protein	0.65 g
Thiamine (vit. B_1)	0.027 mg
Riboflavin (vit. B_2)	0.033 mg
Niacin (vit. B_3)	0.098 mg
Pantothenic acid (B_5)	0.259 mg
Vitamin B_6	0.040 mg
Folate (vit. B_9)	7 μg
Vitamin C	2.8 mg
Calcium	16 mg
Iron	0.28 mg
Magnesium	13 mg
Phosphorus	24 mg
Potassium	147 mg
Zinc	0.20 mg

Figure 39.1: Flow sheet for preparation of cucumber pickle

Cucumbers
↓
Washing
↓
Peeling
↓
Cutting into 5 cm pieces
↓
Mixing with salt
↓
Filling in jars
↓
Standing for 6-8 hrs
↓
Draining off water
↓
Adding spices and vinegar
↓
Keeping in sun for a week
↓
Labeling
↓
Storage
↓
Marketing

Sl.No.	*Item*	*Quantity*	*Sl.No.*	*Item*	*Quantity*
1.	Cucumber	1 kg	5.	Cumin	10 g
2.	Salt	200 g	6.	Cardamom	10 g
3.	Red chilli powder	15 g	7.	Clove headless	6 No
4.	Vinegar	750 ml	8.	Black pepper	10 g

40

Garlic

Allium sativum, commonly known as garlic, is a species in the onion genus, *Allium*. Its close relatives include the onion, shallot, leek, chive, and rakkyo. Dating back over 6,000 years, garlic is native to central Asia, and has long been a staple in the Mediterranean region, as well as a frequent seasoning in Asia, Africa, and Europe. It was known to Ancient Egyptians, and has been used throughout its history for both culinary and medicinal purposes

Origin and Major Types

The ancestry of cultivated garlic is not definitively established. According to Zohary and Hopf, "A difficulty in the identification of its wild progenitor is the sterility of the cultivars", though it is thought to be descendent from the species *Allium longicuspis*, which grows wild in central and southwestern Asia. *Allium sativum* grows in the wild in areas where it has become naturalised. The "wild garlic", "crow garlic", and "field garlic" of Britain are members of the species *Allium ursinum*, *Allium vineale*, and *Allium oleraceum*, respectively. In North America, *Allium vineale* (known as "wild garlic" or "crow garlic") and *Allium canadense*, known as "meadow garlic" or "wild garlic" and "wild onion", are common weeds in fields. One of the best-known "garlic's", the so-called elephant garlic, is actually a wild leek (*Allium ampeloprasum*), and not a true garlic. Single clove garlic (also called pearl or solo garlic) originated in the Yunnan province of China.

European Garlic

There are a number of garlics with Protected Geographical Status in Europe; these include:

- ☆ Aglio Bianco Polesano from Veneto, Italy (PDO)
- ☆ Aglio di Voghiera from Ferrara, Emilia-Romagna, Italy (PDO)
- ☆ Ail blanc de Lomagne from Lomagne in the Gascony area of France (PGI)

- ✰ Ail de la Drôme from Drôme in France (PGI)
- ✰ Ail rose de Lautrec a rose/pink garlic from Lautrec in France (PGI)
- ✰ Ajo Morado de Las Pedroñeras a rose/pink garlic from Las Pedroñeras in Spain (PGI)

Varieties

While botanists classify garlic under the umbrella of the species, *Allium sativum*, there are also two main subspecies.

- ✰ Ophioscorodon, or hard necked garlic, includes porcelain garlics, rocambole garlic, and purple stripe garlics.
- ✰ Sativum, or soft necked garlic, includes artichoke garlic, silver skin garlic, and Creole garlic.

Culinary Uses

Garlic is widely used around the world for its pungent flavor as a seasoning or condiment.

The garlic plant's bulb is the most commonly used part of the plant. With the exception of the single clove types, garlic bulbs are normally divided into numerous fleshy sections called cloves. Garlic cloves are used for consumption (raw or cooked) or for medicinal purposes. They have a characteristic pungent, spicy flavor that mellows and sweetens considerably with cooking.

Other parts of the garlic plant are also edible. The leaves and flowers (bulbils) on the head (spathe) are sometimes eaten. They are milder in flavor than the bulbs, and are most often consumed while immature and still tender. Immature garlic is sometimes pulled, rather like a scallion, and sold as "green garlic". When green garlic is allowed to grow past the "scallion" stage, but not permitted to fully mature, it may produce a garlic "round", a bulb like a boiling onion, but not separated into cloves like a mature bulb. Additionally, the immature flower stalks (scapes) of the hardneck and elephant types are sometimes marketed for uses similar to asparagus in stir-fries.

Inedible or rarely eaten parts of the garlic plant include the "skin" and root cluster. The papery, protective layers of "skin" over various parts of the plant are generally discarded during preparation for most culinary uses, though in Korea immature whole heads are sometimes prepared with the tender skins intact. The root cluster attached to the basal plate of the bulb is the only part not typically considered palatable in any form.

Garlic is a fundamental component in many or most dishes of various regions, including eastern Asia, South Asia, Southeast Asia, the Middle East, northern Africa, southern Europe, and parts of South and Central America. The flavour varies in intensity and aroma with the different cooking methods. It is often paired with onion, tomato, or ginger. The parchment-like skin is much like the skin of an onion, and is typically removed before using in raw or cooked form. An alternative is to cut the top off the bulb, coat the cloves by dribbling olive oil (or other oil-based seasoning) over them, and roast them in an oven. Garlic softens and can be extracted from the cloves by squeezing the (root) end of the bulb, or individually by squeezing one end of the clove. In Korea, heads of garlic are fermented at high temperature; the resulting product, called black garlic, is sweet and syrupy, and is now being sold in the United States, United Kingdom and Australia. Garlic may be applied to breads to create a variety of classic dishes, such as garlic bread, garlic toast, bruschetta, crostini and canapé.

Oils can be flavored with garlic cloves. These infused oils are used to season all categories of vegetables, meats, breads and pasta. In some cuisines, the young bulbs are pickled for three to six weeks in a mixture of sugar, salt, and spices. In Eastern Europe, the shoots are pickled and eaten as an appetizer. Immature scapes are tender and edible. They are also known as "garlic spears", "stems", or "tops". Scapes generally have a milder taste than the cloves. They are often used in stir frying or braised like asparagus. Garlic leaves are a popular vegetable in many parts of Asia. The leaves are cut, cleaned, and then stir-fried with eggs, meat, or vegetables. Mixing garlic with egg yolks and olive oil produces *aioli*. Garlic, oil, and a chunky base produce *skordalia*. Blending garlic, almond, oil, and soaked bread produces *ajoblanco*. Garlic powder has a different taste from fresh garlic. If used as a substitute for fresh garlic, 1/8 teaspoon of garlic powder is equivalent to one clove of garlic.

Storage

Domestically, garlic is stored warm [above 18 °C (64 °F)] and dry to keep it dormant (so it does not sprout). It is traditionally hung; soft neck varieties are often braided in strands called plaits or *grapes*. Peeled cloves may be stored in wine or vinegar in the refrigerator. Commercially, garlic is stored at 0 °C (32 °F), in a dry, low-humidity environment. Garlic will keep longer if the tops remain attached.

Garlic is often kept in oil to produce flavoured oil; however, the practice requires measures to be taken to prevent the garlic from spoiling. Untreated garlic kept in oil can support the growth of *Clostridium botulinum* which causes the deadly botulism illness; refrigeration will not assure the safety of garlic kept in oil. To reduce this risk, the oil should be refrigerated and used within one week. Commercially prepared oils are widely available. Manufacturers add acids and/or other chemicals to eliminate the risk of botulism in their products. Two outbreaks of botulism related to garlic stored in oil have been reported.

Medicinal Use and Health Benefits

In *in-vitro* studies, garlic has been found to have antibacterial, antiviral, and antifungal activity. However, these actions are less clear *in vivo*. Garlic is also claimed to help prevent heart disease (including atherosclerosis, high cholesterol, and high blood pressure) and cancer. Garlic is used to prevent certain types of cancer, including stomach and colon cancers. In fact, countries where garlic is consumed in higher amounts, because of traditional cuisine, have been found to have a lower prevalence of cancer. Animal studies, and some early research studies in humans, have suggested possible cardiovascular benefits of garlic. A Czech study found garlic supplementation reduced accumulation of cholesterol on the vascular walls of animals. Another study had similar results, with garlic supplementation significantly reducing aortic plaque deposits of cholesterol-fed rabbits. Another study showed supplementation with garlic extract inhibited vascular calcification in human patients with high blood cholesterol. The known vasodilative effect of garlic is possibly caused by catabolism of garlic-derived polysulfides to hydrogen sulfide in red blood cells (RBCs), a reaction that is dependent on reduced thiols in or on the RBC membrane. Hydrogen sulfide is an endogenous cardioprotective vascular cell-signaling molecule.

A randomized clinical trial funded by the National Institutes of Health (NIH) in the United States and published in the *Archives of Internal Medicine* in 2007 found the consumption of garlic in any form did not reduce blood cholesterol levels in patients with moderately high baseline cholesterol levels. According to Heart.org, "despite decades of research suggesting that garlic can improve cholesterol profiles, a new NIH-funded trial found absolutely no effects of raw garlic or garlic supplements on LDL, HDL, or triglycerides... The findings underscore the hazards of meta-analyses made up of small, flawed studies and the value of rigorously studying popular herbal remedies". In an editorial regarding the initial report's findings, two physicians from Weill Cornell Medical College of Cornell University, pointed out that there may "be effects of garlic on atherosclerosis specifically that were not picked up in the study".

Allium sativum has been found to reduce platelet aggregation and hyperlipidemia.

In 2007, the BBC reported *Allium sativum* may have other beneficial properties, such as preventing and fighting the common cold. This assertion has the backing of long tradition in herbal medicine, which has used garlic for hoarseness and coughs. The Cherokee also used it as an expectorant for coughs and croup.

Garlic is also alleged to help regulate blood sugar levels. Regular and prolonged use of therapeutic amounts of aged garlic extracts lower blood homocysteine levels and has been shown to prevent some complications of diabetes mellitus. People taking insulin should not consume medicinal amounts of garlic without consulting a physician.

In 1858, Louis Pasteur observed garlic's antibacterial activity, and it was used as an antiseptic to prevent gangrene during World War I and World War II. More recently, it has been found from a clinical trial that a mouthwash containing 2.5 per cent fresh garlic shows good antimicrobial activity, although the majority of the participants reported an unpleasant taste and halitosis.

Garlic cloves are used as a remedy for infections (especially chest problems), digestive disorders, and fungal infections such as thrush. Garlic can be used as a disinfectant because of its bacteriostatic and bacteriocidal properties.

Garlic has been found to enhance thiamin absorption, and therefore reduces the likelihood for developing the thiamin deficiency beriberi.

In 1924, it was found to be an effective way to prevent scurvy, because of its high vitamin C content.

Garlic has been used reasonably successfully in AIDS patients to treat *Cryptosporidium* in an uncontrolled study in China. It has also been used by at least one AIDS patient to treat toxoplasmosis, another protozoal disease. Garlic supplementation has been shown to boost testosterone levels in rats fed a high protein diet.

A 2010 double-blind, parallel, randomized, placebo-controlled trial, involving 50 patients whose routine clinical records in general practice documented treated but uncontrolled hypertension, concluded, "Our trial suggests that aged garlic extract is superior to placebo in lowering systolic blood pressure similarly to current first line medications in patients with treated but uncontrolled hypertension".

Other Uses

The sticky juice within the bulb cloves is used as an adhesive in mending glass and porcelain in China.

Garlic, Raw

Nutritional value per 100 g (3.5 oz)	
Energy	623 kJ (149 kcal)
Carbohydrates	33.06 g
Sugars	1.00g
Dietary fiber	2.1 g
Fat	0.5 g
Protein	6.39 g
β-carotene	5 μg
Thiamine (vit. B_1)	0.2 mg
Riboflavin (vit. B_2)	0.11 mg
Niacin (vit. B_3)	0.7 mg
Pantothenic acid (B_5)	0.596 mg
Vitamin B_6	1.235 mg
Folate (vit. B_9)	3 μg
Vitamin C	31.2 mg
Calcium	181 mg
Iron	1.7 mg
Magnesium	25 mg
Manganese	1.672 mg
Phosphorus	153 mg
Potassium	401 mg
Selenium	14.2 μg
Sodium	17 mg
Zinc	1.16 mg

Figure 40.1: Flow sheet for extraction of garlic oil

Fully Mature Garlic

↓

Removal of undesired material

↓

Washing of tubes

↓

Crushing of bulbs

↓

Steam distillation

↓

Inspection and analyzing

↓

Packing and dispatching

↓

Marketing

Figure 40.2: Flow sheet for preparation of garlic powder

Raw garlic material

↓

Washing

↓

Peeling

↓

(Small amount of sodium sulphate solution spraying)

↓

Dehydration

↓

Pulverization

↓

Sieving

↓

Packaging

↓

Labeling

↓

Storage

41

Ginger

Ginger is the rhizome of the plant *Zingiber officinale,* consumed as a delicacy, medicine, or spice. It lends its name to its genus and family (Zingiberaceae). Other notable members of this plant family are turmeric, cardamom, and galangal. Ginger cultivation began in South Asia and has since spread to East Africa and the Caribbean. It is sometimes called root ginger to distinguish it from other things that share the name *ginger.*

Culinary Use

Ginger produces a hot, fragrant kitchen spice. Young ginger rhizomes are juicy and fleshy with a very mild taste. They are often pickled in vinegar or sherry as a snack or just cooked as an ingredient in many dishes. They can also be steeped in boiling water to make ginger tea, to which honey is often added; sliced orange or lemon fruit may also be added. Ginger can also be made into candy. Mature ginger roots are fibrous and nearly dry. The juice from old ginger roots is extremely potent and is often used as a spice in Indian recipes, and is a quintessential ingredient of Chinese, Japanese and many South Asian cuisines for flavoring dishes such as seafood or goat meat and vegetarian cuisine.

Ginger acts as a useful food preservative. Fresh ginger can be substituted for ground ginger at a ratio of 6 to 1, although the flavors of fresh and dried ginger are somewhat different. Powdered dry ginger root is typically used as a flavoring for recipes such as gingerbread, cookies, crackers and cakes, ginger ale, and ginger beer. Candied ginger is the root cooked in sugar until soft, and is a type of confectionery. Fresh ginger may be peeled before eating. For longer-term storage, the ginger can be placed in a plastic bag and refrigerated or frozen.

Medical Properties and Research

Ginger has been claimed to decrease the pain from arthritis, though studies have been inconsistent. It may also have blood thinning and cholesterol lowering properties that may make it useful for treating heart disease. Preliminary research also indicates that nine compounds found in ginger may bind to human serotonin receptors, possibly helping to affect anxiety.

Advanced glycation end-products are possibly associated in the development of several pathophysiologies, including diabetic cataract for which ginger was effective in preliminary studies, apparently by acting through antiglycating mechanisms.

Ginger compounds are active against a form of diarrhea which is the leading cause of infant death in developing countries. Zingerone is likely to be the active constituent against enterotoxigenic *Escherichia coli* heat-labile enterotoxin-induced diarrhea.

Ginger has been found effective in multiple studies for treating nausea caused by seasickness, morning sickness and chemotherapy though ginger was not found superior over a placebo for pre-emptively treating post-operative nausea. Ginger is a safe remedy for nausea relief during pregnancy. Ginger as a remedy for motion sickness is still a debated issue. The television program *Mythbusters* performed an experiment using one of their staff who suffered from severe motion sickness. The staff member was placed in a moving device which, without treatment, produced severe nausea. Multiple treatments were administered. None, with the exception of the ginger and the two most common drugs, were successful. The staff member preferred the ginger due to lack of side effects. Several studies over the last 20 years were inconclusive with some studies in favor of the herb and some not. A common thread in these studies is the lack of sufficient participants to yield statistical significance. Another issue is the lack of a known chemical pathway for the supposed relief.

Chemistry

The characteristic odor and flavor of ginger is caused by a mixture of zingerone, shogaols and gingerols, volatile oils that compose one to three percent of the weight of fresh ginger. In laboratory animals, the gingerrols increase the motility of the gastrointestinal tract and have analgesic, sedative, antipyretic and antibacterial properties. Ginger oil has been shown to prevent skin cancer in mice and a study at the University of Michigan demonstrated that gingerols can kill ovarian cancer cells. [6]-gingerol (1-[4'-hydroxy-3'-methoxyphenyl]-5-hydroxy-3-decanone) is the major pungent principle of ginger. The chemopreventive potentials of [6]-gingerol present a promising future alternative to expensive and toxic therapeutic agents.

Ginger contains up to three percent of a fragrant essential oil whose main constituents are sesquiterpenoids, with (-)-zingiberene as the main component. Smaller amounts of other sesquiterpenoids (β-sesquiphellandrene, bisabolene and farnesene) and a small monoterpenoid fraction (β-phelladrene, cineol, and citral) have also been identified. The pungent taste of ginger is due to nonvolatile phenylpropanoid-derived compounds, particularly gingerols and shogaols, which form from gingerols when ginger is dried or cooked. Zingerone is also produced from gingerols during this process; this compound is less pungent and has a spicy-sweet aroma. Ginger is also a minor chemical irritant, and because of this was used as a horse suppository by pre-World War I mounted regiments for feaguing. Ginger has a sialagogue action, stimulating the production of saliva, which makes swallowing easier.

Ginger Benefits

For hundreds of years, ginger has been used medicinally and many deem the ginger nutritional value to be wholly associated with medicine. Ginger has frequently being used to treat colic and dyspepsia. And was regularly used to disguise the unpleasant taste of many medicines. It is also now used in many herbal remedies for cleansing the colon, stomach cramps, to help stimulate the circulation, many bowel disorders, muscular aches, indigestion, nausea and vomiting, helps to reduce allergies and because it has anti-microbiotic properties, it can be used to help heal wounds and sores. In the Western Indies, ginger is often used to treat unrinary tract infections. In Nigeria, it is used to treat malaria and yellow fever. It has cholesterol lowering and blood thinning properties – making it useful of heart disease sufferers. It has been associated with alleviating arthritic pain, although results have proved inconsistent to date. In some developing countries, ginger compounds such as zingerone, have been found to be effective and are very much in use to combat *E.coli* induced diarrhea, primarily in children. Ginger has been observed and proved to be effective in treating nausea – the types of nausea caused by morning sickness, chemotherapy, and seasickness. It is entirely safe to be taken throughout pregnancy.

Strangely enough, it helps to stimulate the secretion of saliva from the salivary glands, due to the fact that it has sialagogue action. It is because of this that it is commonly used as an effective aid to digestion. By increasing the amount of saliva production, ginger can help to relieve indigestion, stomach cramps and diarrhea. Because ginger helps to stimulate saliva and thus bile also, it is often referred to patients with a history of gall stones.

Nutritional Value of Ginger

Nutritional Value 10 g	*Amount*
Calories	18
Total Fat	0 gm
Saturated Fat	0 g
Cholesterol	0 mg
Sodium	2 mg
Total Carbohydrate	4 g
Dietary Fiber	1 g
Sugar	0 g
Protein	1 g
Vitamin A	0 per cent
Vitamin C	0 per cent
Calcium	0 per cent
Iron	0 per cent

Candied Ginger

You don't need a candy thermometer to make this. Simply keep an eye on the pot and when the liquid is the consistency of thin honey, it's done and ready to go.

1 pound (500g) fresh ginger, peeled 4 cups (800g) sugar, plus additional sugar for coating the ginger slices, if desired 4 cups (1L) water and pinch of salt

1. Slice the ginger as thinly as possible. It can't be too thin, so use a sharp knife.
2. Put the ginger slices in a non-reactive pot, add enough water to cover the ginger, and bring to a boil. Reduce heat and let ginger simmer for ten minutes. Drain, and repeat, simmering the ginger slices one more time.
3. Mix the sugar and 4 cups (1L) water in the pot, along with a pinch of salt and the ginger slices, and cook until the temperature reaches 225F (106 ºC.)
4. Remove from heat and let stand for at least an hour, although I often let it sit overnight or if you want to coat the slices with sugar, drain very well while the ginger is hot, so the syrup will drain away better.
5. Store ginger slices in its syrup, or toss the drained slices in granulated sugar. Shake off excess sugar, and spread the ginger slices on a cooling rack overnight, until they're somewhat dry. The sugar can be reused in a batter or ice cream base, or for another purpose.

Storage

The ginger, packed in its syrup, can be stored in the refrigerator for up to one year. If you're concerned with it crystallizing, add a tablespoon or two of corn syrup or glucose to the sugar syrup at the beginning of step #3. If tossed in sugar, the pieces can be stored at room temperature for a few months.

Figure 41.1: Flow sheet for preparation of ginger powder

Rhizomes or Green Ginger
↓
Washing
↓
Removal of cork epidermis by scraping
↓
Washing
↓
Scalding in boiling water for 4 min
↓
Drying in sun
↓
Assorting
↓
Grinding
↓
Screening
↓
Packaging
↓
Labeling
↓
Storage
↓
Marketing

Figure 41.2: Flow sheet for preparation of ginger juice

Fully Mature Rhizome of Ginger

↓

Washing

↓

Peeling

↓

Grating

↓

Addition of water (1:2)

↓

Grinding

↓

Straining

↓

Keeping on hour for setting

↓

Syphoning off clear juice

↓

Straining the juice

↓

Pasteurization

↓

Filling cans/bottles

↓

Sealing

↓

Storage

↓

Marketing

Figure 41.3: Flow sheet for preparation of cured dry ginger and glazed ginger

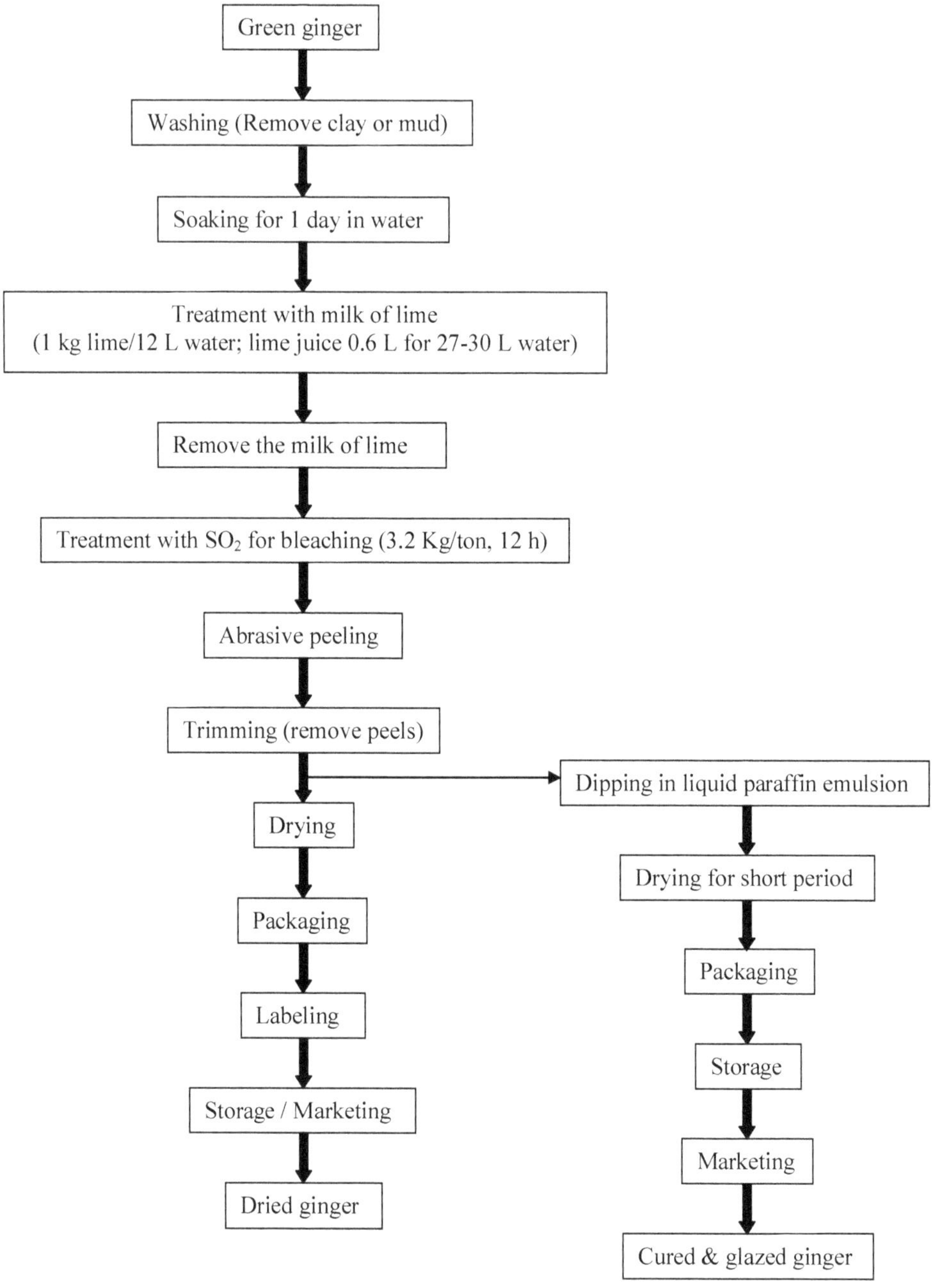

Figure 41.4: Flow sheet for preparation of preserved ginger

Selected rhizomes
↓
Washing and Cleaning
↓
Scrapping
↓
Cutting into desired shape/pieces
↓
Boiling in citric acid solution 0.5 per cent, 30 min
↓
Washing in cold water
↓
Pricking
↓
Keep one night in sugar solution, 30-35 °Brix
↓
Drainout syrup and add more sugar (increase 10 °Brix)
↓
Cool and again pour on ginger and keep overnight
↓
Heat syrup + ginger pieces till to get 70-75 °Brix
↓
Packaging
↓
Labeling
↓
Storage
↓
Marketing

42

Leafy Vegetables

Spinach (*Spinacia oleracea*) is an edible flowering plant in the family of Amaranthaceae. It is native to central and southwestern Asia. It is an annual plant (rarely biennial), which grows to a height of up to 30 cm. Spinach may survive over winter in temperate regions. The leaves are alternate, simple, and ovate to triangular-based, very variable in size from about 2–30 cm long and 1–15 cm broad, with larger leaves at the base of the plant and small leaves higher on the flowering stem. The flowers are inconspicuous, yellow-green, 3–4 mm diameter, maturing into a small, hard, dry, lumpy fruit cluster 5–10 mm across containing several seeds.

Common spinach, *Spinacia oleracea*, was long considered to be in the Chinopodiaceae family, but in 2003 the Chinopodiaceae family was combined with the Amaranthaceae family under the family name 'Amaranthaceae' in the order Caryophyllales. Within the Amaranthaceae family there are now a subfamily Amaranthoideae and a subfamily Chenopodioideae, for the amaranths and the chenopods, respectively.

History

Spinach is thought to have originated in ancient Persia (modern Iran and neighboring countries). Arab traders carried spinach into India, and then the plant was introduced into ancient China, where it was known as "Persian vegetable". The earliest available record of the spinach plant was recorded in Chinese, stating it was introduced into China via Nepal (probably in 647 AD). In AD 827, the Saracens introduced spinach to Sicily. The first written evidence of spinach in the Mediterranean are in three 10th-century works, the medical work by al-Razi (known as Rhazes in the West) and in two agricultural treatises, one by Ibn Wahshiya and the other by Qustus al-Rumi. Spinach became a popular vegetable in the Arab Mediterranean, and arrived in Spain by the latter part of the 12th century, where the great Arab agronomist Ibn al-'Awwam called it the "captain of leafy greens". Spinach was also the subject of a special treatise in the 11th century by Ibn Hajjaj.

The prickly-seeded form of spinach was known in Germany by no later than the 13th century, though the smooth-seeded form was not described till 1552. (The smooth-seeded form is used in modern commercial production). Spinach first appeared in England and France in the 14th century, probably via Spain, and it gained quick popularity because it appeared in early spring, when other vegetables were scarce and when Lenten dietary restrictions discouraged consumption of other foods. Spinach is mentioned in the first known English cookbook, The *Forme of Cury* (1390), where it is referred to as *spinnedge* and/or *spynoches*. Smooth-seeded spinach was described in 1552. In 1533, Catherine de' Medici became queen of France; she so fancied spinach that she insisted it be served at every meal. To this day, dishes made with spinach are known as "Florentine", reflecting Catherine's birth in Florence. During World War I, wine fortified with spinach juice was given to French soldiers weakened by hemorrhage.

Culinary Information/Nutritional Value

Spinach has a high nutritional value and is extremely rich in antioxidants, especially when fresh, steamed, or quickly boiled. It is a rich source of vitamin A (and especially high in lutein), vitamin C, vitamin E, vitamin K, magnesium, manganese, foliate, betaine, iron, vitamin B_2, calcium, potassium, vitamin B_6, folic acid, copper, protein, phosphorus, zinc, niacin, selenium and omega-3 fatty acids. Recently, opioid peptides called rubiscolins have also been found in spinach. Polyglutamyl foliate (vitamin B_9 or folic acid) is a vital constituent of cells, and spinach is a good source of folic acid. Boiling spinach can more than halve the level of foliate left in the spinach, but microwaving does not affect foliate content. Vitamin B_9 was first isolated from spinach in 1941.

Iron

Spinach, along with other green leafy vegetables, is considered to be a rich source of iron. For example, the United States Department of Agriculture states that a 180 g serving of boiled spinach contains 6.43 mg of iron, whereas one 170 g ground hamburger patty contains at most 4.42 mg. The bioavailability of iron is dependent on its absorption, which is influenced by a number of factors. Iron enters the body in two forms: heme iron and nonheme iron. All of the iron in grains and vegetables, and about three-fifths of the iron in animal food sources (meats), is nonheme iron. The remaining portion from meats is heme iron.

The larger portion of dietary iron (nonheme) is absorbed slowly in its many food sources, including spinach. This absorption may vary widely depending on the presence of binders, such as fiber, or enhancers, such as vitamin C. Therefore, the body's absorption of nonheme iron can be improved by consuming foods that are rich in vitamin C. However, spinach contains iron absorption-inhibiting substances, including high levels of oxalate, which can bind to the iron to form ferrous oxalate and render much of the iron in spinach unusable by the body. In addition to preventing absorption and use, high levels of oxalates remove iron from the body. But some studies have found that the addition of oxalic acid to the diet may improve iron absorption in rats over a diet with spinach without additional oxalic acid.

Calcium

Spinach also has a high in calcium content. However, the oxalate content in spinach also binds with calcium, decreasing its absorption. Calcium and zinc also limit iron absorption. The calcium in spinach is the least bioavailable of calcium sources. By way of comparison, the body can absorb about half of the calcium present in broccoli, yet only around 5 per cent of the calcium in spinach.

Types of Spinach

A distinction can be made between older varieties of spinach and more modern ones. Older varieties tend to bolt too early in warm conditions. Newer varieties tend to grow more rapidly, but have less of an inclination to run up to seed. The older varieties have narrower leaves and tend to have a stronger and more bitter taste. Most new varieties have broader leaves and round seeds.

There are three basic types of spinach:

1. *Savoy* has dark green, crinkly and curly leaves. It is the type sold in fresh bunches in most supermarkets. One heirloom variety of savoy is 'Bloomsdale', which is somewhat resistant to bolting. Other common heirloom varieties are 'Merlo Nero' (a mild variety from Italy) and 'Viroflay' very large spinach with great yields).
2. *Flat/smooth* leaf spinach has broad, smooth leaves that are easier to clean than 'Savoy'. This type is often grown for canned and frozen spinach, as well as soups, baby foods, and processed foods.
3. *Semi-savoy* is a hybrid variety with slightly crinkled leaves. It has the same texture as 'Savoy', but it is not as difficult to clean. It is grown for both fresh market and processing. 'Five Stars', a widely grown variety, has good resistance to running up to seed.

Production, Marketing and Storage

Spinach is sold loose, bunched, in prepackaged bags, canned, or frozen. Fresh spinach loses much of its nutritional value with storage of more than a few days. While refrigeration slows this effect to about eight days, spinach will lose most of its foliate and carotenoid content, so for longer storage it is frozen, cooked and frozen, or canned. Storage in the freezer can be for up to eight months.

The Environmental Working Group reported spinach is one of the dozen most heavily pesticide-contaminated produce products. The most common pesticides found on spinach are permethrin, dimethoate, and DDT.

Spinach is packaged in air, or in nitrogen gas to extend shelf life. Some packaged spinach is exposed to radiation to kill any harmful bacteria that may be on the leaves. The Food and Drug Administration approves of irradiation of spinach leaves up to 4.0 kilograys (kGy); however, there is concern that using radiation to sanitize spinach depletes the leaves of their nutritional value. Researchers at the Agricultural Research Service experimentally tested the concentrations of vitamins C, E, K, B_9, and four other carotenoids in packaged spinach following irradiation. They found that with increasing level of irradiation, four nutrients showed little or no change. Those nutrients include vitamins B_9, E, K, and the carotenoid neoxanthin. This study showed the irradiation of packaged spinach to have little or no change to the nutritional value of the crop, and the health benefits of irradiating packed spinach seem to outweigh the health risks of transmitting harmful bacteria.

Spinach, Raw

Nutritional value per 100 g (3.5 oz)	
Energy	97 kJ (23 kcal)
Carbohydrates	3.6 g
Sugars	0.4 g
Dietary fiber	2.2 g
Fat	0.4 g
Protein	2.2 g
Vitamin A equiv.	469 μg
Vitamin A	9400 IU
β-carotene	5626 μg
Lutein and zeaxanthin	12198 μg
Foliate (vit. B_9)	194 μg
Vitamin C	28 mg
Vitamin E	2 mg
Vitamin K	483 μg
Calcium	99 mg
Iron	2.7 mg

Figure 42.1: Flow sheet for preparation of dried or powdered leafy vegetables (Spinach, Mehta, Coriander, Chukka, Shephu)

Mature/tender leafy vegetables

↓

Cleaning/Washing

↓

Grading

↓

Hot water treatment at 40 °C for 30 sec. (1 liter water contain 1g, citric acid, 10g salt, 1g sodium carbonate, 0.1g KMS, 1g Magnesium oxide)

↓

Drying at 45 °C, 12-14 hrs

↓

Sealing in plastic bags or further processing for powder

↓

Grinding for powder (60 sieving)

↓

Packing and labeling

↓

Storage and Marketing

43

Okra

Okra (*Abelmoschus esculentus* Moench, pronounced, known in many English-speaking countries as lady's fingers or gumbo) is a flowering plant in the mallow family. It is valued for its edible green seed pods. The geographical origin of okra is disputed, with supporters of South Asian, Ethiopian and West African origins. The plant is cultivated in tropical, subtropical and warm temperate regions around the world.

Vernacular Names

The name "okra" is most often used in the United States, with a variation of the pronunciation–English Caribbean ("okro") used primarily around the Philippines. "Okra" is of West African origin and is cognate with *kwur* in the Igbo language spoken in Nigeria. Okra is often known as "lady's fingers" outside of the United States, with Singapore being one of the countries and regions. In Persian it is called "Bamiyeh". In various Bantu languages, okra is called *kingombo* or a variant thereof, and this is the origin of its name in Portuguese (*quiabo*), Spanish (*quimbombó* or *guigambó*), Dutch and French, and also of the name "gumbo", used in parts of the United States and English-speaking Caribbean for either the vegetable, or a stew based on it. In India, Pakistan, and often in the United Kingdom, it is called by its Hindi/Urdu name, *bhindi* or *bhendi*. In southern India, it is known as *vendaikkai* in Tamil, *benda Kaya* in Telugu, *bandakka* in Sinhala, and *bende kayi* in Kannada. In Thailand, it is known as *krachiap mon*, which is Thai for "lady's fingers". It is known as *bamya* (*bamiya*) in Arabic-speaking countries. In South and Southeast Europe, it is known as *bamya* (*bamija*). In Romania, it is known as *bamă*. In Albania, it is known as *bamje*. In Israel, it is called *bamia*.

Structure and Physiology

The species is an annual or perennial, growing to 2 m tall. It is related to such species as cotton, cocoa, and hibiscus. The leaves are 10–20 cm long and broad, palmately lobed with 5–7 lobes. The flowers are 4–8 cm in diameter, with five white to yellow petals, often with a red or purple spot at the base of each petal. The fruit is a capsule up to 18 cm long, containing numerous seeds.

Abelmoschus esculentus is cultivated throughout the tropical and warm temperate regions of the world for its fibrous fruits or pods containing round, white seeds. It is among the most heat- and drought-tolerant vegetable species in the world–but severe frost can damage the pods and will tolerate poor soils with heavy clay and intermittent moisture.

In cultivation, the seeds are soaked overnight prior to planting to a depth of 1–2 cm. Germination occurs between six days (soaked seeds) and three weeks. Seedlings require ample water. The seed pods rapidly become fibrous and woody and must be harvested within a week of the fruit being pollinated to be edible. The fruits are harvested when immature and eaten as a vegetable.

Origin and Distribution

Okra is an allopolyploid of uncertain parentage (proposed parents include *Abelmoschus ficulneus*, *A. tuberculatus* and a reported "diploid" form of okra). Truly wild, as opposed to naturalized, populations, are not definitely known, and the species may be a cultigen.

The geographical origin of okra is disputed, with supporters of South Asian, Ethiopian and West African origins. Supporters of a South Asian origin point to the presence of its proposed parents in that region. Opposed to this is the lack of a word for okra in the ancient languages of India, suggesting it arrived there in the Common Era. Supporters of a West African origin point to the greater diversity of okra in that region; however, confusion between okra and *A. caillei* (West African okra) casts doubt on those analyses.

Culinary Uses

The products of the plant are mucilaginous, resulting in the characteristic "goo" or slime when the seed pods are cooked; the mucilage contains a usable form of soluble fiber. Some people cook okra this way, others prefer to minimize sliminess; keeping the pods intact, and brief cooking, for example stir-frying, help to achieve this. Cooking with acidic ingredients such as a few drops of lemon juice, tomatoes, or vinegar may help. Alternatively, the pods can be sliced thinly and cooked for a long time so the mucilage dissolves, as in gumbo. The cooked leaves can also be used as a powerful soup thickener. The immature pods may also be pickled.

In Syria, Tunisia, Egypt, Greece, Iran, Iraq, Jordan, Lebanon, Turkey, and Yemen, and other parts of the eastern Mediterranean, including Cyprus and Israel, okra is widely used in a thick stew made with vegetables and meat. In most of West Asia, okra is known as *bamia* or *bamya*. West Asian cuisine usually uses young okra pods, usually cooked whole. In India, the harvesting is done at a later stage, when the pods and seeds are larger.

It is popular in Indian and Pakistani cuisine, where chopped pieces are stir-fried with spices, pickled, salted or added to gravy-based preparations such as *bhindi ghosts* and *sambar*. In western parts of India (Gujarat, Maharashtra), okra is often stir-fried with some sugar. Okra is also used in Kadhi. In Malaysia okra is commonly a part of *yong tau foo* cuisine, typically stuffed with processed fish paste (surimi) and boiled with a selection of vegetables and tofu, and served in a soup with noodles.

Okra leaves may be cooked in a similar way to the greens of beets or dandelions. The leaves are also eaten raw in salads. Okra seeds may be roasted and ground to form a caffeine-free substitute for coffee. When importation of coffee was disrupted by the American Civil War in 1861, the *Austin State Gazette* said "An acre of okra will produce seed enough to furnish a plantation of fifty negroes with coffee in every way equal to that imported from Rio."

Greenish-yellow edible okra oil is pressed from okra seeds; it has a pleasant taste and odor, and is high in unsaturated fats such as oleic acid and linoleic acid. The oil content of some varieties of the seed can be quite high, about 40 per cent . Oil yields from okra crops are also high. At 794 kg/ha, the yield was exceeded only by that of sunflower oil in one trial. A 1920 study found that a sample contained 15 per cent oil. A 2009 study found okra oil suitable for use as a biofuel.

Medicinal Properties

Okra is low in calories and is a good source of many nutrients including vitamin B_6 and C, fiber, calcium, and folic acid. It is effective for the prevention of neural tube defects in developing fetuses mainly due to its high content of vitamin B_6, calcium, fiber, and folic acid.

Unspecified parts of the plant were reported in 1898 to possess diuretic properties; this is cited (or simply stated) in many sources associated with herbal and traditional medicine. Okra (and rhubarb, beets, spinach, Swiss chard, sweet potatoes, tea, chocolate and soy products) is rich in oxalates; the Mayo clinic recommends that people who tend to form calcium oxalate kidney stones may benefit from restricting such foods.

Okra, Raw

Nutritional value per 100 g (3.5 oz)	
Energy	129 kJ (31 kcal)
Carbohydrates	7.03 g
Sugars	1.20 g
Dietary fiber	3.2 g
Fat	0.10 g
Protein	2.00 g
Water	90.17 g

Figure 43.1: Flow sheet for preparation of canned okra

Tender okra fruits
↓
Washing
↓
Grading
↓
Cutting similar size pieces or use whole fruits
↓
Blanching for 2-3 min
↓
Cooling in 1.5 per cent salt solution
↓
Filling cans with okra in 2 per cent salt solution
↓
Pasteurization at 79-82 °C for 30-50 min
↓
Sealing cans and labeling
↓
Storage and marketing

Figure 43.2: Flow sheet for preparation of dried okra

Tender okra fruits
↓
Cleaning and washing
↓
Grading
↓
Whole fruits or similar size cut pieces
↓
Blanching for 5 min
↓
Drying at 55 °C 14 hrs
↓
Packing in plastic bags
↓
Sealing and labeling
↓
Storage and marketing

44

Onion

The onion (*Allium cepa*), also known as the bulb onion, common onion and garden onion, is the most widely cultivated species of the genus *Allium*. The genus *Allium* also contains a number of other species variously referred to as onions and cultivated for food, such as the Japanese bunching onion (*A. fistulosum*), Egyptian onion (*A.* ×*proliferum*), and Canada onion (*A. canadense*). The name "wild onion" is applied to a number of *Allium* species.

The vast majority of cultivars of *A. cepa* belong to the 'common onion group' (*A. cepa* var. *cepa*) and are usually referred to simply as 'onions'. The 'Aggregatum group' of cultivars (*A. cepa* var. *aggregatum*) includes both shallots and potato onions. *Allium cepa* is known only in cultivation, but related wild species occur in Central Asia.

Uses

Culinary Uses

Onions are often chopped and used an ingredient in various hearty warm dishes, although onions can also be a more prominent ingredient, for example in French onion soup, or be used raw in cold salads. Onions pickled in vinegar are eaten as a snack. These are often served as a side serving in fish and chip shops throughout the United Kingdom and Australia, often served with cheese in the United Kingdom, and are referred to simply as "pickled onions" in Eastern Europe.

Onion Types and Products

Common onions are normally available in three colors: yellow, red, and white. Yellow onions are full-flavored and are a reliable standby for cooking almost anything. Yellow onions turn a rich, dark brown when cooked and give French onion soup its tangy sweet flavor. The red onion is a good choice for fresh uses or in grilling and char-broiling. White onions are the traditional onion used in classic Mexican cuisine. They have a golden color and sweet flavor when sautéed.

While the large mature onion bulb is the onion most often eaten, onions can be eaten at immature stages. Young plants may be harvested before bulbing occurs and used whole as scallions. When an onion is harvested after bulbing has begun but the onion is not yet mature, the plants are sometimes referred to as summer onions.

Additionally, onions may be bred and grown to mature at smaller sizes. Depending on the mature size and the purpose for which the onion is used, these may be referred to as pearl, boiler, or pickler onions. (However, true pearl onions are a different species.) Pearl and boiler onions may be cooked as a vegetable rather than an ingredient. Pickler onions are, unsurprisingly, often pickled.

Onion seed may be "sprouted", and the resulting sprouts used in salads, sandwiches, and other dishes. (See sprouting). Onions are available in fresh, frozen, canned, caramelized, pickled, powdered, chopped, and dehydrated forms.

Onion powder is a spice used for seasoning in cooking. It is made from finely ground, dehydrated onions, mainly the pungent varieties of bulb onions, which causes the powder to have a very strong odor. Onion powder comes in a few varieties: white, yellow, red and toasted.

Medicinal Properties and Health Effects of Onions

Wide-ranging claims have been made for the effectiveness of onions against conditions ranging from the common cold to heart disease, diabetes, osteoporosis, and other diseases. They contain chemical compounds believed to have anti-inflammatory, anticholesterol, anticancer, and antioxidant properties, such as quercetin. Preliminary studies have shown increased consumption of onions reduces the risk of head and neck cancers.

Among all varieties, Asian white onions have the most eye irritating chemical reaction. Regular use of white onion, if eaten raw, is claimed to be good due to its antioxidant and anti-inflammatory properties.

In India some sects do not eat onions as they believe them to be an aphrodisiac; various schools of Buddhism also advise against eating onions and other vegetables of the Allium family. In many parts of the undeveloped world, onions are used to heal blisters and boils. A traditional Maltese remedy for sea urchin wounds is to tie half a baked onion to the afflicted area overnight. A similar traditional cure is known in Bulgaria. Half-baked onion with sugar is placed over the finger and fingernail in case of inflammation.

An application of raw onion is also said to be helpful in reducing swelling from bee stings. In the United States, products that contain onion extract are used in the treatment of topical scars; some studies have found their action to be ineffective, while others found that they may act as an anti-inflammatory or bacteriostatic and can improve collagen organization in rabbits.

Onions may be beneficial for women, who are at increased risk for osteoporosis as they go through menopause, by destroying osteoclasts so they do not break down bone.

An American chemist has stated the pleiomeric chemicals in onions have the potential to alleviate or prevent sore throat. Onion in combination with jaggery has been widely used as a traditional household remedy for sore throat in India.

Shallots have the most phenols, six times the amount found in Vidalia onion, the variety with the lowest phenolic content. Shallots also have the most antioxidant activity, followed by Western Yellow, pungent yellow (New York Bold), Northern Red, Mexico, Empire Sweet, Western White, Peruvian Sweet, Texas 1015, Imperial Valley Sweet, and Vidalia. Western Yellow onions have the most flavonoids, eleven times the amount found in Western White, the variety with the lowest flavonoid content.

For all varieties of onions, the more phenols and flavonoids they contain, the more reputed antioxidant and anticancer activity they provide. When tested against liver and colon cancer cells in laboratory studies, 'Western Yellow', pungent yellow (New York Bold) and shallots were most effective in inhibiting their growth. The milder-tasting cultivars (*i.e.*, 'Western White,' 'Peruvian Sweet,' 'Empire Sweet,' 'Mexico,' 'Texas 1015,' 'Imperial Valley Sweet' and 'Vidalia') showed little cancer-fighting ability.

Shallots and ten other onion (*Allium cepa* L.) varieties commonly available in the United States were evaluated: Western Yellow, Northern Red, pungent yellow (New York Bold), Western White, Peruvian Sweet, Empire Sweet, Mexico, Texas 1015, Imperial Valley Sweet, and Vidalia. In general, the most pungent onions delivered many times the effects of their milder cousins.

The 3-mercapto-2-methylpentan-1-ol in onion was found to inhibit peroxynitrite-induced mechanisms *in vitro*. While members of the onion family appear to have medicinal properties for humans, they can be deadly for dogs, cats, and guinea pigs.

Eye Irritation

As onions are sliced or eaten, cells are broken, allowing enzymes called alliinases to break down amino acid sulphoxides and generate sulphenic acids. A specific sulfenic acid, 1-propenesulfenic acid, formed when onions are cut, is rapidly rearranged by a second enzyme, called the lachrymatory factor synthase or LFS, giving syn-propanethial-S-oxide, a volatile gas known as the onion lachrymatory factor or LF. The LF gas diffuses through the air and eventually reaches the eye, where it activates sensory neurons, creating a stinging sensation. Tear glands produce tears to dilute and flush out the irritant. Chemicals that exhibit such an effect on the eyes are known as lachrymatory agents.

Supplying ample water to the reaction while peeling onions prevents the gas from reaching the eyes. Eye irritation can, therefore, be avoided by cutting onions under running water or submerged in a basin of water. Another way to reduce irritation is by chilling, or by not cutting off the root of the onion (or by doing it last), as the root of the onion has a higher concentration of enzymes. Using a sharp blade to chop onions will limit the cell damage and the release of enzymes that drive the irritation response. Chilling or freezing onions prevents the enzymes from activating, limiting the amount of gas generated.

Storage

Green onion and leeks are optimally stored refrigerated. Cooking onions and sweet onions, on the other hand, can be stored at room temperature, optimally in a single layer, in mesh bag in a dry, cool, dark, well ventilated location. In this environment, cooking onions have a shelf life of 3 to 4 weeks, and sweet onions 1 to 2 weeks. Cooking onions will absorb odours from apples and pears. Also, they draw moisture from vegetables they are stored with which may cause them to decay. Sweet onions have a greater water and sugar content than cooking onions. This makes them sweeter and milder tasting, but also reduces their shelf life. Sweet onions can also be stored refrigerated; they have a shelf life of approximately 1 month, optimally uncovered. Irrespective of type, any cut pieces of onion are optimally tightly wrapped, stored away from other produce, and used within 2 to 3 days.

Raw Onions

Nutritional value per 100 g (3.5 oz)	
Energy	166 kJ (40 kcal)
Carbohydrates	9.34 g
Sugars	4.24 g
Dietary fiber	1.7 g
Fat	0.1 g
Saturated	0.042 g
Monounsaturated	0.013 g
Polyunsaturated	0.017 g
Protein	1.1 g
Water	89.11 g
Vitamin A equiv.	0 μg
Thiamine (vit. B_1)	0.046 mg
Riboflavin (vit. B_2)	0.027 mg
Niacin (vit. B_3)	0.116 mg
Vitamin B_6	0.12 mg
Folate (vit. B_9)	19 μg
Vitamin B_{12}	0 μg
Vitamin C	7.4 mg
Vitamin E	0.02 mg
Vitamin K	0.4 μg
Calcium	23 mg
Iron	0.21 mg
Magnesium	0.129 mg
Phosphorus	29 mg
Potassium	146 mg
Sodium	4 mg
Zinc	0.17 mg

Figure 44.1: Flow sheet for preparation of dried onion shreds or powder

Fully Mature onion

↓

Cleaning and Size Grading

↓

Cutting 0.5 to 1.0 mm size slices (machine or manually)

↓

Drying in sun or drying machine at 55 °C

↓

For colored onion use 2 per cent salt in 2 per cent calcium chloride solution for 10-15 min dipping shreds before drying

↓

Dried onion shreds

↓

Grinding with pulverizer

↓

Sieving and bagging

↓

Sealing and labeling

↓

Onion Powder

↓

Storage and marketing

Figure 44.2: Flow sheet for preparation of onion pickle

Small onions

↓

Peeling

↓

Blanching for 5 min

↓

Filling blanched onions in jar

↓

Addition of salt

↓

Standing for 24-48 hrs

↓

Draining off water

↓

Addition of vinegar and spices

↓

Storage

↓

Marketing

Sl.No.	*Item*	*Quantity*	*Sl.No.*	*Item*	*Quantity*
1.	Onions	1 kg	5.	Cumin	10 g
2.	Salt	250 g	6.	Cardamom	10 g
3.	Red chilli powder	10 g	7.	Clove headless	5 Nos
4.	Vinegar	1000 ml	8.	Black pepper	10 g

45

Peas

A pea is most commonly the small spherical seed or the seed-pod of the pod fruit *Pisum sativum*. Each pod contains several peas. Peapods are botanically a fruit, since they contain seeds developed from the ovary of a (pea) flower. However, peas are considered to be a vegetable in cooking. The name is also used to describe other edible seeds from the Fabaceae such as the pigeon pea (*Cajanus cajan*), the cowpea (*Vigna unguiculata*), and the seeds from several species of *Lathyrus*. *P. sativum* is an annual plant, with a life cycle of one year. It is a cool season crop grown in many parts of the world; planting can take place from winter to early summer depending on location. The average pea weighs between 0.1 and 0.36 grams. The species is used as a vegetable, fresh, frozen or canned, and is also grown to produce dry peas like the split pea. These varieties are typically called field peas.

The wild pea is restricted to the Mediterranean basin and the Near East. The earliest archaeological finds of peas come from Neolithic Syria, Turkey and Jordan. In Egypt, early finds date from *ca.* 4800–4400 BC in the Nile delta area, and from *ca.* 3800–3600 BC in Upper Egypt. The pea was also present in Georgia in the 5th millennium BC. Farther east, the finds are younger. Peas were present in Afghanistan *ca.* 2000 BC, in Harappa, Pakistan, and in northwest India in 2250–1750 BC. In the second half of the 2nd millennium BC this pulse crop appears in the Gangetic basin and southern India.

The pea is a most commonly green, occasionally purple or golden yellow, pod-shaped vegetable, widely grown as a cool season vegetable crop. The seeds may be planted as soon as the soil temperature reaches 10 °C (50 °F), with the plants growing best at temperatures of 13 to 18 °C (55 to 64 °F). They do not thrive in the summer heat of warmer temperate and lowland tropical climates but do grow well in cooler high altitude tropical areas. Many cultivars reach maturity about 60 days after planting.

Peas have both low-growing and vining cultivars. The vining cultivars grow thin tendrils from leaves that coil around any available support and can climb to be

1–2 m high. A traditional approach to supporting climbing peas is to thrust branches pruned from trees or other woody plants upright into the soil, providing a lattice for the peas to climb. Branches used in this fashion are sometimes called *pea brush.* Metal fences, twine, or netting supported by a frame are used for the same purpose. In dense plantings, peas give each other some measure of mutual support. Pea plants can self-pollinate.

Uses

Culinary Uses

In early times, peas were grown mostly for their dry seeds. In modern times, however, peas are usually boiled or steamed, which breaks down the cell walls and makes the taste sweeter and the nutrients more bio-available. Along with broad beans and lentils, these formed an important part of the diet of most people in the Middle East, North Africa and Europe during the middle Ages. By the 17th and 18th centuries it had become popular to eat peas "green", that is, while they are immature and right after they are picked. This was especially true in France and England, where the eating of green peas was said to be "both a fashion and madness". New cultivars of peas were developed by the English during this time which became known as garden peas and English peas. The popularity of green peas spread to North America. Thomas Jefferson grew more than 30 cultivars of peas on his estate. With the invention of canning and freezing of foods, green peas became available year-round, and not just in the spring as before.

Fresh peas are often eaten boiled and flavored with butter and/or spearmint as a side dish vegetable. Salt and pepper are also commonly added to peas when served. Fresh peas are also used in pot pies, salads and casseroles. Pod peas (particularly sweet cultivars called mange tout and *sugar peas,* or the flatter "snow peas," are used in stir-fried dishes, particularly those in American Chinese cuisine. Pea pods do not keep well once picked, and if not used quickly are best preserved by drying, canning or freezing within a few hours of harvest.

In India, fresh peas are used in various dishes such as *aloo matar* (curried potatoes with peas) or *matar paneer* (paneer cheese with peas), though they can be substituted with frozen peas as well. Peas are also eaten raw, as they are sweet when fresh off the bush. Split peas are also used to make dhal, particularly in Guyana, and Trinidad, where there is a significant population of Indians.

Dried peas are often made into a soup or simply eaten on their own. In Japan, China, Taiwan and some Southeast Asian countries, including Thailand and Malaysia, peas are roasted and salted, and eaten as snacks. In the UK, dried yellow split peas are used to make pease pudding (or "pease porridge"), a traditional dish. In North America, a similarly traditional dish is split pea soup.

Pea soup is eaten in many other parts of the world, including northern Europe, parts of middle Europe, Russia, Iran, Iraq and India. In Sweden it is called *ärtsoppa,* and is eaten as a traditional Swedish food which predates the Viking era. This food was made from a fast-growing pea that would mature in a short growing season. *Ärtsoppa* was especially popular among the many poor who traditionally only had one pot and everything was cooked together for a dinner using a tripod to hold the pot over the fire.

In Chinese cuisine, pea sprouts are commonly used in stir-fries. Pea leaves are often considered a delicacy as well.

In Greece, Turkey, Cyprus, and other parts of the Mediterranean, peas are made into a stew with meat and potatoes. In Hungary and Serbia, pea soup is often served with dumplings and spiced with hot paprika. In the United Kingdom, dried, rehydrated and mashed marrowfat peas, known by the public as mushy peas, are popular, originally in the north of England but now ubiquitously, and especially as an accompaniment to fish and chips or meat pies, particularly in fish and chip shops. Sodium bicarbonate is sometimes added to soften the peas. In 2005, a poll of 2,000 people revealed the pea to be Britain's 7th favorite culinary vegetable. Processed peas are mature peas which have been dried, soaked and then heat treated (processed) to prevent spoilage in the same manner as pasteurizing. Cooked peas are sometimes sold dried and coated with wasabi, salt, or other spices.

Nutritional Value

Green peas, botanically classified as a fruit, have been used as vegetable, for cooking purposes, since ages. They can be described as the small, spherical seeds or the pods of the legume Pisum sativum. Green peas are mainly cultivated during the cool season *i.e.* from the start of winters to early-summers. Apart from being very rich in protein, they hold a high nutritional value in the form of vitamins and minerals too. Owing to this, they have been associated with a large number of nutrition benefits.

Peas are high in fiber, protein, vitamins, minerals, and lutein. Dry weight is about one-quarter protein and one-quarter sugar. Pea seed peptide fractions have less ability to scavenge free radicals than glutathione, but greater ability to chelate metals and inhibit linoleic acid oxidation.

1 cup of raw green peas (145g) contains:

- ✰ Calories: 117
- ✰ Fat (g): 0.58
- ✰ Carbohydrates (g): 20.95
- ✰ Fibers (g): 7.4
- ✰ Protein (g): 7.86

Raw Green Pea

Nutritional value per 100 g (3.5 oz)	
Energy	339 kJ (81 kcal)
Carbohydrates	14.5 g
Sugars	5.7 g
Dietary fiber	5.1 g
Fat	0.4 g
Protein	5.4 g

Vitamin A equiv.	38 μg
β-carotene	449 μg
lutein and zeaxanthin	2593 μg
Thiamine (vit. B_1)	0.3 mg
Riboflavin (vit. B_2)	0.1 mg
Niacin (vit. B_3)	2.1 mg
Pantothenic acid (B_5)	0.1 mg
Vitamin B_6	0.2 mg
Folate (vit. B_9)	65 μg
Vitamin C	40.0 mg
Calcium	25.0 mg
Iron	1.5 mg
Magnesium	33.0 mg
Phosphorus	108 mg
Potassium	244 mg
Zinc	1.2 mg

Health and Nutrition Benefits of Eating Green Peas

- Being low in calories, green peas are good for those who are trying to lose weight.
- Green peas are rich in dietary fiber, making them good for those suffering from constipation.
- Studies have indicated that green peas might prove beneficial for those suffering from the problem of high cholesterol.
- The high amount of iron and vitamin C in green peas has been found to help strengthen the immune system.
- The lutein present in green peas helps reduce the risk of age-related mascular degeneration and cataracts.
- Green peas slow down the appearance of glucose in the blood and thus, help keep the energy levels steady.
- Green peas have been found to aid energy production, nerve function and carbohydrate metabolism.
- Green peas provide the body with those nutrients that are important for maintaining bone health.
- The folic acid and vitamin B_6 in green peas are good for promoting the cardiovascular health of an individual.
- Being rich in antioxidants, like vitamin C, green peas can help keep cancer at bay.

Figure 45.1: Flow sheet for preparation of frozen peas

Mature green peas
↓
Cleaning and sorting
↓
Separation of seeds and peel
↓
Blanching at 85 °C for 2 min
↓
Draining
↓
Filling cans/Pouches
↓
Sealing cans/Pouches
↓
Pasteurization
↓
Cooling and labeling
↓
Storage at –20 to –40 °C
↓
Marketing

Figure 45.2: Flow sheet for preparation of dried green peas

Fresh mature green pea pods
↓
Shelling
↓
Grading
↓
Pricking
↓
Blanching (96 °C, 5 min)
↓
Drying in tray dryer (60 °C)
↓
Dried product (4.9 per cent M. C.)
↓
Sensory evaluation
↓
Storage

46

Potato

The potato is a starchy, tuberous crop from the perennial *Solanum tuberosum* of the Solanaceae family (also known as the nightshades). The word potato may refer to the plant itself as well as the edible tuber. In the region of the Andes, there are some other closely related cultivated potato species. Potatoes were first introduced outside the Andes region four centuries ago, and have become an integral part of much of the world's cuisine. It is the world's fourth-largest food crop, following rice, wheat, and maize. Long-term storage of potatoes requires specialised care in cold warehouses.

Wild potato species occur throughout the Americas, from the United States to Uruguay. The potato was originally believed to have been domesticated independently in multiple locations, but later genetic testing of the wide variety of cultivars and wild species proved a single origin for potatoes in the area of present-day southern Peru (from a species in the *Solanum brevicaule* complex), where they were domesticated 7,000–10,000 years ago. Following centuries of selective breeding, there are now over a thousand different types of potatoes. Of these subspecies, a variety that at one point grew in the Chiloé Archipelago (the potato's south-central Chilean sub-center of origin) left its germplasm on over 99 per cent of the cultivated potatoes worldwide.

Following the Spanish conquest of the Inca Empire, the Spanish introduced the potato to Europe in the second half of the 16th century. The staple was subsequently conveyed by European mariners to territories and ports throughout the world. The potato was slow to be adopted by distrustful European farmers, but soon enough it became an important food staple and field crop that played a major role in the European 19th century population boom. However, lack of genetic diversity, due to the very limited number of varieties initially introduced, left the crop vulnerable to disease. In 1845, a plant disease known as late blight, caused by the fungus-like oomycete *Phytophthora infestans*, spread rapidly through the poorer communities of western Ireland, resulting in the crop failures that led to the Great Irish Famine.

Nonetheless, thousands of varieties persist in the Andes, where over 100 cultivars might be found in a single valley and a dozen or more might be maintained by a single agricultural household.

The annual diet of an average global citizen in the first decade of the 21st century included about 33 kg (73 lb) of potato. However, the local importance of potato is extremely variable and rapidly changing. It remains an essential crop in Europe (especially eastern and central Europe), where per capita production is still the highest in the world, but the most rapid expansion over the past few decades has occurred in southern and eastern Asia. China is now the world's largest potato-producing country, and nearly a third of the world's potatoes are harvested in China and India.

The English word *potato* comes from Spanish *patata* (the name used in Spain). The Spanish Royal Academy says the Spanish word is a compound of the Taino *batata* (sweet potato) and the Quechua *papa* (potato). The name potato originally referred to a type of sweet potato rather than the other way around, although there is actually no close relationship between the two plants. The English confused the two plants one for the other. In many of the chronicles detailing agriculture and plants, no distinction is made between the two. The 16th-century English herbalist John Gerard used the terms "bastard potatoes" and "Virginia potatoes" for this species, and referred to sweet potatoes as "common potatoes". Potatoes are occasionally referred to as "Irish potatoes" or "white potatoes" in the United States, to distinguish them from sweet potatoes.

The name spud for a small potato comes from the digging of soil (or a hole) prior to the planting of potatoes. The word has an unknown origin and was originally (c. 1440) used as a term for a short knife or dagger, probably related to Dutch *spyd* and/ or the Latin "spad-" root meaning "sword"; cf. Spanish "espada", English "spade" and "spadroon". The word spud traces back to the 16th century. It subsequently transferred over to a variety of digging tools. Around 1845 it transferred over to the tuber itself. The origin of "spud" has erroneously been attributed to a 19th century activist group dedicated to keeping the potato out of Britain, calling itself. The Society for the Prevention of an Unwholesome Diet. It was Mario Pei's 1949 *The Story of Language* that can be blamed for the false origin. Pei writes, "The potato, for its part, was in disrepute some centuries ago. Some Englishmen who did not fancy potatoes formed a Society for the Prevention of Unwholesome Diet. The initials of the main words in this title gave rise to spud." Like most other pre-20th century acronymic origins, this one is false.

Asia

The potato diffused widely after 1600, becoming a major food resource in Europe and East Asia. Following its introduction into China toward the end of the Ming dynasty, the potato immediately became a delicacy of the imperial family. After the middle period of the Qianlong reign (1735–96), population increases and a subsequent need to increase grain yields coupled with greater peasant geographic mobility led to the rapid spread of potato cultivation throughout China, and it was acclimated to local natural conditions. Boomgaard (2003) looks at the adoption of various root and tuber crops in Indonesia throughout the colonial period and examines the chronology and reasons for progressive adoption of foreign crops: sweet potato, Irish potato, bengkuang (yam beans), and cassava. The potato was introduced in the Philippines during the late 16th century and to Java and China during the 17th century. It was well established as a crop in India by the late 18th century and in Africa by the mid-20th century.

Nutrition

The potato contains vitamins and minerals, as well as an assortment of phytochemicals, such as carotenoids and natural phenols. Chlorogenic acid constitutes up to 90 per cent of the potato tuber natural phenols. Others found in potatoes are 4-O-caffeoylquinic acid (crypto-chlorogenic acid), 5-O-caffeoylquinic (neo-chlorogenic acid), 3,4-dicaffeoylquinic and 3,5-dicaffeoylquinic acids. A medium-size 150 g (5.3 oz) potato with the skin provides 27 mg of vitamin C (45 per cent of the Daily Value (DV)), 620 mg of potassium (18 per cent of DV), 0.2 mg vitamin B_6 (10 per cent of DV) and trace amounts of thiamin, riboflavin, folate, niacin, magnesium, phosphorus, iron, and zinc. The fiber content of a potato with skin (2 g) is equivalent to that of many whole grain breads, pastas, and cereals.

In terms of nutrition, the potato is best known for its carbohydrate content (approximately 26 grams in a medium potato). The predominant form of this carbohydrate is starch. A small but significant portion of this starch is resistant to digestion by enzymes in the stomach and small intestine, and so reaches the large intestine essentially intact. This resistant starch is considered to have similar physiological effects and health benefits as fiber: It provides bulk, offers protection against colon cancer, improves glucose tolerance and insulin sensitivity, lowers plasma cholesterol and triglyceride concentrations, increases satiety, and possibly even reduces fat storage. The amount of resistant starch in potatoes depends much on preparation methods. Cooking and then cooling potatoes significantly increases resistant starch. For example, cooked potato starch contains about 7 per cent resistant starch, which increases to about 13 per cent upon cooling.

The cooking method used can significantly impact the nutrient availability of the potato.

Potatoes are often broadly classified as high on the glycemic index (GI) and so are often excluded from the diets of individuals trying to follow a low-GI diet. In fact, the GI of potatoes can vary considerably depending on type (such as red, russet, white, or Prince Edward), origin (where it was grown), preparation methods (*i.e.*, cooking method, whether it is eaten hot or cold, whether it is mashed or cubed or consumed whole, etc.), and with what it is consumed (*i.e.*, the addition of various high-fat or high-protein toppings).

Potatoes are not considered by the NHS as counting towards the five portions of fruit and vegetables diet.

Toxicity

Potatoes contain toxic compounds known as glycoalkaloids, of which the most prevalent are solanine and chaconine. Solanine is also found in other plants in the family Solanaceae, which includes such plants as the deadly nightshade (*Atropa belladonna*), henbane (*Hyoscyamus niger*) and tobacco (*Nicotiana*) as well as the potato, eggplant, and tomato. This toxin affects the nervous system, causing weakness and confusion.

These compounds, which protect the plant from its predators, are, in general, concentrated in its leaves, stems, sprouts, and fruits. Exposure to light, physical damage and age increase glycoalkaloid content within the tuber; the highest concentrations occur just underneath the skin. Cooking at high temperatures (over 170 °C or 340 °F) partly destroys these. The concentration of glycoalkaloid in wild potatoes suffices to produce toxic effects in humans. Glycoalkaloids may cause headaches, diarrhea, cramps, and in severe cases coma and death; however, poisoning from potatoes occurs very rarely. Light exposure causes greening from chlorophyll synthesis, thus giving a visual clue as to areas of the tuber that may have become more toxic; however, this does not provide a definitive guide, as greening and glycoalkaloid accumulation can occur independently of each other. Some varieties of potato contain greater glycoalkaloid concentrations than others; breeders developing new varieties test for this, and sometimes have to discard an otherwise promising cultivar.

Breeders try to keep solanine levels below 200 mg/kg (200 ppm). However, when these commercial varieties turn green, even they can approach concentrations of solanine of 1000 mg/kg (1000 ppm). In normal potatoes, analysis has shown solanine levels may be as little as 3.5 per cent of the breeders' maximum, with 7–187 mg/kg being found. While a normal potato has 12-20 mg/kg of glycoalkaloid content, a green tuber contains 250-280 mg/kg, and green skin 1500-2200 mg/kg.

The U.S. National Toxicology Program suggests that the average American consume at most 12.5 mg/day of solanine from potatoes (the toxic dose is actually several times this, depending on body weight). Douglas L. Holt, the State Extension Specialist for Food Safety at the University of Missouri, notes that no reported cases of potato-source solanine poisoning have occurred in the U.S. in the last 50 years, and most cases involved eating green potatoes or drinking potato-leaf tea.

Potato, Raw, with Skin

Nutritional value per 100 g (3.5 oz)	
Energy	321 kJ (77 kcal)
Carbohydrates	19 g
Starch	15 g
Dietary fiber	2.2 g
Fat	0.1 g
Protein	2 g
Water	75 g
Thiamine (vit. B_1)	0.08 mg
Riboflavin (vit. B_2)	0.03 mg
Niacin (vit. B_3)	1.1 mg

Vitamin B_6	0.25 mg
Vitamin C	20 mg
Calcium	12 mg
Iron	1.8 mg
Magnesium	23 mg
Phosphorus	57 mg
Potassium	421 mg
Sodium	6 mg

Storage

Storage facilities need to be carefully designed to keep the potatoes alive and slow the natural process of decomposition, which involves the breakdown of starch. It is crucial that the storage area is dark, well ventilated and for long-term storage maintained at temperatures near 4 °C (39 °F). For short-term storage before cooking, temperatures of about 7 °C (45 °F) to 10 °C (50 °F) are preferred.

On the other hand, temperatures below 4 °C (39 °F) convert potatoes' starch into sugar, which alters their taste and cooking qualities and leads to higher acrylamide levels in the cooked product, especially in deep-fried dishes the discovery of acrylamides in starchy foods in 2002 has led to many international health concerns as they are believed to be possible carcinogens and their occurrence in cooked foods are currently under study as possible influences in potential health problems.

Under optimum conditions possible in commercial warehouses, potatoes can be stored for up to ten to twelve months. When stored in homes, the shelf life is usually only a few weeks. If potatoes develop green areas or start to sprout, these areas should be trimmed before using. Trimming or peeling green areas is inadequate to remove co present toxins, and such potatoes are no longer suitable as animal food. Commercial storage of potatoes involves several phases: drying of surface moisture; a wound healing phase at 85 per cent to 95 per cent relative humidity and temperatures below 25 °C (77 °F); a staged cooling phase; a holding phase; and a reconditioning phase, during which the tubers are slowly warmed. Mechanical ventilation is used at various points during the process to prevent condensation and accumulation of carbon dioxide.

When stored in the home, mature potatoes are optimally kept at room temperature, where they last 1 to 2 weeks in a paper bag, in a dry, cool, dark, well ventilated location. If mature potatoes are refrigerated, dark spots can occur and conversion of starch into sugar can give rise to an unpleasant sweet flavour when cooked. Only new potatoes can be refrigerated, and should be kept so, where they have a shelf life of 1 week. If kept in too warm a temperature, both mature and new potatoes will sprout and shrivel. Exposure to light causes them to turn green. Also, potatoes absorb odours produced by pears.

Varieties

While there are close to 4000 different varieties of potato, it has been bred into many standard or well-known varieties, each of which has particular agricultural or culinary attributes. In general, varieties are categorized into a few main groups, such as russets, reds, whites, yellows (also called Yukons) and purples–based on common characteristics. Around 80 varieties are commercially available in the UK. For culinary purposes, varieties are often described in terms of their waxiness. Floury or mealy (baking) potatoes have more starch (20–22 per cent) than waxy (boiling) potatoes (16–18 per cent). The distinction may also arise from variation in the comparative ratio of two potato starch compounds: amylose and amylopectin. Amylose, a long-chain molecule, diffuses out of the starch granule when cooked in water, and lends itself to dishes where the potato is mashed. Varieties that contain slightly higher amylopectin content, a highly branched molecule, help the potato retain its shape when boiled.

The European Cultivated Potato Database (ECPD) is an online collaborative database of potato variety descriptions, updated and maintained by the Scottish Agricultural Science Agency within the framework of the European Cooperative Programme for Crop Genetic Resources Networks (ECP/GR) which is organised by the International Plant Genetic Resources Institute (IPGRI).

Popular Varieties (Cultivars) include

- Adirondack Blue
- Adirondack Red
- Agata
- Almond
- Amandine
- Anya
- Arran Victory
- Atlantic
- Avalanche
- Bamberg
- Belle de Fontenay
- BF-15
- Bildtstar
- Bintje
- Blue Congo
- Bonnotte
- British Queens
- Cabritas
- Camota
- Cara
- Chelina
- Chiloé
- Cielo
- Clavela Blanca
- Désirée
- Estima
- Fianna
- Fingerling
- Flava
- Golden Wonder
- Home Guard
- Innovator
- Jersey Royal
- Kerr's Pink
- Kestrel
- King Edward
- Kipfler
- Lady Balfour
- Linda
- Marfona
- Maris Piper
- Marquis
- Nicola
- Pachacoña
- Pink Eye
- Pink Fir Apple
- Primura
- Ratte
- Record
- Red Norland
- Red Pontiac
- Rooster
- Russet Burbank
- Russet Norkotah
- Selma
- Shepody
- Sieglinde
- Sirco
- Spunta
- Stobrawa
- Vivaldi
- Vitelotte
- Yellow Finn
- Yukon Gold

Blue Varieties

The *blue potato* originated in South America. It has purple skin and flesh, which becomes blue once cooked. It has a slight whitish scab that seems to be present in all samples. The variety, called "Cream of the Crop", has been introduced into Ireland and has proved popular.

A mutation in the varieties' P locus causes production of the antioxidant anthocyanin.

Genetically-Modified Varieties

Genetic research has produced several genetically modified varieties. 'New Leaf', owned by Monsanto Company, incorporates genes from *Bacillus thuringiensis*, which confers resistance to the Colorado potato beetle; 'New Leaf Plus' and 'New Leaf Y', approved by US regulatory agencies during the 1990s, also include resistance to viruses. McDonald's, Burger King, Frito-Lay, and Procter and Gamble announced they would not use genetically modified potatoes, and Monsanto published its intent to discontinue the line in March 2001. The starch content of 'Amflora', a waxy potato variety from the German chemical company BASF, has been modified to contain only amylopectin, making it inedible but more useful for industrial purposes. In 2010, the European Commission cleared the way for 'Amflora' to be grown in the European Union. Nevertheless, under EU rules, individual countries have the right to decide whether they will allow this potato to be grown on their territory. Commercial planting of 'Amflora' was expected in the Czech Republic and Germany in the spring of 2010, and Sweden and the Netherlands in subsequent years. Another GM potato variety developed by BASF is 'Fortuna'. In 2010, a team of Indian scientists announced they had developed a genetically modified potato with 35 to 60 per cent more protein than non-modified potatoes. Protein content was boosted by adding the gene AmA1 from the grain amaranth. They also found 15 to 25 per cent greater crop yields with these potatoes. The researchers expected that a key market for the GM potato would be the developing world, where more than a billion people are chronically undernourished.

New Non-GM Varieties

Other new potato cultivars are conventionally bred. For example, on 22 September 2007, Benguet State University (BSU) announced that four potato varieties–'Igorota', 'Solibao', 'Ganza' and one not yet officially named possess more than 18 per cent dry matter content required by fast-food chains to make crispy and sturdy French fries. Since 2005, a natural 100 per cent amylopectin waxy potato variety called 'Eliane' is being cultivated by the starch company AVEBE.

Some horticulturists sell chimeras, made by grafting a tomato plant onto a potato plant, producing both edible tomatoes and potatoes. This practice is not very widespread.

Uses

- Potatoes are used to brew alcoholic beverages such as vodka, potcheen, or akvavit.

- They are also used as food for domestic animals.
- Potato starch is used in the food industry as, for example, thickeners and binders of soups and sauces, in the textile industry, as adhesives, and for the manufacturing of papers and boards.
- Maine companies are exploring the possibilities of using waste potatoes to obtain polylactic acid for use in plastic products; other research projects seek ways to use the starch as a base for biodegradable packaging.

Culinary Uses

Potatoes are prepared in many ways: skin-on or peeled, whole or cut up, with seasonings or without. The only requirement involves cooking to swell the starch granules. Most potato dishes are served hot, but some are first cooked, and then served cold, notably potato salad and potato chips/crisps.

Common dishes are: mashed potatoes, which are first boiled (usually peeled), and then mashed with milk or yogurt and butter; whole baked potatoes; boiled or steamed potatoes; French-fried potatoes or chips; cut into cubes and roasted; scalloped, diced, or sliced and fried (home fries); grated into small thin strips and fried (hash browns); grated and formed into dumplings, Rösti or potato pancakes. Unlike many foods, potatoes can also be easily cooked in a microwave oven and still retain nearly all of their nutritional value, provided they are covered in ventilated plastic wrap to prevent moisture from escaping; this method produces a meal very similar to a steamed potato, while retaining the appearance of a conventionally baked potato. Potato chunks also commonly appear as a stew ingredient. Potatoes are boiled between 10 and 25 minutes, depending on size and type, to become soft.

In India, the most popular potato dishes are *aloo ki sabzi*, batata vada, and samosa, which is spicy mashed potato mixed with a small amount of vegetable stuffed in conical dough, and deep fried. Potatoes are also a major ingredient as fast food items, such as aloo chaat, where they are deep fried and served with chutney. In Northern India, alu dum and alu paratha are a favorite part of the diet; the first is a spicy curry of boiled potato, the second is a type of stuffed chapatti. A dish called masala dosa from South India is very notable all over India. It is a thin pancake of rice and pulse paste rolled over spicy smashed potato and eaten with sambhar and chutney. Poori in sound India in particular in Tamil Nadu is almost always taken with smashed potato masal. Other favorite dishes are alu tikki and pakoda items. Vada pav is a popular vegetarian fast food dish in Mumbai and other regions in the Maharashtra in India. Aloo posto (a curry with potatoes and poppy seeds) is immensely popular in East India, especially Bengal.

Figure 46.1: Process for preparation of potato chips

Potatoes (fully mature)

↓

Washing

↓

Sorting (removal of defected and small tubers)

↓

Peeling (by abrasion peelers)

↓

Trimming (removal of residual skin, discolored area, green portion)

↓

Slicing (1/15 – 1/25 inch thick slices by rotary slicer)

↓

Drying (optional process may be done by drum driers)

↓

Frying (in continuous or batch fryers using refined oil)

↓

Salting and addition of flavouring materials

↓

Cooling and inspection

↓

Packaging

↓

Storage

Figure 46.2: Flow sheet for preparation of potato powder

Potatoes (1kg)

↓

Peeling

↓

Cooking in cooker with 2 Lit. Water, 5g KMS

↓

Mashing cooked potatoes

↓

Spread on a tray

↓

Dry at 60-80 °C in oven

↓

Grinding dry mash

↓

Sieve with 60 mesh (get 200g powder)

↓

Packaging air tight

↓

Storage

Figure 46.3: Flow sheet for preparation of potato starch

Potatoes (1 kg)
↓
Peeling
↓
Cut into small pieces
↓
Add 1 lit water, 2.5g KMS
↓
Grind the pieces in a mixer-grinder
↓
Add 2-3 lit water
↓
Filter through muslin cloth
↓
Decant the clear water
↓
Let the settle the starch at the bottom of the container
↓
Repeat the same procedure 2-3 times to get clear water
↓
Lastly remove water and spread starch in the tray
↓
Dry in an oven at 60 °C
↓
Grind dried starch to fine powder (get 100g starch)
↓
Seal in airtight container
↓
Storage

Figure 46.4: Flow sheet for preparation of osmotic dehydrated or chilled potato

Fresh potatoes
↓
Pass through grader
↓
Grading of potatoes
↓
Stone removal by pass through screen
↓
Stone free potatoes
↓
Washing with water
↓
Dust and Clay free potatoes
↓
Pass through alkali and solar fire
↓
Alkali treated potatoes
↓
Pass through abrasion machine
↓
De skinned potatoes
↓
Slicing of potatoes
↓
Treatment with sodium metabisulphate solution
↓
Osmotic dehydration product
↓
Washing
↓
Chilling of potatoes
↓
Freeze dried potatoes
↓
Packaging
↓
Storage at low temperature
↓
Marketing

47

Pumpkin

A pumpkin is a gourd-like squash of the genus *Cucurbita* and the family Cucurbitaceae (which also includes gourds). It commonly refers to cultivars of any one of the species *Cucurbita pepo, Cucurbita mixta, Cucurbita maxima,* and *Cucurbita moschata,* and is native to North America. They typically have a thick, orange or yellow shell, creased from the stem to the bottom, containing the seeds and pulp. In Australian English, the name 'pumpkin' generally refers to the broader category called winter squash in North America.

Description

The word pumpkin originates from the word *pepon* which is Greek for "large melon". The French adapted this word to *pompon,* which the British changed to *pumpion* and later American colonists changed that to the word we use today, "pumpkin". The origin of pumpkins is not definitively known, although they are thought to have originated in North America. The oldest evidence, pumpkin-related seeds dating between 7000 and 5500 BC, were found in Mexico. Pumpkins are a squash-like fruit that range in size from less than 1 pound (0.45 kilograms) to over 1,000 pounds (450 kilograms). Since some squash share the same botanical classifications as pumpkins, the names are frequently used interchangeably. In general, pumpkin stems are more rigid, prickly, and angular (with an approximate five-degree angle) than squash stems, which are generally softer, more rounded, and more flared where joined to the fruit.

Pumpkins generally weigh 9–18 lbs (4–8 kg) with the largest (of the species *C. maxima*) capable of reaching a weight of over 75 lbs (34 kg). The pumpkin varies greatly in shape, ranging from oblate to oblong. The rind is smooth and usually lightly ribbed. Although pumpkins are usually orange or yellow, some fruits are dark green, pale green, orange-yellow, white, red and gray. Pumpkins are monoecious, having both male and female flowers on the same plant. The female flower is distinguished by the small ovary at the base of the petals. These bright and colorful flowers have extremely short life spans and may only open for as short a time as one day. The color of pumpkins is derived from the orange pigments abundant in them. The main nutrients are lutein and both alpha and beta carotene, the latter of which generates vitamin A in the body.

Distribution and Habitation

Pumpkins are grown all around the world for a variety of reasons ranging from agricultural purposes (such as animal feed) to commercial and ornamental sales. Of the seven continents, only Antarctica is unable to produce pumpkins; the biggest international producers of pumpkins include the United States, Canada, Mexico, India, and China. The traditional American pumpkin is the Connecticut Field variety.

Uses

Pumpkins are very versatile in their uses for cooking. Most parts of the pumpkin are edible, including the fleshy shell, the seeds, the leaves, and even the flowers. In the United States and Canada, pumpkin is a popular Halloween and Thanksgiving staple. Homemade pumpkin purée can serve the same purpose.

When ripe, the pumpkin can be boiled, baked, steamed, or roasted. In its native North America, it is a very important, traditional part of the autumn harvest, eaten mashed and making its way into soups and purees. Often, it is made into pie, various kinds of which are a traditional staple of the Canadian and American Thanksgiving holidays. In Canada, Mexico, the U.S., Europe and China, the seeds are often roasted and eaten as a snack.

Pumpkins that are still small and green may be eaten in the same way as squash or zucchini. In the Middle East, pumpkin is used for sweet dishes; a well-known sweet delicacy is called *halawa yaqtin*. In South Asian countries such as India, pumpkin is cooked with butter, sugar, and spices in a dish called *kadu ka halwa*. Pumpkin is used to make *sambar* in Udupi cuisine. In Guangxi province, China, the leaves of the pumpkin plant are consumed as a cooked vegetable or in soups. In Australia and New Zealand, pumpkin is often roasted in conjunction with other vegetables. In Japan, small pumpkins are served in savory dishes, including tempura. In Myanmar, pumpkins are used in both cooking and desserts (candied). The seeds are a popular sunflower seed substitute. In Thailand, small pumpkins are steamed with custard inside and served as a dessert. In Italy, it can be used with cheeses as a savory stuffing for ravioli. Also, pumpkin can be used to flavor both alcoholic and nonalcoholic beverages.

In the southwestern United States and Mexico, pumpkin and squash flowers are a popular and widely available food item. They may be used to garnish dishes, and they may be dredged in a batter then fried in oil. Pumpkin leaves are a popular vegetable in the Western and central regions of Kenya; they are called *seveve*, and are an ingredient of *mukimo*, respectively, whereas the pumpkin itself is usually boiled or steamed. The seeds are popular with children who roast them on a pan before eating them.

Extract

East China Normal University research on type-1 diabetic rats, published in July 2007, suggests that chemical compounds found in pumpkin promote regeneration of damaged pancreatic cells, resulting in increased bloodstream insulin levels. According to the research team leader, pumpkin extract may be "a very good product for pre-diabetic people, as well as those who already have diabetes," possibly reducing or eliminating the need for insulin injections for some type-1 diabetics. It is unknown whether pumpkin extract has any effect on diabetes mellitus type-2, as it was not the subject of the study.

Seeds

Pumpkin seeds, also known as pepitas, are small, flat, green, edible seeds. Most pumpkin seeds are covered by a white husk, although some pumpkin varieties produce seeds without them. Pumpkin seeds are a popular snack that can be found hulled or semi-hulled at most grocery stores. However, roasting pumpkin seeds (usually scooped out of jack-o-lanterns) is a popular Halloween treat. Pumpkin seeds have many health benefits, as they are a good source of protein, zinc, and other vitamins, and they are even said to lower cholesterol. One gram of pumpkin seed protein contains as much tryptophan as a full glass of milk. Pumpkin seeds are a good source of magnesium, manganese, phosphorus, and phytosterols.

Pumpkin Seed Oil

Pumpkin seed oil is thick, green-red oil that is produced from roasted pumpkin seeds. When used for cooking or as a salad dressing, pumpkin seed oil is generally mixed with other oils because of its robust flavor. It is used in cooking in central and Eastern Europe. It is considered a delicacy in Austria, where a little is often added in traditional local cuisine on pumpkin soup and on potato salad. In some restaurants in Vienna, they propose even to add a few drops on vanilla ice cream. Long believed to be a folk remedy for prostate problems, it has been claimed to combat benign prostatic hyperplasia. Pumpkin seed oil contains essential fatty acids that help maintain healthy blood vessels, nerves and tissues.

Other Uses

Canned pumpkin is often recommended by veterinarians as a dietary supplement for dogs and cats that are experiencing certain digestive ailments such as constipation, diarrhea, or hairballs. The high fiber content helps to aid proper digestion. Raw pumpkin can be fed to poultry, as a supplement to regular feed, during the winter to help maintain egg production, which usually drops off during the cold months. The medicinal properties of pumpkin include anti-diabetic, antioxidant, anti-carcinogenic, and anti-inflammatory.

Pumpkin, raw

Nutritional value per 100 g (3.5 oz)	
Energy	56 kJ (13 kcal)
Carbohydrates	6.5 g
Sugars	1.36 g
Dietary fiber	0.5 g
Fat	0.1 g
Saturated	0.05 g
Monounsaturated	0.01 g
Polyunsaturated	0.01 g
Protein	1.0 g
Vitamin A equiv.	369 μg
β-carotene	3100 μg
Thiamine (vit. B_1)	0.05 mg
Riboflavin (vit. B_2)	0.110 mg
Niacin (vit. B_3)	0.6 mg
Pantothenic acid (B_5)	0.298 mg
Vitamin B_6	0.061 mg
Foliate (vit. B_9)	16 μg
Vitamin C	9 mg
Vitamin E	1.06 mg
Calcium	21 mg
Iron	0.8 mg
Magnesium	12 mg
Phosphorus	44 mg
Potassium	340 mg
Sodium	1 mg
Zinc	0.32 mg

Figure 47.1: Flow sheet for preparation of pumpkin tutti-frutti

Selection of mature pumpkin fruits

↓

Washing; removal of skin

↓

Cleaning

↓

Removing seeds

↓

Cutting into small cubes

↓

Blanching (2-3 min)

↓

Syruping

↓

Steeping in 40 per cent syrup + 1 per cent citric acid + edible color for 24 h

↓

Steeping in 50 per cent syrup for 24 h

↓

Steeping in 60 per cent syrup for 24 h

↓

Steeping in 70 per cent syrup for 7 days

↓

Quickly washing

↓

Drying in shade (16-18 per cent moisture)

↓

Packaging

↓

Storage

Figure 47.2: Flow sheet for preparation of pumpkin pulp

Selection of mature pumpkin fruits

↓

Washing; removal of skin

↓

Cleaning

↓

Removing seeds

↓

Cutting into small cubes

↓

Pulping by pulping machine

↓

Pasteurization at 82-85 °C for 5 min

↓

Filling hot in cans and again pasteurization

↓

Sealing and labeling

↓

Storage and marketing

48

Ridge Gourd

Ridge gourd scientifically known as *Luffa acutangula* and also known as turiya or turai or beerakai or dodka in different languages in India is another variety of gourd and a versatile vegetable which adds a facet to the vegetarian cuisine. This is also known as Chinese Okra or Sponge gourd.

This vegetable is an extremely popular vegetable in the Asian, African and Arabic countries. Also known by other name like loofah, luffa or tori etc this vegetable is believed to be originated in the Arabic desserts and spread throughout the world. It is a dark green vegetable on the exterior side with white pulp and white seeds embedded within in a spongy flesh. All the species of the ridge gourd or loofah are edible and must be consumed before they mature or else they are too woody and fibrous to eat.

Luffa or ridge gourds are tropical and subtropical vines of the plant family Cucurbitaceous. Luffa acutangula and Luffa aegyptiaca (luffa cylindrical) are at least two species that are grown and harvested before maturity and eaten as vegetable in Asia and Africa. In India, a variety of dishes are prepared with the ridge gourd where they prepare the chutney, add them to the dal or make a curry with this vegetable. In Maharashtra, India, dodka (ridged gourd/luffa) and ghosavala (smooth luffa) are common vegetables prepared with either crushed dried peanuts or with beans.

Luffa or ridge gourd grows arrow straight, slightly curved, or much curved. One variety of this vegetable grows very short and wide sponges. These are typically used for pot scrubbers. Another type, grown more often for food, produces extremely thin long fruits. When small, the fruits are very flexible and will conform to whatever shape they are against. The ridged or angled luffa (*Luffa acutangula*) is a different species of luffa. It produces fiber but is often used as a food crop. It can be identified by the pronounced ridges that run along the skin lengthwise. It tends to be thinner, bent or curved, and more elongated than the Luffa cylindrica. Edible luffa found in markets is probably more often *acutangula* than the *cylindrica* species. *Luffa acutangula* is also commonly known as Patola in the Tagalog language of the Philippines.

The edible size fruits taste something like a cross between a zucchini and a cucumber. Some luffa varieties may produce fruits that are too bitter to eat. Peeling the skin off reduces bitterness but is probably not usually necessary to do so. If it tastes bad, don't eat it. Many commonly cultivated varieties are edible. The ridge gourd could be either stir fried or make into a nice spicy curry. Ridge gourd could also be cut into slices and fired in batter like our chilli bajjis. A healthier option would be to make gravy out of it. Fry some onions, tomatoes, garlic with some salt and chilli powder. Add some cubed ridge gourd and let it cook on a simmering flame. Add half a cup of milk and let the gravy thicken. After removing from flame, season and serve hot.

Ridge gourd is low in saturated fat and cholesterol, high in dietary fiber, vitamin C, riboflavin, zinc, thiamin, iron, magnesium and manganese. The nutritional value of gourd makes it suitable for maintaining optimum health, weight lose. It has excellent cooling properties. Ridge gourd is a powerhouse of nutrients and are rich is dietary fiber, vitamin C, riboflavin, zinc, thiamine, iron and magnesium. It is also low in fat and calories. Being low in saturated fat and cholesterol make it the ideal choice for those who wish to lose weight. It also has very high water content, making it a very low calorie vegetable (only 60 calories in 100 gm of cooked ridge gourd).

This vegetable has excellent blood purifying properties which helps to purify, restore and nourish liver protecting it from alcohol intoxication. Cellulose in ridge gourd helps to overcome constipation and also aids in curing piles. A glass of Ridge gourd juice mixed with other healthy vegetables in the morning can help to strengthen your immune system against any infection.

Health Benefits of Ridge Gourd

Ridge gourd is low in saturated fat and cholesterol, high in dietary fiber, vitamin C, riboflavin, zinc, thiamin, iron, magnesium and manganese. The nutritional value of gourd makes it suitable for maintaining optimum health, weight lose. It has excellent cooling properties.

Ridge gourd can be cut into slices and fired in batter like our chilli bajjies. A healthier option would be to make gravy out of it. Fry some onions, tomatoes, garlic with some salt and chilli powder. Add some cubed ridge gourd and let it cook on a simmering flame. Add half a cup of milk and let the gravy thicken. After removing from flame, season and serve hot.

The nutritional value per 100 gm of edible portion of ridge gourd is:

Nutrient	*Content*
Energy	17 (KCal)
Carbohydrate	3.4 (g)
Protein	0.5 (g)
Fat	0.1 (g)
Calcium	18 (mg)
Iron	0.39 (mg)

Figure 48.1: Flow sheet for preparation of canned ridge gourd

Tender ridge gourd fruits
↓
Washing
↓
Grading
↓
Cutting similar size pieces
↓
Blanching 1-2 min.
↓
Cooling in 1.5 per cent salt solution
↓
Filling cans with ridge gourd in 2 per cent salt solution
↓
Pasteurization at 79-82 °C for 30-50 min
↓
Sealing cans and labeling
↓
Storage and marketing

Figure 48.2: Flow sheet for preparation of dried ridge gourd

Tender ridge gourd fruits
↓
Cleaning and washing
↓
Grading
↓
Similar size cut pieces
↓
Blanching for 5 min
↓
Drying at 55 °C 14 hrs
↓
Packing in plastic bags
↓
Sealing and labeling
↓
Storage and marketing

49

Sweet Potato

The sweet potato (*Ipomoea batatas*) is a dicotyledonous plant that belongs to the family Convolvulaceae. Its large, starchy, sweet-tasting, tuberous roots are an important root vegetable. The young leaves and shoots are sometimes eaten as greens. Of the approximately 50 genera and more than 1,000 species of Convolvulaceae, *I. batatas* is the only crop plant of major importance some others are used locally, but many are actually poisonous. The sweet potato is only distantly related to the potato (*Solanum tuberosum*).

Although the softer, orange variety is often called a yam in parts of North America, the sweet potato is botanically very distinct from the other vegetable called a yam, which is native to Africa and Asia and belongs to the monocot family Dioscoreaceae. To prevent confusion, the United States Department of Agriculture requires sweet potatoes labeled as "yams" to be labeled also as "sweet potatoes". In New Zealand English, the Mâori term *kûmara* is commonly used.

The genus *Ipomoea* that contains the sweet potato also includes several garden flowers called morning glories, though that term is not usually extended to *Ipomoea batatas*. Some cultivars of *Ipomoea batatas* are grown as ornamental plants; the name "tuberous morning glory" may be used in a horticultural context.

The plant is a herbaceous perennial vine, bearing alternate heart-shaped or palmately lobed leaves and medium-sized sympetalous flowers. The edible tuberous root is long and tapered, with a smooth skin whose color ranges between purple, red, brown, and beige. Its flesh ranges from beige through yellow, orange, and purple.

Origin and Distribution

The center of origin and domestication of sweet potato is thought to be either in Central America or South America. In Central America, sweet potatoes were domesticated at least 5,000 years ago.

Austin (1988) postulated that the center of origin of *I. batatas* was between the Yucatán Peninsula of Mexico and the mouth of the Orinoco River in Venezuela. The 'cultigen' had most likely been spread by local people to the Caribbean and South America by 2500 BC. Zhang *et al.,* (1998) provided strong supporting evidence that the geographical zone postulated by Austin is the primary center of diversity. The much lower molecular diversity found in Peru Ecuador suggests this region should be considered as secondary center of sweet potato diversity.

The sweet potato was also grown before western exploration in Polynesia. Sweet potato has been radiocarbon-dated in the Cook Islands to 1000 AD, and current thinking is that it was brought to central Polynesia around 700 AD, possibly by Polynesians who had traveled to South America and back, and spread across Polynesia to Hawaii and New Zealand from there. It is possible; however, that South Americans brought it to the Pacific, although this is unlikely as it was the Polynesians who had a strong maritime tradition and not the native South Americans. The theory that the plant could spread by floating seeds across the ocean is not supported by evidence. Another point is that the sweet potato in Polynesia is the cultivated *Ipomoea batatas,* which is generally spread by vine cuttings and not by seeds. Sweet potatoes are now cultivated throughout tropical and warm temperate regions wherever there is sufficient water to support their growth.

According to the Food and Agriculture Organization (FAO) statistics, world production in 2004 was 127 million tonnes. The majority comes from China, with a production of 105 million tonnes from 49,000 km^2. About half of the Chinese crop is used for livestock feed. Per capita production is greatest in countries where sweet potatoes are a staple of human consumption, led by Papua New Guinea at 550 kg per person per year, the Solomon Islands at 160 kg, Burundi and Rwanda at 130 kg and Uganda at 100 kg.

Cultivation

The plant does not tolerate frost. It grows best at an average temperature of 24 °C (75 °F), abundant sunshine and warm nights. Annual rainfalls of 750–1,000 mm (30–39 in) are considered most suitable, with a minimum of 500 mm (20 in) in the growing season. The crop is sensitive to drought at the tuber initiation stage 50–60 days after planting, and it is not tolerant to water-logging, as it may cause tuber rots and reduce growth of storage roots if aeration is poor.

Depending on the cultivar and conditions, tuberous roots mature in two to nine months. With care, early-maturing cultivars can be grown as an annual summer crop in temperate areas, such as the northern United States. Sweet potatoes rarely flower when the daylight is longer than 11 hours, as is normal outside of the tropics. They are mostly propagated by stem or root cuttings or by adventitious roots called "slips" that grow out from the tuberous roots during storage. True seeds are used for breeding only.

They grow well in many farming conditions and have few natural enemies; pesticides are rarely needed. Sweet potatoes are grown on a variety of soils, but well-drained, light- and medium-textured soils with a pH range of 4.5-7.0 are more favorable for the plant. They can be grown in poor soils with little fertilizer. However, sweet potatoes are very sensitive to aluminum toxicity and will die about six weeks after planting if lime is not applied at planting in this type of soil. Because they are sown by vine cuttings rather than seeds, sweet potatoes are relatively easy to plant. Because the rapidly growing vines shade out weeds, little weeding is needed. In the tropics, the crop can be maintained in the ground and harvested as needed for market or home consumption. In temperate regions, sweet potatoes are most often grown on larger farms and are harvested before first frosts.

China is the largest grower of sweet potatoes, providing about 80 per cent of the world's supply; 130 million tons were produced in 1990; about half that of common potatoes. Historically, most of China's sweet potatoes were grown for human consumption, but now most (60 per cent) are grown to feed pigs. The rest are grown for human food and for other products. Some are grown for export, mainly to Japan. China grows over 100 varieties of sweet potato.

Sweet potatoes very early became popular in the islands of the Pacific Ocean, spreading from Polynesia to Japan and the Philippines. One reason is that they were a reliable crop in cases of crop failure of other staple foods because of typhoon flooding. They are featured in many favorite dishes in Japan, Taiwan, the Philippines, and other island nations. Indonesia, Vietnam, India, and some other Asian countries are also large sweet potato growers. Sweet potato, also known as *kelang* in Tulu is part of Udupi cusine. Uganda (the third largest grower after Indonesia), Rwanda, and some other African countries also grow a large crop which is an important part of their peoples' diets. North and South America, the original home of the sweet potato, together grow less than three percent of the world's supply. Europe has only a very small sweet potato production, mostly in Portugal. In the Caribbean, a variety of the sweet potato called the *boniato* is popular. The flesh of the *boniato* is cream-colored, unlike the more popular orange hue seen in other varieties. *Boniatos* are not as sweet and moist as other sweet potatoes, but many people prefer their fluffier consistency and more delicate flavor.

Sweet potatoes have been an important part of the diet in the United States for most of its history, especially in the Southeast. From the middle of the 20th century, however, they have become less popular. The average per capita consumption of sweet potatoes in the United States is only about 1.5–2 kg (3.3–4.4 lb) per year, down from 13 kg (29 lb) in 1920. Southerner Kent Wrench writes: "The Sweet Potato became associated with hard times in the minds of our ancestors and when they became affluent enough to change their menu, the potato was served less often."

New Zealanders grow enough *kûmara* to provide each person with 7 kg (15 lb) per year, and they also import substantially more than this from China.

In the Southeastern United States, sweet potatoes are traditionally cured to improve storage, flavor, and nutrition, and to allow wounds on the periderm of the harvested root to heal. Proper curing requires drying the freshly dug roots on the ground for two to three hours, then storage at 85–90 °F (29–32 °C) and 90 to 95 per cent relative humidity from five to fourteen days. Cured sweet potatoes can keep for thirteen months when kept at 55–59 °F (13–15 °C) and > 90 per cent relative humidity. Colder temperatures injure the roots.

Nutritional Values and Uses

Besides simple starches, sweet potatoes are rich in complex carbohydrates, dietary fiber, beta carotene (a vitamin A equivalent nutrient), vitamin C, and vitamin B_6. Pink, yellow and green varieties are high in carotene, the precursor of vitamin A.

In 1992, the Center for Science in the Public Interest compared the nutritional value of sweet potatoes to other vegetables. Considering fiber content, complex carbohydrates, protein, vitamins A and C, iron, and calcium, the sweet potato ranked highest in nutritional value. According to these criteria, sweet potatoes earned 184 points, 100 points over the next on the list, the common potato. Sweet potato varieties with dark orange flesh have more beta carotene than those with light-colored flesh, and their increased cultivation is being encouraged in Africa, where vitamin A deficiency is a serious health problem. Despite the name "sweet", it may be a beneficial food for diabetics, as preliminary studies on animals have revealed it helps to stabilize blood sugar levels and to lower insulin resistance.

Culinary Uses

Although the leaves and shoots are also edible, the starchy tuberous roots are by far the most important product. In some tropical areas, they are a staple food crop.

Asia

In China, sweet potatoes, typically of the yellow variety, are baked in a large iron drum, and sold as street food during winter. In Japan, this is called *yaki-imo* (roasted sweet potato), which typically uses either the yellow-fleshed Japanese sweet potato or the purple-fleshed (Okinawan) sweet potato. Sweet potato soup, a Chinese *tong sui* (sweet soup) served during winter, consists of boiling sweet potato in water with rock sugar and ginger. Sweet potato greens are a common side dish in Taiwanese cuisine, often boiled or sautéed and served with a garlic and soy sauce mixture, or simply salted before serving. They, as well as dishes featuring the sweet potato root, are commonly found at bento (POJ: *piân-tong*) restaurants. In northeastern Chinese cuisine, sweet potatoes are often cut into chunks and fried, before being drenched into a pan of boiling syrup.

In Korean cuisine, sweet potato starch is used to produce *dangmyeon* (cellophane noodles). Sweet potatoes are also boiled, steamed, or roasted, and young stems are eaten as *namul*. Pizza restaurants such as Pizza Hut and Domino's in Korea are using sweet potatoes as a popular topping.

In Japan, boiled sweet potato is the most common way to eat it at home. Also, the use in vegetable tempura is common. *Daigaku-imo* is a baked sweet potato dessert. Because it is sweet and starchy, it is used in *imo-kinton* and some other *wagashi* (Japanese sweets), such as *ofukuimo*. *Shôchû*, a Japanese spirit normally made from the fermentation of rice, can also be made from sweet potato, in which case it is called *imo-jôchû*. *Imo-gohan*, sweet potato cooked with rice, is popular in Guangdong, Taiwan and Japan. It is also served in *nimono* or *nitsuke*, boiled and typically flavored with soy sauce, *mirin* and *dashi*. In Malaysia and Singapore, sweet potato is often cut into small cubes and cooked with yam and coconut milk (*santan*) to make a sweet dessert called *bubur caca.* A favorite way of cooking sweet potato is deep frying slices of sweet potato in batter, and served as a tea-time snack. In homes, sweet potatoes are usually boiled. The leaves of sweet potatoes are usually stir-fried with only garlic or with *sambal belacan* and dried shrimp by the Malaysian Chinese.

In some regions of India, fasts of religious nature are an occasion for a change in normal diet, and a total absence from cooking or eating is held as elective while a normal diet for a fasting day is a light feast consisting of different foods from usual, amongst which sweet potato is one of the prime sources of sustenance. Sweet potato is eaten otherwise, too, and a popular variety of preparation in most parts is roasted slow over kitchen coals at night and eaten with some dressing–primarily salt, possibly yogurt–while the easier way in the south is simply boiling or pressure-cooking before peeling, cubing and seasoning for a vegetable dish as part of the meal. Usually the preparations for sweet potato are similar to the blander ones for potato, while the sharper versions–with greener chili or red pepper–are reserved for potato.

In the Philippines, sweet potatoes (locally known as *camote* or *kamote*) are an important food crop in rural areas. They are often a staple among impoverished families in provinces, as they are easier to cultivate and cost less than rice. The tubers are boiled or baked in coals and may be dipped in sugar or syrup. Young leaves and shoots (locally known as *talbos ng kamote* or *camote* tops) are eaten fresh in salads with shrimp paste (*bagoong alamang*) or fish sauce. They can be cooked in vinegar and soy sauce and served with fried fish (a dish known as *adobong talbosng kamote*), or with recipes such as *sinigang*. The stew obtained from boiling *camote* tops is purple-colored, and is often mixed with lemon as juice. Sweet potatoes are also sold as street food in suburban and rural areas. Fried sweet potatoes coated with caramelized sugar and served in skewers (*camote cue*) are popular afternoon snacks. Sweet potatoes are also used in a variant of *halo-halo* called *ginatan*, where they are cooked in coconut milk and sugar and mixed with a variety of rootcrops, sago, jackfruit and *bilu-bilo* (glutinous rice balls). Bread made from sweet potato flour is also gaining popularity. Sweet potato is relatively easy to propagate, and in rural areas that can be seen abundantly at canals and dikes. The uncultivated plant is usually fed to pigs.

In the mountainous regions of West Papua, sweet potatoes are the staple food among the natives there. Using the *bakar batu* way of cooking (free translation: burning rocks), rocks that have been burned in a nearby bonfire are thrown into a pit lined with leaves. Layers of sweet potatoes, an assortment of vegetables, and pork are piled on top of the rocks. The top of the pile then is insulated with more leaves, creating a pressure of heat and steam inside which cooks all food within the pile after a several hours. In most parts of Indonesia, sweet potatoes are frequently fried with batter and served as snacks.

Non-culinary Uses

In South America, the juice of red sweet potatoes is combined with lime juice to make a dye for cloth. By varying the proportions of the juices, every shade from pink to black can be obtained. All parts of the plant are used for animal fodder. Sweet potatoes or *camotes* are often found in Moche ceramics. Several selections are cultivated in gardens as ornamental plants for their attractive foliage, including the dark-leafed cultivars 'Blackie' and 'Ace of Spades' and the chartreuse-foliaged 'Margarita'. The species called wild sweet potato vine, man root, or man-of-the-earth, is not edible, but it is cultivated as an ornamental vine. Researchers at North Carolina State University are breeding sweet potato varieties that would be grown primarily for biofuel production.

Health and Nutrition Benefits of Eating Sweet Potatoes

- ☆ Being rich in dietary fiber, sweet potato lowers the risk of constipation, diverticulitis and colon and rectal cancer.
- ☆ Sweet potatoes have been found to helpful in minimizing the risk of heart disease, diabetes and obesity.
- ☆ Sweet potato is a good snack for those trying to lose weight. It induces a feeling of fullness soon and thus, helps control food intake.
- ☆ Consumption of sweet potatoes has been known to help avoid stroke, by bringing down the harmful effects of low-density cholesterol and preventing blood clots.
- ☆ The presence of β-carotene in sweet potato helps the body fight against free radicals and thus, prevents cancer.
- ☆ Since sweet potatoes have a low glycemic index, they can be consumed by those suffering from diabetes, without any apprehensions.
- ☆ The high amount of potassium in sweet potato helps the body in maintaining fluid and electrolyte balance as well as cell integrity.
- ☆ Sweet potatoes have been found to be beneficial for blood purification as well as lowering of blood pressure.
- ☆ Owing to the presence of iron and calcium, sweet potatoes ensure proper blood flow and also improve bone density.
- ☆ Regular consumption of sweet potato is good for stomach ulcers and inflamed conditions of the colon.
- ☆ Being rich in vitamins and minerals, sweet potato is good for people involved in heavy muscular work.

Nutrient Content

Raw Sweet Potato

Nutritional value per 100 g (3.5 oz)	
Energy	360 kJ (86 kcal)
Carbohydrates	20.1 g
Starch	12.7 g
Sugars	4.2 g
Dietary fiber	3.0 g
Fat	0.1 g
Protein	1.6 g
Vitamin A equiv.	709 μg
β-carotene	8509 μg
Lutein and zeaxanthin	0 μg
Thiamine (vit. B_1)	0.1 mg
Riboflavin (vit. B_2)	0.1 mg
Niacin (vit. B_3)	0.61 mg
Pantothenic acid (B_5)	0.8 mg
Vitamin B_6	0.2 mg
Foliate (vit. B_9)	11 μg
Vitamin C	2.4 mg
Vitamin E	0.26 mg
Calcium	30.0 mg
Iron	0.6 mg
Magnesium	25.0 mg
Phosphorus	47.0 mg
Potassium	337 mg
Sodium	55 mg
Zinc	0.3 mg

Figure 49.1: Process for preparation of sweet potato chips

Sweet Potatoes (fully mature)

↓

Washing

↓

Sorting (removal of defected and small tubers)

↓

Peeling (by abrasion peelers)

↓

Trimming (removal of residual skin, discolored area, green portion)

↓

Slicing (1/15 – 1/25 inch thick slices by rotary slicer)

↓

Drying (optional process may be done by drum driers)

↓

Frying (in continuous or batch fryers using refined oil)

↓

Salting and addition of flavouring materials

↓

Cooling and inspection

↓

Packaging

↓

Storage

Figure 49.2: Flow sheet for preparation of sweet potato powder

Sweet potatoes (1kg)

↓

Peeling

↓

Cooking in cooker with 2 Lit. Water, 5g KMS

↓

Mashing cooked potatoes

↓

Spread on a tray

↓

Dry at 60-80 °C in oven

↓

Grinding dry mash

↓

Sieve with 60 mesh (get 200g powder)

↓

Packaging air tight

↓

Storage

Figure 49.3: Flow sheet for preparation of sweet potato starch

Sweet potatoes (1 kg)
↓
Peeling
↓
Cut into small pieces
↓
Add 1 lit water, 2.5g KMS
↓
Grind the pieces in a mixer-grinder
↓
Add 2-3 lit water
↓
Filter through muslin cloth
↓
Decant the clear water
↓
Let the settle the starch at the bottom of the container
↓
Repeat the same procedure 2-3 times to get clear water
↓
Lastly remove water and spread starch in the tray
↓
Dry in an oven at 60 °C
↓
Grind dried starch to fine powder (get 100g starch)
↓
Seal in airtight container
↓
Storage

50

Sweet Turnip

The turnip or white turnip (*Brassica rapa* var. *rapa*) is a root vegetable commonly grown in temperate climates worldwide for its white, bulbous taproot. Small, tender varieties are grown for human consumption, while larger varieties are grown as feed for livestock.

The most common type is mostly white-skinned apart from the upper 1–6 centimeters, which protrude above the ground and are purple, red, or greenish wherever sunlight has fallen. This above-ground part develops from stem tissue, but is fused with the root. The interior flesh is entirely white. The entire root is roughly conical, but can be occasionally global, about 5–20 centimeters in diameter, and lacks side roots. The taproot (the normal root below the swollen storage root) is thin and 10 centimeters or more in length; it is trimmed off before marketing. The leaves grow directly from the above-ground shoulder of the root, with little or no visible crown or neck (as found in rutabagas).

Turnip leaves are sometimes eaten as "turnip greens" ("turnip tops" in the UK), and they resemble mustard greens in flavor. Turnip greens are a common side dish in southeastern US cooking, primarily during late fall and winter. Smaller leaves are preferred; however, any bitter taste of larger leaves can be reduced by pouring off the water from initial boiling and replacing it with fresh water. Varieties specifically grown for the leaves resemble mustard greens more than those grown for the roots, with small or no storage roots. Varieties of *B. rapa* that have been developed only for use as leaves are called Chinese cabbage. Both leaves and root have a pungent flavor similar to raw cabbage or radishes that becomes mild after cooking.

Turnip roots weigh up to about one kilogram, although they can be harvested when smaller. Size is partly a function of variety and partly a function of the length of time the turnip has grown. Most very small turnips (also called baby turnips) are specialty varieties. These are only available when freshly harvested and do not keep well. Most baby turnips can be eaten whole, including their leaves. Baby turnips come in yellow-, orange-, and red-fleshed varieties as well as white-fleshed. Their flavor is mild, so they can be eaten raw in salads like radishes.

Nutrition

The turnip's root is high only in vitamin C. The green leaves of the turnip top ("turnip greens") are a good source of vitamin A, foliate, vitamin C, vitamin K and calcium. Turnip greens are high in lutein (8.5 mg/100g).

Like rutabaga, turnip contains bitter cyanoglucosides that release small amounts of cyanide. Sensitivity to the bitterness of these cyanoglucosides is controlled by a paired gene. Subjects who have inherited two copies of the "sensitive" gene find turnips twice as bitter as those who have two "insensitive" genes, and thus may find turnips and other cyanoglucoside containing foods intolerably bitter.

Health Benefits of Turnips

- Turnips are very low calorie root vegetables; contains only 28 calories per 100 g. However, they are very good source of anti-oxidants, minerals, vitamins and dietary fiber.
- Fresh roots are indeed one of the vegetables rich in vitamin C; provide about 21mg or 35 per cent of DRA of vitamin C per 100 g. Vitamin-C is a powerful water-soluble anti-oxidant required by the body for synthesis of collagen. It also helps body scavenge harmful free radicals, prevents from cancers, inflammation, and helps boost immunity.
- Turnip greens are the storehouse of many vital nutrients, in fact several times than the roots. The greens are very rich in antioxidants like vitamin A, vitamin C, carotenoids xanthins and lutein. In addition, the greens are excellent source of vitamin K.
- Top greens are also very good source of B-complex group of vitamins such as foliates riboflavin, pyridoxine, pantothenic acid and thiamin.
- Fresh greens are also excellent sources of important minerals like calcium, copper, iron and manganese.

Turnip Greens

Nutritional value per 100 g (3.5 oz)	
Energy	84 kJ (20 kcal)
Carbohydrates	4.4 g
Dietary fiber	3.5 g
Fat	0.2 g
Protein	1.1 g
Vitamin A equiv.	381 μg
Folate (vit. B_9)	118 μg
Vitamin C	27 mg
Vitamin K	368 μg
Calcium	137 mg

Turnips (*Brassica rapa (Rapifera Group))*, Fresh, raw, Nutrition Value per 100 g, (*Source*: USDA National Nutrient data base)

Principle	*Nutrient Value*	*Percentage of RDA*
Energy	28 Kcal	1.5 per cent
Carbohydrates	6.43 g	5 per cent
Protein	0.90 g	1.5 per cent
Total Fat	0.10 g	<1 per cent
Cholesterol	0 mg	0 per cent
Dietary Fiber	1.8 g	5 per cent
Vitamins		
Foliates	15 mcg	4 per cent
Niacin	0.400 mg	2.5 per cent
Pantothenic acid	0.200 mg	4 per cent
Pyridoxine	0.090 mg	7 per cent
Riboflavin	0.030 mg	2.5 per cent
Thiamin	0.040 mg	4 per cent
Vitamin A	0 IU	0 per cent
Vitamin C	21 mg	35 per cent
Vitamin E	0.03 mg	<1 per cent
Vitamin K	0.1 mcg	<1 per cent
Electrolytes		
Sodium	39 mg	2.5 per cent
Potassium	233 mg	5 per cent
Minerals		
Calcium	30 mg	3 per cent
Copper	0.085 mg	9 per cent
Iron	0.30 mg	4 per cent
Magnesium	11 mg	2.5 per cent
Manganese	0.134 mg	6 per cent
Zinc	0.27 mg	2 per cent
Phyto-nutrients		
Carotene-β	0 mcg	–
Carotene-α	0 mcg	–
Lutein-zeaxanthin	0 mcg	–

Figure 50.1: Flow sheet for preparation of sweet turnip pickle

Turnips (sound and tender)

↓

Washing

↓

Trimming

↓

Peeling (rough and thick skin)

↓

Cutting round slices (1-1.5 cm thick)

↓

Blanching for 5 min

↓

Draining off water

↓

Frying spices in a little oil (except tamarind pulp)

↓

Drying in shade for an hour

↓

Mixing pieces with fried spices

↓

Cooling

↓

Making paste of tamarind pulp, jiggery and vinegar

↓

Adding paste to pieces

↓

Filling in jars

↓

Keeping in sun for a week

↓

Addition of oil after heating and cooling it

↓

Sealing and labeling

↓

Storage

↓

Marketing

Sl.No.	*Item*	*Quantity*	*Sl.No.*	*Item*	*Quantity*
1.	Turnip	1 kg	9.	Cumin	10 g
2.	Salt	100 g	10.	Cardamom (large)	10 g
3.	Ginger chopped	20 g	11.	Aniseed powder	10 g
4.	Onion chopped	50 g	12.	Jaggery	200 g
5.	Garlic chopped	10 g	13.	Black pepper	10 g
6.	Red chilli powder	10 g	14.	Vinegar	100 ml
7.	Tamarind pulp	100 g	15.	Mustard ground	20 g
8.	Cinnamon	10 g	16.	Mustard oil	200 ml

51

Tomato

Tomato may refer to both the plant (*Solanum lycopersicum*) and the edible, typically red, fruit which it bears. Originating in South America, the tomato was spread around the world following the Spanish colonization of the Americas, and its many varieties are now widely grown, often in greenhouses in cooler climates.

The tomato fruit is consumed in diverse ways, including raw, as an ingredient in many dishes and sauces, and in drinks. While it is botanically a fruit, it is considered a vegetable for culinary purposes (as well as by the United States Supreme Court, see *Nix v. Hedden*), which has caused some confusion. The fruit is rich in lycopene, which may have beneficial health effects. The tomato belongs to the nightshade family. The plants typically grow to 1–3 meters (3–10 ft) in height and have a weak stem that often sprawls over the ground and vines over other plants. It is a perennial in its native habitat, although often grown outdoors in temperate climates as an annual.

Tomatoes that have been modified using genetic engineering have been developed, and although none are commercially available now, they have been in the past. The first commercially available genetically modified food was a variety of tomato named (the Flavr Savr), which was engineered to have a longer shelf life. Scientists are continuing to develop tomatoes with new traits not found in natural crops, such as increased resistance to pests or environmental stresses. Other projects aim to enrich tomatoes with substances that may offer health benefits or provide better nutrition. The tomato is now grown and eaten around the world. It is used in diverse ways, including raw in salads, and processed into ketchup or tomato soup. Unripe green tomatoes can also be breaded and fried, used to make salsa, or pickled. Tomato juice is sold as a drink, and is used in cocktails such as the Bloody Mary.

Tomatoes are acidic, making them especially easy to preserve in home canning whole, in pieces, as tomato sauce or paste. The fruit is also preserved by drying, often in the sun, and sold either in bags or in jars with oil. Tomatoes are used extensively in Mediterranean cuisine, especially Italian and Middle Eastern cuisines. They are a key ingredient in pizza, and are commonly used in pasta sauces. They are also used in gazpacho (Spanish cuisine) and *pa amb tomàquet* (Catalan cuisine).

Though it is botanically a berry, a subset of fruit, the tomato is a vegetable for culinary purposes, because of Tomatoes are now eaten freely throughout the world, and their consumption is believed to benefit the heart, among other organs. They contain the carotene lycopene, one of the most powerful natural antioxidants. In some studies, lycopene, especially in cooked tomatoes, has been found to help prevent prostate cancer, but other research contradicts this claim. Lycopene has also been shown to improve the skin's ability to protect against harmful UV rays. Natural genetic variation in tomatoes and their wild relatives has given a genetic plethora of genes that produce lycopene, carotene, anthocyanin, and other antioxidants. Tomato varieties are available with double the normal vitamin C (Doublerich), 40 times normal vitamin A, high levels of anthocyanin (resulting in blue tomatoes), and two to four times the normal amount of lycopene (numerous available cultivars with the high crimson gene).

Lycopene has also been shown to protect against oxidative damage in many epidemiological and experimental studies. In addition to its antioxidant activity, other metabolic effects of lycopene have also been demonstrated. The richest source of lycopene in the diet is tomato and tomato derived products. Tomato consumption has been associated with decreased risk of breast cancer, head and neck cancers and might be strongly protective against neurodegenerative diseases. Tomatoes and tomato sauces and puree are said to help lower urinary tract symptoms (BPH) and may have anticancer properties. Tomato consumption might be beneficial for reducing cardiovascular risk associated with types two diabetes.

Red Tomatoes, Raw

Nutritional value per 100 g (3.5 oz)	
Energy	74 kJ (18 kcal)
Carbohydrates	3.9 g
Sugars	2.6 g
Dietary fiber	1.2 g
Fat	0.2 g
Protein	0.9 g
Water	94.5 g
Vitamin A equiv.	42 μg
lutein and zeaxanthin	123 μg
Vitamin C	14 mg
Vitamin E	0.54 mg
Potassium	237 mg

Figure 51.1: Flow sheet for preparation of tomato juice

Fully ripe tomatoes

↓

Washing

↓

Sorting and trimming (removing green portion and stem)

↓

Cutting and chopping into 4-6 pieces

↓

Heating at 70-90 °C, 3-5 min to soften

↓

Pulping or extraction of juice/pulp (mechanically or sieving)

↓

Addition of salt, sugar and citric acid

↓

Homogenization

↓

Heating at 82-88 °C, 1 min

↓

Filling hot into bottles or cans

↓

Sterilization in boiling water for 30 min

↓

Cooling and Labeling

↓

Storage and Marketing

Figure 51.2: Flow sheet for preparation of tomato puree or paste

Tomato juice
↓
Cooking to desired consistency
↓
Judging of end point for puree or paste
(12 per cent solids for puree and 25 per cent solids for paste)
↓
Filling into bottles or cans (82–88°C)
↓
Sealing
↓
Sterilization in boiling water for 30 min
↓
Cooling and Labeling
↓
Storage and Marketing

Figure 51.3: Flow sheet for preparation of tomato soup

Tomato pulp/juice
↓
Boiling to concentrate
↓
Putting spice bag in cooking pan (as a sauce)
↓
Preparing paste of butter and flour/starch
↓
Addition to concentrated juice
↓
Cooking to desired consistency with continues stirring
(to prevent coagulation of starch)
↓
Addition of sugar and salt and cooking for 2 min
↓
Removal of spice bag after squeezing
↓
Filling into cans (plain cans)
↓
Sterilization at 115 °C for 40-45 min
↓
Cooling and Labeling
↓
Storage and Marketing

Sl.No.	Item	Quantity	Sl.No.	Item	Quantity
1.	Tomato pulp	1 kg	6.	Cumin	1 g
2.	Salt	20 g	7.	Black pepper	1g
3.	Sugar	20 g	8.	Cinnamon (powder)	1g
4.	Butter/cream	20 g	9.	Cardamom	1g
5.	Flour/Starch	10 g	10.	Water	350 ml

Figure 51.4: Flow sheet for preparation of tomato chilli sauce

Fully ripe red tomatoes

⬇

Washing

⬇

Sorting and trimming

⬇

Blanching for 2 min

⬇

Placing quickly in cold water to crack skin

⬇

Peeling

⬇

Crushing

⬇

Cooking pulp with 1/3 sugar to thick mass and seed

⬇

Rest of the process is similar to tomato sauce

Sl.No.	Item	Quantity	Sl.No.	Item	Quantity
1.	Tomato pulp with seeds	1 kg	6.	Cumin	1 g
2.	Salt	20 g	7.	Black pepper	1g
3.	Sugar	20 g	8.	Cinnamon (powder)	1g
4.	Butter/cream	20 g	9.	Cardamom	1g
5.	Flour/Starch	10 g	10.	Water	350 ml

Figure 51.5: Flow sheet for preparation of tomato cocktail

Tomato juice

↓

Putting spice bag in juice and cooking gently for 20 min

↓

Addition of strained lime juice, vinegar and salt, if desired little sugar

↓

Heating at 82-88 °C

↓

Filling hot in bottles

↓

Crown corking

↓

Sterilization at 85-90 °C, 30 min

↓

Cooling and Labeling

↓

Storage and Marketing

Sl.No.	*Item*	*Quantity*	*Sl.No.*	*Item*	*Quantity*
1.	Tomato juice	5 Liters	7.	Black pepper	1.5 g
2.	Salt	45 g	8.	Coriander seed	1.5 g
3.	Lime juice	60 ml	9.	Cardamom	1.5 g
4.	Red chilli powder	0.25 g	10.	Cinnamon (powder)	1.5 g
5.	Clove (headless)	5 Nos	11.	Vinegar/acetic acid	300 ml
6.	Cumin	1.5 g			

Figure 51.6: Flow sheet for preparation of tomato chutney

Tomatoes (Fully ripe, red)
↓
Washing
↓
Sorting
↓
Blanching for 2 min
↓
Putting quickly in cold water
↓
Peeling and Crushing
↓
Addition of ingredients except salt and vinegar and cooking to desired consistency
↓
Addition of salt and vinegar and cooking for 5 min
↓
Addition hot into sterilized bottles
↓
Sealing and Labeling
↓
Storage and Marketing

Sl.No.	*Item*	*Quantity*	*Sl.No.*	*Item*	*Quantity*
1.	Tomato	1 Kg	8.	Cinnamon	10 g
2.	Salt	25 g	9.	Cardamom	10 g
3.	Sugar	500 g	10.	Aniseed	10 g
4.	Onion (chopped)	100 g	11.	Black pepper (powder)	10 g
5.	Ginger (chopped)	10 g	12.	Cumin	10 g
6.	Garlic (chopped)	5 g	13.	Sodium benzoate	0.5 g
7.	Red chilli powder	10 g	14.	Vinegar	100 ml

Figure 51.7: Flow sheet for preparation of tomato sauce/ketchup

Fully mature, red Tomatoes
↓
Washing
↓
Sorting and trimming
↓
Cutting and chopping
↓
Heating at 70 – 90 °C for 3-5 min to soften
↓
Pulping or extraction of juice/pulp mechanically/sieving
↓
Straining tomato pulp/juice
↓
Cooking pulp with one third quantity of sugar
↓
Putting spice bag in pulp and pressing occasionally
↓
Cooking to one-third of original volume of pulp/juice
↓
Addition of remaining sugar and salt
↓
Cooking
↓
Addition of vinegar and preservative
↓
Removal of spice bag (after squeezing in pulp)
↓
Judging of end point (TSS 25°)
↓
Filling hot into bottles at 88 °C
↓
Crown corking
↓
Pasteurization (at 85-90 °C; 30 min)
↓
Cooling and Labeling
↓
Storage and Marketing

Sl.No.	Item	Quantity	Sl.No.	Item	Quantity
1.	Tomato pulp	1Kg	8.	Cinnamon	10 g
2.	Salt	10 g	9.	Cardamom	10 g
3.	Sugar	75 g	10.	Aniseed (powder)	10 g
4.	Onion (chopped)	50 g	11.	Black pepper (powder)	10 g
5.	Ginger (chopped)	10 g	12.	Cumin	10 g
6.	Garlic (chopped)	5 g	13.	Clove	5 Nos
7.	Red chilli powder	5 g	14.	Vinegar	25 ml
			15.	Sodium benzoate	0.25 g

Figure 51.8: Flow sheet for preparation of tomato pickle

Tomatoes (Ripe firm, pulpy)
↓
Washing
↓
Blanching for 5 min
↓
Cooling immediately in water
↓
Peeling
↓
Cutting in to 4-6 pieces
↓
Frying all ingredients in a little oil (except vinegar)
↓
Mixing with pieces
↓
Heating for 2 min
↓
Cooling
↓
Addition of vinegar and remaining oil
↓
Filling in jar
↓
Sealing and labeling
↓
Storage and Marketing

Sl.No.	*Item*	*Quantity*	*Sl.No.*	*Item*	*Quantity*
1.	Tomatoes	1 kg	8.	Cardamom (large)	10 g
2.	Salt	75 g	9.	Cinnamon	10 g
3.	Garlic (chopped)	10 g	10.	Turmeric	10 g
4.	Ginger (chopped)	50 g	11.	Fenugreek	10 g
5.	Red chilli	10 g	12.	Clove (headless)	5 Nos
6.	Cumin	10 g	13.	Vinegar	250 ml
7.	Aniseed (powder)	10 g	14.	Mustard oil	300 ml

52

Turmeric

Turmeric (*Curcuma longa*) is a rhizomatous herbaceous perennial plant of the ginger family, Zingiberaceae. It is native to tropical South Asia and needs temperatures between 20 °C and 30 °C and a considerable amount of annual rainfall to thrive. Plants are gathered annually for their rhizomes, and propagated from some of those rhizomes in the following season.

When not used fresh, the rhizomes are boiled for several hours and then dried in hot ovens, after which they are ground into a deep orange-yellow powder commonly used as a spice in curries and other South Asian and Middle Eastern cuisine, for dyeing, and to impart color to mustard condiments. Its active ingredient is curcumin and it has a distinctly earthy, slightly bitter, slightly hot peppery flavor and a mustardy smell. In medieval Europe, turmeric became known as Indian saffron, since it was widely used as an alternative to the far more expensive saffron spice.

Nizamabad, a city in the south Indian state of Andhra pradesh, is the world's largest producer and most important trading center of turmeric in Asia. For these reasons, Nizamabad in history is also known as "Turmeric City" Sangli, a town in the southern part of the Indian western state of Maharashtra, is the second largest and most important trading center for turmeric in Asia. Kasur district of Pakistan is the largest producer of turmeric in Pakistan. Mayo cultivators introduced different varieties of turmeric in Kasur. Turmeric is commonly called "Pasupu" in Telugu, Manjal in Tamil and haridra, haldar or Haldi in Hindi. Attempts to patent turmeric have been defeated.

Uses

1. *Medicinal use*: Turmeric is widely known for curing a number of diseases, *viz.* preventing various sexual transmitted diseases. It has anti-tumoral properties and is considered to be a possible treatment for cancer.

2. *Food additive*: Turmeric is a widely used food additive. It is also used in mustard, pickles and other consumable items.
3. *Food Industry*: Turmeric has found extensive usage in the food industry. It is widely used in a number of products including bakery products, dairy products, cereals, sauces and ketchups, biscuits, cakes, etc.
4. *Cosmetics*: Turmeric is used in various cosmetic products to manufacture sunscreen, fairness creams and lotions. The anti-oxidant properties help to lighten the skin and in treating skin inflammations.

Culinary Uses

Turmeric grows wild in the forests of South and Southeast Asia. It is one of the key ingredients for many Indian, Persian and Thai dishes such as in curry and many more. Ancient Indian medicine, Ayurveda has recommended its use in food for its medicinal value, much of which is now being researched in the modern day. Its use a coloring agent is not of primary value in South Asian cuisine.

In Indonesia, the turmeric leaves are used for Minangese or Padangese curry base of Sumatra, such as *rendang*, *sate padang* and many other varieties.

Although most usage of turmeric is in the form of root powder, in some regions (especially in Maharashtra, Goa and Konkan), leaves of turmeric are used to wrap and cook food. This usually takes place in areas where turmeric is grown locally, since the leaves used are freshly picked. This imparts a distinct flavor. In recipes outside South Asia, turmeric is sometimes used as an agent to impart a rich, custard-like yellow color. It is used in canned beverages and baked products, dairy products, ice cream, yogurt, yellow cakes, orange juice, biscuits, popcorn color, sweets, cake icings, cereals, sauces, gelatins, etc. It is a significant ingredient in most commercial curry powders. Turmeric is mostly used in savory dishes, as well as some sweet dishes, such as the cake *sfouf*. Although usually used in its dried, powdered form, turmeric is also used fresh, much like ginger. It has numerous uses in Far Eastern recipes, such as fresh turmeric pickle, which contains large chunks of soft turmeric.

Turmeric (coded as E100 when used as a food additive), indicating how it is used as a food coloring (it normally gives food slightly yellow color) is used to protect food products from sunlight. The oleoresin is used for oil-containing products. The curcumin/polysorbate solution or curcumin powder dissolved in alcohol is used for water-containing products. Over coloring, such as in pickles, relishes, and mustard, are sometimes used to compensate for fading.

In combination with annatto (E160b), turmeric has been used to color cheeses, yogurt, dry mixes, salad dressings, winter butter and margarine. Turmeric is also used to give a yellow color to some prepared mustards, canned chicken broths and other foods (often as a much cheaper replacement for saffron).

Turmeric is widely used as a spice in South Asian and Middle Eastern cooking. Many Persian dishes use turmeric as a starter ingredient for almost all Iranian fry ups (which typically consist of oil, onions and turmeric followed by any other ingredients that are to be included). In Nepal, turmeric is widely grown and is extensively used in almost every vegetable and meat dish in the country for its color, as well as for its medicinal value. In South Africa, turmeric is traditionally used to give boiled white rice a golden color.

In Goa and Dakshina Kannada (Karnataka state, India), turmeric plant leaf is used to prepare special sweet dishes, *patoleo,* by layering on the leaf rice flour, and coconut-jaggery mixture, and then closing and steaming in a special copper steamer (goa). In Tamil Nadu, an Indian State, it is called "Manjal", which is extensively used for its aroma, color and as a disinfectant.

Preliminary Medical Research

Turmeric is currently being investigated for possible benefits in Alzheimer's disease, cancer, arthritis, and other clinical disorders. As an example of preliminary laboratory research, turmeric ameliorated the severity of pancreatitis-associated lung injury in mice. According to a 2005 article in the *Wall Street Journal,* research activity into curcumin and turmeric is increasing. The U.S. National Institutes of Health currently has registered 61 clinical trials completed or underway to study use of dietary curcumin for a variety of clinical disorders.

Cosmetics

Turmeric paste is traditionally used by Indian women to keep them free of superfluous hair and as an antimicrobial. Turmeric paste, as part of both home remedies and Ayurveda, is also said to improve the skin and is touted as an anti-aging agent. Turmeric figures prominently in the bridal beautification ceremonies of India, Bangladesh, and Pakistan. Staining oneself with turmeric is believed to improve the skin tone and tan. Turmeric is currently used in the formulation of some sunscreens. The government of Thailand is funding a project to extract and isolate tetrahydrocurcuminoids (THC) from turmeric. Tetrahydrocurcuminoids are colorless compounds that might have antioxidant and skin-lightening properties, and might be used to treat skin inflammations, making these compounds useful in cosmetics formulations.

Dye

Turmeric makes a poor fabric dye, as it is not very light fast (it fades with exposure to sunlight). However, turmeric is commonly used in Indian clothing, such as saris.

Ceremonial Uses

Turmeric is considered highly auspicious in India and has been used extensively in various Indian ceremonies for millennia. Even today it is used in every part of India during wedding ceremonies and religious ceremonies. It is used in Pujas to make a form of Hindu god Ganesha. Lord Ganesha, the remover of obstacles, is invoked at the beginning of almost any ceremony and a form of Ganesha for this purpose is made by mixing turmeric with water and forming it into a cone-like shape.

Gaye holud (literally "yellow on the body") is a ceremony observed mostly in the region of Bengal (comprising Bangladesh and Indian West Bengal). The Gaye holud takes place one or two days prior to the religious and legal Bengali wedding ceremonies. The turmeric paste is applied by friends to the bodies of the couple. This is said to soften the skin, but also colors them with the distinctive yellow hue that gives its name to this ceremony. It may be a joint event for the bride and groom's families, or it may consist of separate events for the bride's family and the groom's family. During the south Indian festival Pongal, a whole turmeric plant with fresh rhizomes is offered as a thanksgiving offering to Surya, the Sun god. Also, the fresh plant sometimes is tied around the sacred Pongal pot in which an offering of pongal is prepared.

In southern India, as a part of the marriage ritual, dried turmeric tuber tied with string is used to replace the Mangalsutra temporarily or permanently. The Hindu Marriage act recognizes this custom. Thali necklace is the equivalent of marriage rings of west. In western and coastal India, during weddings of the Marathi and Konkani people turmeric tubers are tied with strings by the couple to their wrists during a ceremony called *Kankanabandhana*. Modern Neopagans list it with the quality of fire, and it is used for power and purification rites. Friedrich Ratzel in *The History of Mankind* reported in 1896 that in Micronesia the preparation of turmeric powder for embellishment of body, clothing and utensils had a highly ceremonial character. He quotes an example of the roots being ground by four to six women in special public buildings and then allowed to stand in water. The following morning, three young coconuts and three old soma nuts are offered by a priestess with prayer, after which the dye which has settled down in the water is collected, baked into cakes in coconut molds, wrapped in banana leaves, and hung up in the huts till required for use.

Composition/Nutritional Value

Turmeric contains up to 5 per cent essential oils and up to 5 per cent curcumin, a polyphenol. Curcumin is the active substance of turmeric and curcumin is known as C.I. 75300, or Natural Yellow 3. The systematic chemical name is (1*E*, 6*E*)-1, 7-bis (4-hydroxy-3-methoxyphenyl)-1,6-heptadiene-3,5-dione. It can exist at least in two tautomeric forms, keto and enol. The keto form is preferred in solid phase and the enol form in solution. Curcumin is a pH indicator. In acidic solutions (pH < 7.4) it turns yellow, whereas in basic (pH > 8.6) solutions it turns bright red.

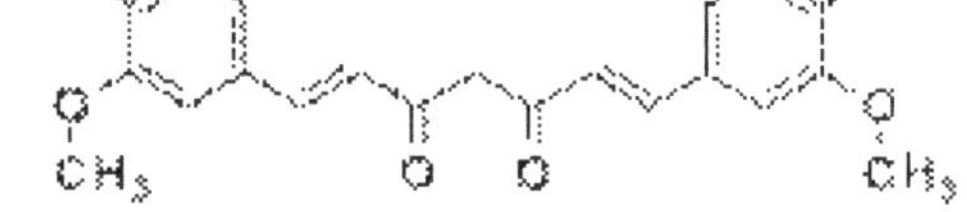

Curcumin keto form

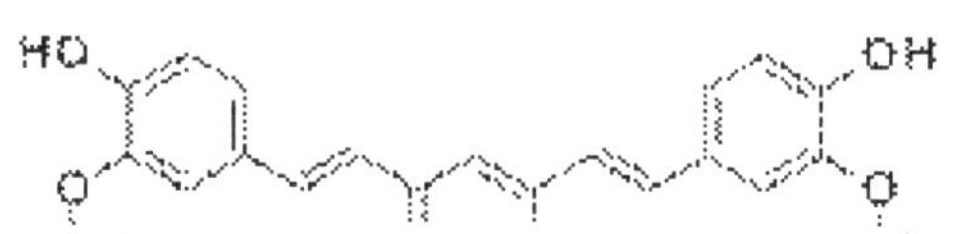

Curcumin enol form

Health Benefits of Turmeric

- It is a natural antiseptic and antibacterial agent, useful in disinfecting cuts and burns.
- When combined with cauliflower, it has shown to prevent prostate cancer and stop the growth of existing prostate cancer.
- Prevented breast cancer from spreading to the lungs in mice.
- May prevent melanoma and cause existing melanoma cells to commit suicide.
- Reduces the risk of childhood leukemia.
- Is a natural liver detoxifier.
- May prevent and slow the progression of Alzheimer's disease by removing amyloyd plaque buildup in the brain.
- May prevent metastases from occurring in many different forms of cancer.
- It is a potent natural anti-inflammatory that works as well as many anti-inflammatory drugs but without the side effects.
- Has shown promise in slowing the progression of multiple sclerosis in mice.
- Is a natural painkiller and cox-2 inhibitor.
- May aid in fat metabolism and help in weight management.
- Has long been used in Chinese medicine as a treatment for depression.
- Because of its anti-inflammatory properties, it is a natural treatment for arthritis and rheumatoid arthritis.
- Boosts the effects of chemo drug paclitaxel and reduces its side effects.
- Promising studies are underway on the effects of turmeric on pancreatic cancer.
- Studies are ongoing in the positive effects of turmeric on multiple myeloma.
- Has been shown to stop the growth of new blood vessels in tumors.
- Speeds up wound healing and assists in remodeling of damaged skin.
- May help in the treatment of psoriasis and other inflammatory skin conditions.

Figure 52.1: Flow sheet for preparation of dry turmeric powder

Harvesting of fully mature Turmeric Rhizomes

↓

Sweating (1 day under leaves)

↓

Curing (Boiling)

↓

Drying

↓

Polishing

↓

Grading

↓

Grinding

↓

Sieving (60-80 mesh)

↓

Packaging/Sealing

↓

Labeling

↓

Storage

↓

Marketing

Figure 52.2: Flow sheet for preparation of curcumin from turmeric

Harvesting of fully mature Turmeric Rhizomes

↓

Sweating (1 day under leaves)

↓

Curing (Boiling)

↓

Drying

↓

Polishing

↓

Grading

↓

Grinding

↓

Sieving (60-80 mesh)

↓

Drying of powder (moisture free)

↓

Dissolve in absolute ethanol

↓

Reflux 3-5 hrs

↓

Evaporate ethanol and get curcumin

↓

Packaging/Labeling/Storage

↓

Marketing

53

Fruit Products

The term has different meanings dependent on context. In non-technical usage, such as food preparation, fruit normally means the fleshy seed-associated structures of certain plants that are sweet and edible in the raw state, such as apples, oranges, grapes, strawberries, juniper berries and bananas. Seed-associated structures that do not fit these informal criteria are usually called by other names, such as vegetables, pods, nut, ears and cones.

In biology (botany), a "fruit" is a part of a flowering plant that derives from specific tissues of the flower, mainly one or more ovaries. Taken strictly, this definition excludes many structures that are "fruits" in the common sense of the term, such as those produced by non-flowering plants (like juniper berries, which are the seed-containing female cones of conifers.) On the other hand, the botanical sense includes many structures that are not commonly called "fruits", such as bean pods, corn kernels, wheat grains, tomatoes, the section of a fungus that produces spores, and many more. However, there are several variants of the biological definition of fruit that emphasize different aspects of the enormous variety that is found among plant fruits.

Fruits (in either sense of the word) are the means by which many plants disseminate seeds. Most plants bearing edible fruits, in particular, coevolved with animals in a symbiotic relationship as a means for seed dispersal and nutrition, respectively; in fact, many animals (including humans to some extent) have become dependent on fruits as a source of food. Fruits account for a substantial fraction of world's agricultural output, and some (such as the apple and the pomegranate) have acquired extensive cultural and symbolic meanings.

Many fruits that, in a botanical sense, are true fruits are actually treated as vegetables in cooking and food preparation, because they are not particularly sweet. These culinary vegetables include cucurbits (*e.g.*, squash, pumpkin, and cucumber), tomatoes, peas, beans, corn, eggplant, and sweet pepper. In addition, some spices, such as allspice and chilies, are fruits, botanically speaking. In contrast, occasionally a culinary "fruit" is not a true fruit in the botanical sense. For example, rhubarb is often referred to as a fruit, because it is used to make sweet desserts such as pies, though only the petiole of the rhubarb plant is edible. In the culinary sense of these words, a fruit is usually any sweet-tasting plant product, especially those associated with seed(s), a vegetable is any savoury or less sweet plant product, and a nut is any hard, oily, and shelled plant product.

Technically, a cereal grain is also a kind of fruit, a kind which is termed a caryopsis. However, the fruit wall is very thin, and is fused to the seed coat, so almost all of the edible grain is actually a seed. Therefore, cereal grains, such as corn, wheat and rice are better considered as edible seeds, although some references do list them as fruits. Edible gymnosperm seeds are often misleadingly given fruit names, *e.g.*, pine nuts, ginkgo nuts, and juniper berries.

Uses

Many hundreds of fruits, including fleshy fruits like apple, peach, pear, kiwifruit, watermelon and mango are commercially valuable as human food, eaten both fresh and as jams, marmalade and other preserves. Fruits are also in manufactured foods like cookies, muffins, yoghurt, ice cream, cakes, and many more. Many fruits are used to make beverages, such as fruit juices (orange juice, apple juice, grape juice, etc.) or alcoholic beverages, such as wine or brandy. Apples are often used to make vinegar. Fruits are also used for gift giving; Fruit Basket and Fruit Bouquet are some common forms of fruit gifts.

Many vegetables are botanical fruits, including tomato, bell pepper, eggplant, okra, squash, pumpkin, green bean, cucumber and zucchini. Olive fruit is pressed for olive oil. Spices like vanilla, paprika, allspice and black pepper are derived from berries.

Nutritional Value

Fruits are generally high in fiber, water, vitamin C and sugars, although this latter varies widely from traces as in lime, to 61 per cent of the fresh weight of the date. Fruits also contain various phytochemicals that do not yet have an RDA/RDI listing under most nutritional factsheets, and which research indicates are required for proper long-term cellular health and disease prevention. Regular consumption of fruit is associated with reduced risks of cancer, cardiovascular disease (especially coronary heart disease), stroke, Alzheimer disease, cataracts, and some of the functional declines associated with aging.

Diets that include a sufficient amount of potassium from fruits and vegetables also help reduce the chance of developing kidney stones and may help reduce the effects of bone-loss. Fruits are also low in calories which would help lower one's calorie intake as part of a weight-loss diet.

Nonfood Uses

Because fruits have been such a major part of the human diet, different cultures have developed many different uses for various fruits that they do not depend on as being edible. Many dry fruits are used as decorations or in dried flower arrangements, such as unicorn plant, lotus, wheat, annual honesty and milkweed. Ornamental trees and shrubs are often cultivated for their colorful fruits, including holly, pyracantha, viburnum, skimmia, beautyberry and cotoneaster.

Fruits of opium poppy are the source of opium which contains the drugs morphine and codeine, as well as the biologically inactive chemical theabaine from which the drug oxycodone is synthysized. Osage orange fruits are used to repel cockroaches. Bayberry fruits provide a wax often used to make candles. Many fruits provide natural dyes, *e.g.* walnut, sumac, cherry and mulberry. Dried gourds are used as decorations, water jugs, bird houses, musical instruments, cups and dishes. Pumpkins are carved into Jack-o'-lanterns for Halloween. The spiny fruit of burdock or cocklebur were the inspiration for the invention of Velcro.

Coir is a fiber from the fruit of coconut that is used for doormats, brushes, mattresses, floortiles, sacking, insulation and as a growing medium for container plants. The shell of the coconut fruit is used to make souvenir heads, cups, bowls, musical instruments and bird houses. Fruit is often used as a subject of still life paintings.

Safety

For food safety, the CDC recommends proper fruit handling and preparation to reduce the risk of food contamination and food borne illness. Fresh fruits and vegetables should carefully be selected. At the store, they should not be damaged or bruised and pre-cut pieces should be refrigerated or surrounded by ice. All fruits and vegetables should be rinsed before eating. This recommendation also applies to produce with rinds or skins that are not eaten. It should be done just before preparing or eating to avoid premature spoilage. Fruits and vegetables should be kept separate from raw foods like meat, poultry, and seafood, as well as utensils that have come in contact with raw foods. Fruits and vegetables, if they are not going to be cooked, should be thrown away if they have touched raw meat, poultry, seafood or eggs. All cut, peeled, or cooked fruits and vegetables should be refrigerated within two hours. After a certain time, harmful bacteria may grow on them and increase the risk of food borne illness.

Figure 53.1: Flow sheet for preparation of mixed fruit juice beverage

Mixed fruit juice (10 per cent)

↓

Addition of water

↓

Addition of sugar, citric acid

↓

Adjust °Brix at 15, acidity 0.25 per cent

↓

Cool and bottling

↓

RTS beverage

Some time two or more juices are mixed to yield a well-balanced, rightly flavoured, highly palatable and refreshing drink. Juices are blended so as to utilize a too sweet fruit (grape), a bitter (grape fruit), too acidic or tart fruits (sour lime, sour plum, galgal sour cherry etc.) blended and insipid tasting fruits like pear or apple and strongly flavoured fruits (guava and banana). Some of the common commercial blends of juice are: Grape (97 per cent) and lime (3 per cent), Grape (50 per cent) and orange (50 per cent), Orange (50-75 per cent) and grape fruit (25-50 per cent), Apple (97 per cent) and lime (3 per cent), Apple (74 per cent) and grape fruit (25 per cent) + 1 per cent sugar, Apple (37 per cent) and plum (62 per cent) + 1 per cent sugar.

This is a type of fruit beverage which contains at least 10 per cent fruit juice and 10-15 per cent soluble solids besides about 0.3 per cent acidity. It is not diluted before serving; hence it is known as ready-to-serve (RTS).

Figure 53.2: Flow sheet for preparation of fruit juice concentrate

Fully Mature Fruits

↓

Washing

↓

Sorting and trimming

↓

Juice extraction

↓

Evaporation

↓

Pasteurization

↓

Canning

↓

Labeling

↓

Storage

↓

Marketing of fruit concentrate

Figure 53.3: Flow sheet for preparation of fruit cheese

Fruits (Firm and ripe)
↓
Washing
↓
Cutting into thin slices
↓
Boiling with equal quantity of water (to soften pulp)
↓
Sieving and making fine pulp (to remove seed and skin)
↓
Adding sugar, citric acid and butter to pulp
↓
Mixing thoroughly
↓
Cooking till sufficiently thick
↓
Adding salt and color
↓
Removing from fire (when the mass starts leaving side of pan)
↓
Spreading hot cheese in 0.6 cm thick layer on tray smeared with butter
↓
Allowing cooling and setting
↓
Cutting into small pieces of suitable size
↓
Wrapping in butter paper or polythene sheet
↓
Packing in dry jar
↓
Storage and Marketing
↓
Sealing and Labeling

Figure 53.4: Flow sheet for preparation of fruit butter

Fruits (Firm and Ripe)
↓
Washing
↓
Cutting into thin pieces
↓
Sieving and making fine pulp (remove seed and skin)
↓
Mixing pulp with sugar (750 g sugar for 1 kg pulp)
↓
Cooking to consistency of jam
↓
Adding finely ground spices and mixing
(small quantity of nutmeg, cinnamon, clove etc.)
↓
Removing from fire
↓
Filling hot into clean sterilized bottles
↓
Sealing and Labeling
↓
Storage and Marketing

Figure 53.5: Flow sheet for preparation of cordial

Fully Mature Fruits
↓
Washing
↓
Cutting into halves
↓
Juice extraction
↓
Straining of juice
↓
Addition of preservative (KMS 1 g/liter juice)
↓
Storing in glass container for 10-15 days for clarification
↓
Suspended material settles down
↓
Syphoning off the supernatant clear juice
↓
Straining and measuring
↓
Preparation of syrup
↓
Straining
↓
Mixing of juice with syrup
↓
Addition of preservative (KMS 0.6g/Liter product)
↓
Bottling
↓
Capping
↓
Storage at cool and dry place
↓
Marketing

Figure 53.6: Flow sheet for preparation of squash

Fully Mature Fruits

↓

Washing

↓

Trimming

↓

Cutting or grating

↓

Juice extraction

↓

Straining

↓

Juice measuring

↓

Preparation of syrup by adding sugar, water, acid and heating etc.

↓

Straining

↓

Mixing with juice

↓

Addition of preservative (0.6g KMS or 1.0 Sodium benzoate/Liter squash)

↓

Bottling

↓

Capping

↓

Storage

↓

Marketing

Figure 53.7: Flow sheet for preparation of syrups

Fruit juice/Pulp/Extract

↓

Mixing with strained syrup solution

↓

Sugar + Water + Citric acid and heating to dissolve these materials

↓

Bottling

↓

Capping

↓

Labeling

↓

Storage

↓

Marketing

Figure 53.8: Flow sheet for preparation of cider

Fully Ripe Fruits
↓
Washing
↓
Crushing and pressing
↓
Extraction of juice
↓
Addition of sugar (22 per cent TSS)
↓
Addition of preservative
↓
Addition of pure culture of wine yeast
↓
Fermentation
↓
Racking
↓
Filtration
↓
Aging
↓
Heating at 82 °C
↓
Filtration
↓
Bottling
↓
Crown corking
↓
Pasteurization
↓
Cooling
↓
Storage
↓
Marketing

54

Weaning Foods

Weaning foods are generally introduced between the ages of six months to three years old as breastfeeding is discontinued. While breastfed infants are often able to maintain adequate growth through their sixth month, additional nutrients are required to complement or, in some cases, replace breastfeeding completely. The main concern is making sure that there is no gap between nutrient requirements and what a child is able to consume, absorb, and utilize. Nutritional status in children is most vulnerable during the weaning stages when both macro and micronutrients may be insufficient to maintain growth and development. Protein energy malnutrition and micronutrient under-nutrition occur together. It is an important part of weaning strategies to optimize nutritional status and to tackle under-nutrition-related problems as a group for maximum effectiveness. Traditionally, weaning foods are liquids and semisolids which are later replaced by foods eaten by older family members. In some cases these types of foods can be filling and yet not meet the child's nutrient needs. Establishing appropriate characteristics for nutrients and other aspects of weaning foods that assist in motor skills and mental development will be important to assure the appropriateness of targeting foods to this age group.

Policy Considerations

In addition to the characteristics, policy guidelines and directives should be considered in the development, distribution, and commercialization of weaning foods. Several policy recommendations and regulations are available online. Guidelines are not always clear to the end user and definitions that pertain to other populations may not transfer well to weaning foods. Such is the issue of "fiber" and "non-digestible carbohydrates", of which the former can be problematic in feeding young children and the latter has been suggested as helpful in supporting the colonization of the gut with beneficial bacteria. Policy development requires research on which to build evidence-based practice guidelines. Outcomes from processing to health impact require documentation for the recommendation of feasible methods and materials for weaning foods. Examples of guidelines include the World Health Organization's International Code of Marketing of Breast Milk Substitutes, Baby Friendly Hospital Initiative, the Innocent Declaration from the "Convention of the Rights of the Child", and Codex Alimentarius, among others. In addition, each country is likely to have governmental regulations that pertain to the development, distribution, and marketing of infant weaning products.

Dietary Reference Intake Recommendations

Nutrient	*RDA 7-12 months*
Energy (kcal)	743
Carbohydrate	95 g/d
Protein	11 g/kg/d
Fat	30 g/d
Protein	13.5 g
Vitamin A	500 ug/d
Vitamin C	50 mg/d
Vitamin D	5 mg/d
Vitamin E	5 mg/d
Vitamin K	2.5 mg/d
Thiamin	0.3 mg/d
Riboflavin	0.4 mg/d
Niacin	4 mg/d
Vitamin B_6	0.3 mg/d
Foliate	80 ug/d
Vitamin B_{12}	0.5 ug/d
Pantothenic acid	1.8 mg/d
Biotin	6 ug/d
Choline	150 mg/d
Fluoride	0.7 mg/d
Iron	40 mg/d
Selenium	45 ug/d
Zinc	5 mg/d

Table 54.1: Composition of malted weaning food (MWF) as compared with proprietary weaning foods.

Constituent	*MWF*	*Farex*	*Nestum*	*Cerelac*	*Balamul*
Moisture, per cent	6.0	3.5	5.0	2.2	4.0
Protein, per cent	11.5	12.0	7.5	11.0	22.0
Fat, per cent	1.5	3.0	-	7.8	3.5
Crude fiber, per cent	1.8	0.9	-	-	-

Contd...

Table 54.1–*Contd...*

Constituent	*MWF*	*Farex*	*Nestum*	*Cerelac*	*Balamul*
Total ash, per cent	2.3	3.5	3.0	2.0	-
Calcium, mg	240	750	690	275	800
Phosphorus, mg	210	400	570	225	690
Calories	396	350	366	422	380
PER	2.4	2.3	-	-	-
Cooked paste viscosity, CPU; 15 per cent slurry	250	9000	12000	3500	900
Market price, Rs/Kg	15.00	33.75	28.50	32.25	17.10

Figure 54.1: Flow sheet for preparation of malted weaning food

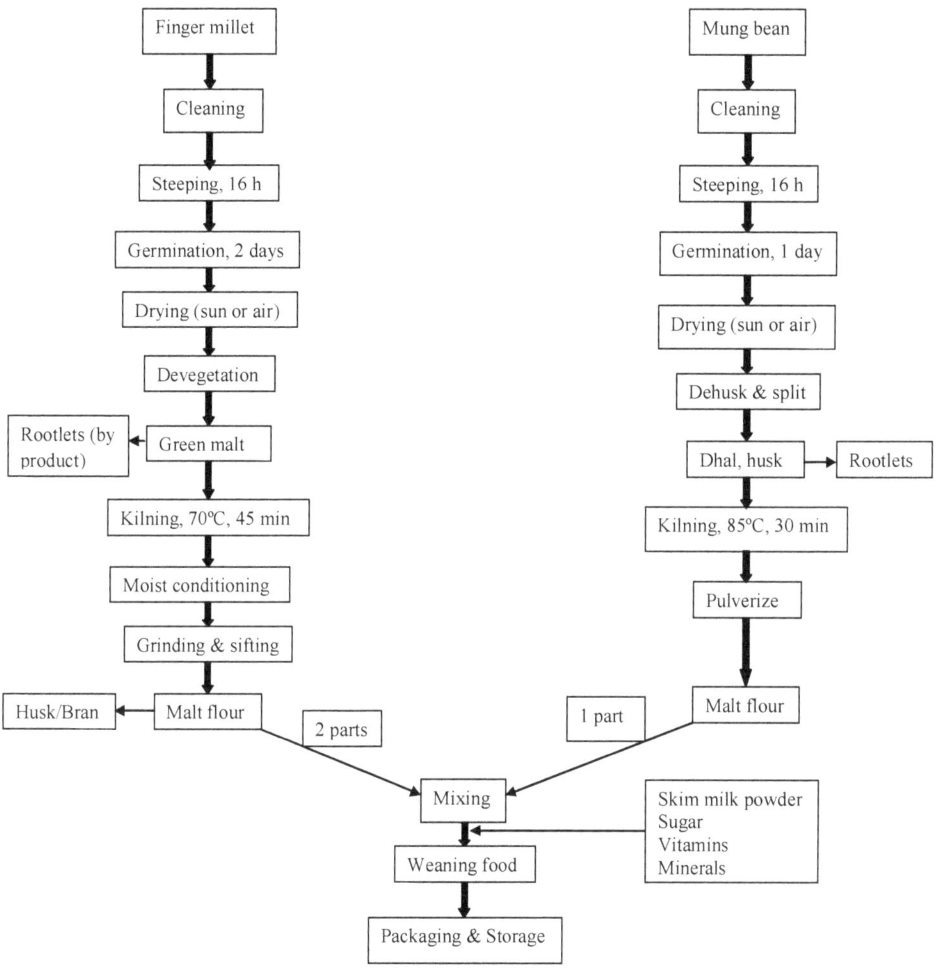

55

Mineral and Vitamins

1. Minerals

Minerals are elements that originate in the soil and cannot be created by living things, such as plants and animals. Yet plants, animals and humans need minerals in order to be healthy. Plants absorb minerals from the soil, and animals get their minerals from the plants or other animals they eat. Most of the minerals in the human diet come directly from plants, such as fruits and vegetables, or indirectly from animal sources. Minerals may also be present in your drinking water, but this depends on where you live, and what kind of water you drink (bottled, tap). Minerals from plant sources may also vary from place to place, because the mineral content of the soil varies according to the location in which the plant was grown.

Nutrient - Daily Amount Needed	*Information*	*Fruit Sources*	*Vegetable Sources*
Calcium Adults need 1000 mg/ day. Children need 800 to 1300 mg/day. Recommended supplement: Coral Calcium Supreme	Calcium eases insomnia and helps regulate the passage of nutrients through cell walls. Without calcium, your muscles wouldn't contract correctly, your blood wouldn't clot and your nerves wouldn't carry messages. If you don't get enough calcium from the food you eat, your body automatically takes the calcium needed from your bones. If your body continues to tear down more bone than it replaces over a period of years in order to get sufficient calcium, your bones will become weak and break easily.	Most fruits contain some calcium, these have a bit more than usual: Blackberries Blackcurrants Dates Grapefruit Mulberries Orange Pomegranate Prickly Pears	Most vegetables contain some calcium, these have a bit more than usual: Amaranth leaves Bok Choy Brussels Sprouts Butternut Squash Celery Chinese Broccoli French Beans Kale Okra

Nutrient - Daily Amount Needed	Information	Fruit Sources	Vegetable Sources
	Deficiency may result in muscle spasms and cramps in the short term and osteoporosis.		Parsnip Spirulina Swiss Chard Turnip
Copper The estimated safe and adequate intake for copper is 1.5 - 3.0 mg/day. Many survey studies show that Americans consume about 1.0 mg or less of copper per day	Copper is involved in the absorption, storage and metabolism of iron and the formation of red blood cells. It also helps supply oxygen to the body. The symptoms of a copper deficiency are similar to iron-deficiency anemia.	Most fruits contain a small amount of copper, but kiwi fruit has a significant amount. Avocado Blackberries Dates Guava Kiwi Fruit Lychee Mango Passionfruit Pomegranate	Most vegetables have some copper, but Lima Beans have a significant amount. Amaranth leaves Artichoke French Beans Kale Lima Beans Parsnip Peas Potatoes Pumpkin Spirulina Squash - Winter Sweet Potato Swiss Chard Taro
Iodine Adults should get 150 mcgs per day.The children's recommendation for iodine is 70 to 150 mcg (that is micrograms).	Iodine helps regulate the rate of energy production and body weight and promotes proper growth. It also promotes healthy hair, nails, skin and teeth. In countries where iodine is deficient in the soil, rates of hypothyroidism, goiter and retarded growth from iodine deficiency are very high. In developed countries, however, because iodine is added to table salt, iodine deficiencies are rare.	Fruits grown in iodine-rich soils contain iodine.	Vegetables grown in iodine-rich soils contain iodine.
Iron Women and teenage girls need at least 15 mg a day, whereas men can get by on 10. It is important that children get about 10 to 12 mg of iron per day, preferably from their diet. Breast-	Most at risk of iron deficiency are infants, adolescent girls and pregnant women. Iron deficiency in infants can result in impaired learning ability and behavioral problems. It can also affect the immune system and cause weakness and fatigue. To aid in the absorption of iron, eat foods rich in vitamin C at the same	While most fruits have some iron, probably the best source of iron for children is raisins, which are rich in iron. Other fruits which have a good amount of iron are:	Vegetables: Amaranth leaves Bok Choy Brussels Sprouts Butternut squash French Beans Kale Leeks Lima Beans Peas

Nutrient - Daily Amount Needed	*Information*	*Fruit Sources*	*Vegetable Sources*
feeding is the best insurance against iron deficiency in babies.	time you eat the food containing iron. The tannin in non-herbal tea can hinder absorption of iron. Take iron supplements and your vitamin E at different times of the day, as the iron supplements will tend to neutralize the vitamin E. Vegetarians need to get twice as much dietary iron as meat eaters.	Avocado Blackberries Blackcurrant Boysenberries Breadfruit Cherries Dates Figs Grapes Kiwi Lemon Loganberries Lychee Mulberries Passion Fruit Persimmon Pomegranate Raspberries Strawberry Watermelon	Potatoes Pumpkin Spirulina Swiss Chard
Magnesium Adults need 310 to 420 mg/ day. Children need 130 to 240 mg/day.	Magnesium is needed for bone, protein, making new cells, activating B vitamins, relaxing nerves and muscles, clotting blood, and in energy production. Insulin secretion and function also requires magnesium. Magnesium also assists in the absorption of calcium, vitamin C and potassium. Deficiency may result in fatigue, nervousness, insomnia, heart problems, high blood pressure, osteoporosis, muscle weakness and cramps.	Fruits: Avocado Banana Blackberries Blackcurrants Breadfruit Cherimoya Dates Guava Kiwi Loganberries Mulberries Passion Fruit Pomegranate Prickly Pear Raspberries Watermelon	Vegetables: Amaranth leaves Artichoke Butternut squash French Beans Lima Beans Okra Peas Spirulina Swiss Chard
Manganese 2.0-5.0 mg/day for adults 2.0-3.0 mg for children 7-10 yr 1.5-2.0 mg for children 4-6 yr 1.0-1.5 mg for children 1-3 yr 0.6-1.0 mg for children 6 mo-1 yr	The functions of this mineral are not specific since other minerals can perform in its place. Manganese does function in enzyme reactions concerning blood sugar, metabolism, and thyroid hormone function. Deficiency is rare in humans.	Most fruits contain manganese, but the following fruits have a significant amount: Avocado Banana Blackberries Blackcurrants Blueberries Boysenberries Cranberries Dates Gooseberries Grapefruit	Vegetables: Amaranth leaves Brussels Sprouts Butternut squash French Beans Kale Leeks Lima Beans Okra Parsnip Peas Potatoes Spirulina Squash - Winter Sweet Potato

Nutrient - Daily Amount Needed	*Information*	*Fruit Sources*	*Vegetable Sources*
0.3-0.6 mg for infants 0-6 months		Guava Loganberries Pineapple Pomegranate Raspberries Strawberry	Swiss Chard Taro
Phosphorus Adults need 700 mg/ day. Children need 500 to 1250 mg/day.	In combination with calcium, phosphorus is necessary for the formation of bones and teeth and of the nerve cells. Phosphorus is second to calcium in abundance in the body. It is very widely distributed in both plant and animal foods so it is unlikely that deficiency would be a problem.	Fruits: Avocado Blackcurrants Breadfruit Dates Guava Kiwi Lychee Mulberries Passionfruit Pomegranate	Vegetables: Amaranth leaves Artichoke Brussels Sprouts Celeriac Corn French Beans Lima Beans Parsnip Peas Potatoes Pumpkin Spirulina Taro
Potassium Estimated Minimum Requirements 2000 mg/day for adults and adolescents.	Potassium is essential for the body's growth and maintenance. It is necessary to keep a normal water balance between the cells and body fluids. Potassium plays an essential role in proper heart function. Deficiency may cause muscular cramps, twitching and weakness, irregular heartbeat, insomnia, kidney and lung failure.	Fruits: Avocado Bananas Blackcurrants Breadfruit Cherimoya Cherries Chinesepear Dates Grapefruit Guava Kiwi Lychee Papaya Passionfruit Pomegranate Prickly pear Watermelon	Vegetables: Amaranth leaves Bamboo Shoots Bok Choy Butternut squash French Beans Lima Beans Parsnips Potatoes Pumpkin Spirulina Sweet Potatoes Swiss Chard
Selenium Men need 70 µg/day. Women need 55 mcgs/ day.	Selenium is a part of several enzymes necessary for the body to properly function. Generally, selenium functions as an antioxidant that works in conjunction with vitamin E. Selenium deficiency is rare in humans.	Most fruits contain a small amount of selenium, but dates have a significant amount. Bananas Breadfruit Guava Lychee Mango Passionfruit Pomegranate Watermelon	Vegetables: Asparagus Brussels Sprouts French Beans Lima Beans Mushrooms Parsnip Peas Spirulina

Nutrient - Daily Amount Needed	*Information*	*Fruit Sources*	*Vegetable Sources*
Sodium			
500 mg/day for adults 120 mg for infants Daily value recommendation - no more than 2,400 to 3,000 mg/day	Sodium is required by the body to regulate blood pressure and blood volume. It helps regulate the fluid balance in your body. Sodium also helps in the proper functioning of muscles and nerves. Many people get far more sodium than they need, which tends to cause health problems. Different body types need different amounts of sodium.	Sodium occurs naturally in almost all fresh, whole fruits but passion fruit has a significant amount.	Sodium occurs naturally in almost all fresh, whole vegetables, these have significant amounts: Amaranth leaves Artichoke Broccoli Beetroot Bok Choy Brussels Sprouts Celeriac Celery Fennel Kale Spirulina Spaghetti squash Sweet Potatoes Swiss Chard
Zinc			
Men need 15 mgs/day. Women should get 12 mg/day. Children need 10 to 15 mg/day. Vegetarians need about 50 percent more zinc in their diet than meat eaters.	This metal is important in a number of key activities, ranging from protein and carbohydrate metabolism to the immune system, wound healing, growth and vision. Severe deficiency can contribute to stunted growth. Deficiency can sometimes be seen in white spots on the fingernails.	Most fruits contain a small amount of zinc, but the following have a significant amount: Avocado Blackberries Dates Loganberries Pomegranate Raspberries	Vegetables: Amaranth leaves Asparagus Bamboo Shoots Brussels Sprouts Corn French Beans Lima Beans Okra Peas Potatoes Pumpkin Spirulina Swiss Chard

2. Vitamins

Vitamins are organic food substances found only in living things, *i.e.* plants and animals. They are essential for our bodies to function properly, for growth, energy and for our general well-being. With very few exceptions the human body cannot manufacture or synthesize vitamins. They must be supplied in our diet or in man-made dietary supplements. Some people believe that vitamins can replace food, but that is incorrect. In fact, vitamins cannot be assimilated without also ingesting food.

That is why it is best to take them with a meal. Synthetic vitamin supplements can be of varying quality, so it is a good idea to get your supplements from a reliable source.

Nutrient - Daily Amount Needed	*Information*	*Fruit Sources*	*Vegetable Sources*
Vitamin A 10,000 IU/day (plant-derived) for adult males. 8,000 for adult females - 12,000 if lactating. 4,000 for children ages 1-3 5,000 for children ages 4-6 7,000 for children ages 7-10	Vitamin A helps cell reproduction. It also stimulates immunity and is needed for formation of some hormones. Vitamin A helps vision and promotes bone growth, tooth development, and helps maintain healthy skin, hair, and mucous membranes. It has been shown to be an effective preventive against measles. Deficiency can cause night blindness, dry skin, poor bone growth, and weak tooth enamel. Alpha-carotene, beta-carotene and retinol are all versions of Vitamin A.	Most fruits contain vitamin A, but the following fruits have a significant amount: Cantaloupes Grapefruit Guava Mango Papaya Passionfruit Tomatoes Watermelon	Amaranth Leaves Bok Choy Broccoli Brussels Sprouts Butternut Squash Carrots Chinese Broccoli Chinese Cabbage Kale Leeks Peas Pumpkin Rapini Spinach Squash summer Squash - winter Sweet Potato Swiss Chard
Vitamin B_1 (Thiamine) 1.2 mg for adult males and 1.1 mg for women - 1.5 mg if lactating. Children need .6 to .9 mg of B_1/thiamine per day.	Vitamin B_1/thiamine is important in the production of energy. It helps the body cells convert carbohydrates into energy. It is also essential for the functioning of the heart, muscles, and nervous system. Not getting enough thiamine can leave one fatigued and weak. Note: Most fruits and vegetables are not a significant source of thiamine.	Avocado Boysenberries Breadfruit Cherimoya Dates Grapes Grapefruit Guava Loganberries Mango Orange Pineapple Pomegranate Watermelon	Asparagus Brussels Sprouts Butternut Squash Corn French Beans Lima Beans Okra Parsnips Peas Potatoes Spirulina Sweet Potato
Vitamin B_2 (Riboflavin) 1.3 mg for adult males and 1.1 mg for women - 1.5 mg if pregnant/ lactating. Children need 0.6 to 0.9 mg of B_2/riboflavin per day.	Vitamin B_2 or riboflavin is important for body growth, reproduction and red cell production. It also helps in releasing energy from carbohydrates.	Avocado Banana Cherimoya Dates Grapes Lychee Mango Mulberries Passion Fruit Pomegranate Prickly Pear	Amaranth Leaves Artichoke Asparagus Bok Choy Brussels Sprouts Chinese Broccoli French Beans Lima Beans

Nutrient - Daily Amount Needed	*Information*	*Fruit Sources*	*Vegetable Sources*
			Mushrooms Peas Pumpkin Spirulina quash - winter Sweet Potato Swiss Chard
Vitamin B_3 (Niacin) 16 mg for adult males and 14 mg for women - 17-18 mg if pregnant/ lactating. Children need 9 - 16 mg of niacin per day.	Niacin assists in the functioning of the digestive system, skin, and nerves. It is also important for the conversion of food to energy.	Avocado Boysenberries Breadfruit Cherimoya Dates Guava Loganberries Lychee Mango Nectarine Passion Fruit Peach	Artichoke Butternut Squash Corn Mushrooms Okra Parsnip Peas Potatoes Pumpkin Spirulina Spaghetti Squash Squash - winter Sweet Potato
Vitamin B_5 (Pantothenic acid) 5 mg for adults and 6 - 7 mg for women who are pregnant or lactating. Children need 2 - 4 mg of pantothenic acid per day.	Pantothenic acid is essential for the metabolism of food as well as in the formation of hormones and (good) cholesterol.	Avocado Black Currants Breadfruit Cherimoya Dates Gooseberries Grapefruit Guava Pomegranate Raspberries Starfruit Watermelon	Broccoli Brussels Sprouts Butternut Squash Corn French Beans Mushrooms Okra Parsnip Potatoes Pumpkin Spirulina Spaghetti Squash Squash summer Squash - winter Sweet Potato
Vitamin B_6 (Pyridoxine) 1.3 to 1.7 mg for adults - 2 mg for women who are pregnant or lactating. Children need between 0.6 to 1.3 mg B_6 per day.	B6 plays a role in the creation of antibodies in the immune system. It helps maintain normal nerve function and acts in the formation of red blood cells. It is also required for the chemical reactions of proteins. The higher the protein intake, the more need there is for vitamin B_6. Too little B_6 in the diet can cause dizziness, nausea, confusion, irritability and convulsions.	Avocado Banana Breadfruit Cherimoya Dates Gooseberries Grapes Guava Lychee Mango Passion Fruit Pineapple	Amaranth Leaves Bok Choy Broccoli Brussels Sprouts Butternut Squash Celeriac Corn French Beans Green Pepper

Nutrient - Daily Amount Needed	*Information*	*Fruit Sources*	*Vegetable Sources*
		Pomegranate Watermelon	Kale Lima Beans Okra Peas Potatoes Spirulina Spaghetti Squash Squash - winter Sweet Potato Taro
Vitamin B_9 (Foliate) At least 400 mcgs for most adults - pregnant women 600 mcgs and breastfeeding women should get at least 500 mcgs. Children need between 150 to 300 mcg per day.	Foliate and folic acid is both forms of B_9. Foliate occurs naturally in fresh foods, whereas folic acid is the synthetic form found in supplements. Your body needs foliate to produce red blood cells, as well as components of the nervous system. It helps in the formation and creation of DNA and maintaining normal brain function, and is a critical part of spinal fluid. It has also been proven to reduce the risk for an NTD-affected (neural tube defect) pregnancy by 50 to 70 percent. Folic acid is vital for proper cell growth and development of the embryo. That is why it is important for a woman to have enough foliate/folic acid in her body both before and during pregnancy.	Avocado Blackberries Boysenberries Breadfruit Cherimoya Dates Guava Loganberries Lychee Mango Orange Papaya Passionfruit Pineapple Pomegranate Raspberries Strawberries	Amaranth Leaves Artichoke Asparagus Beetroot Bok Choy Broccoli Brussels Sprouts Chinese Broc-coli Chinese Cab-bage French Beans Lima Beans Okra Parsnip Peas Potatoes Spinach Spirulina Squash summer Squash - winter
Vitamin B_{12} (Cobalamin) 2.4 mcg for adults and 2.6 - 2.8 mcg for women who are pregnant or lactating. Children need 0.9 - 2.4 mcg per day.	Like the other B vitamins, vitamin B_{12} is important for metabolism. It helps in the formation of red blood cells and in the maintenance of the central nervous system. Vitamin B_{12} is the one vitamin that is available only from fish, poultry, meat or dairy sources in food.	None	None
Vitamin C 60 mg for adults - 70 mg for women who are pregnant and 95 for those lactating.	Vitamin C is one of the most important of all vitamins. It plays a significant role as an antioxidant, thereby protecting body tissue from the damage of oxidation. Antioxidants act to protect your cells against the effects of free	Black Currants Breadfruit Grapefruit Guava Kiwi Lychee Mango	Amaranth Leaves Bok Choy Broccoli Brussels Sprouts Butternut

Nutrient - Daily Amount Needed	*Information*	*Fruit Sources*	*Vegetable Sources*
Children need between 45 and 50 mg	radicals, which are potentially damaging by-products of the body's metabolism. Free radicals can cause cell damage that may contribute to the development of cardiovascular disease and cancer. Vitamin C has also been found by scientists to be an effective antiviral agent.	Mulberries Orange Papaya Passionfruit Pineapple Strawberries	Squash Green Pepper Kale Swiss Chard
Vitamin D 5 mcg (200 IU) for most adults. Between 50 - 70 yrs 10 mcg (400 IU), and after 70 15 mcg (600 IU). Children need about 5 mcg (200 IU)/day.	Vitamin D is known as the "sunshine vitamin" since it is manufactured by the body after being exposed to sunshine. Ten to fifteen minutes of good sunshine three times weekly is adequate to produce the body's requirement of vitamin D. This means that we don't need to obtain vitamin D from our diet unless we get very little sunlight – usually not a problem for children. Vitamin D is vital to the human body as it promotes absorption of calcium and magnesium, which are essential for the normal development of healthy teeth and bones. It also helps maintain adequate levels of calcium and phosphorus in the blood.	None	Mushrooms
Vitamin E 30 IU for most adults. Children need between 6-11 mg/day. (1 IU is equal to approximately 0.75 mg) Note: some researchers and medical experts believe that with all of the positive studies using higher doses of vitamin E, this daily recommended intake is not high enough.	Like vitamin C, vitamin E plays a significant role as an antioxidant, thereby protecting body tissue from the damage of oxidation. It is important in the formation of red blood cells and the use of vitamin K. Many women also use it to help minimize the appearance of wrinkles, and mothers use it to help heal minor wounds without scarring, as it is valued for its ability to soothe and heal broken or stressed skin tissue.	Avocado Blackberries Black Currants Blueberries Boysenberries Breadfruit Cranberries Guava Kiwi Loganberries Mango Mulberries Nectarine Papaya Peach Pomegranate Raspberries	Butternut Squash Parsnip Potatoes Pumpkin Spirulina Swiss Chard Taro
Vitamin K 70-80 micrograms/ day for adult males,	Vitamin K is fat soluble and plays a critical role in blood clotting. It regulates blood calcium levels and	Avocado Blackberries Blueberries	Alfalfa, sprouted Artichoke Asparagus

Nutrient - Daily Amount Needed	*Information*	*Fruit Sources*	*Vegetable Sources*
60-65 micrograms per day for adult females. Children need about half the amount, depending on age.	activates at least 3 proteins involved in bone health.	Boysenberries Chinese Pear Cranberries Grapes Kiwi Loganberries Mango Mulberries Pear Plum Pomegranate Raspberries Tomatoes	Bok Choy Broccoli Brussels Sprouts Cabbage Carrots Cauliflower Celery Chinese Broccoli Cucumber Kale Leeks Okra Peas Rapini Spinach Spirulina Squash - winter Swiss Chard

56

Flavouring Compounds

An aroma compound, also known as odorant, aroma, fragrance or flavor, is a chemical compound that has a smell or odor. A chemical compound has a smell or odor when two conditions are met: the compound needs to be volatile, so it can be transported to the olfactory system in the upper part of the nose, and it needs to be in a sufficiently high concentration to be able to interact with one or more of the olfactory receptors.

Aroma compounds can be found in food, wine, spices, perfumes, fragrance oils, and essential oils. For example, many form biochemically during ripening of fruits and other crops. In wines, most form as byproducts of fermentation. Odorants can also be added to a dangerous odorless substance, like propane, natural gas, or hydrogen, as a warning. Also, many of the aroma compounds play a significant role in the production of flavorants, which are used in the food service industry to flavor, improve, and generally increase the appeal of their products.

Flavor or flavour is the sensory impression of a food or other substance, and is determined mainly by the chemical senses of taste and smell. The "trigeminal senses", which detect chemical irritants in the mouth and throat as well as temperature and texture, are also very important to the overall Gestalt of flavor perception. The flavor of the food, as such, can be altered with natural or artificial flavorants, which affect these senses.

Flavorant is defined as a substance that gives another substance flavor, altering the characteristics of the solute, causing it to become sweet, sour, tangy, etc.

Of the three chemical senses, smell is the main determinant of a food item's flavor. While the taste of food is limited to sweet, sour, bitter, salty, and savory (umami) – the basic tastes – the smells of a food are potentially limitless. A food's flavor, therefore, can be easily altered by changing its smell while keeping its taste similar. Nowhere is this better exemplified than in artificially flavored jellies, soft drinks and candies, which, while made of bases with a similar taste, have dramatically different flavors due to the use of different scents or fragrances. The flavorings of commercially produced food products are typically created by flavorists. Although the terms "*flavoring*" or "*flavorant*" in common language denote the combined chemical sensations of taste and smell, the same terms are usually used in the fragrance and flavors industry to refer to edible chemicals and extracts that alter the flavor of food and food products through the sense of smell. Due to the high cost or unavailability of natural flavor extracts, most commercial flavorants are *nature-identical*, which means that they are the chemical equivalent of natural flavors but chemically synthesized rather than being extracted from the source materials. Identification of nature-identical flavorants is done using technology such as headspace techniques.

Flavorings are focused on altering or enhancing the flavors of natural food product such as meats and vegetables, or creating flavor for food products that do not have the desired flavors such as candies and other snacks. Most types of flavorings are focused on scent and taste. Few commercial products exist to stimulate the trigeminal senses, since these are sharp, astringent, and typically unpleasant flavors.

Table 56.1: Various types of flavouring compounds.

Type	*Description*
Natural flavoring substances	Flavoring substances obtained from plant or animal raw materials, by physical, microbiological or enzymatic processes. They can be either used in their natural state or processed for human consumption, but cannot contain any nature-identical or artificial flavoring substances.
Nature-identical flavoring substances	Flavoring substances that are obtained by synthesis or isolated through chemical processes, which are chemically and organoleptically identical to flavoring substances naturally present in products intended for human consumption. They cannot contain any artificial flavoring substances.
Artificial flavoring substances	Flavoring substances not identified in a natural product intended for human consumption, whether or not the product is processed. These are typically produced by fractional distillation and additional chemical manipulation of naturally sourced chemicals, crude oil or coal tar. Although they are chemically different, in sensory characteristics are the same as natural ones.

Table 56.2: Some important or abundant flavor compounds in selected fruits and vegetables.

Fruit	*Major Sugars*	*Major Acids*	*Important Aroma Compounds*
Apple	sucrose glucose fructose	malic citric	β-damascenone butyl hexanoate isoamyl hexanoate hexyl hexanoate ethyl butanoate propyl butanoate hexyl butanoate butylacetate 2-ethyl-1-butyl acetate ethyl acetate butanol

Contd...

Table 56.2–*Contd...*

Fruit	*Major Sugars*	*Major Acids*	*Important Aroma Compounds*
Peach	sucrose	malic	benzaldehyde
	glucose	citric	benzyl alcohol
	fructose		nonanol
	sorbitol		linalool
			ethyl hexanoate
			3-methylbutanoate
			α-terpineol
			γ-hexalactone
			δ-decalactone
			γ-undecalactone
			δ-undecalactone
			γ-dodecalactone
			δ-dodecalactone
			α-pyrone
			6-pentyl-α-pyrone
Strawberry	sucrose,	citric	hexanal
	glucose,		*cis*-3-hexanal
	fructose		*trans*-2-hexanal
			furaneol
			mesifuran
			ethyl hexanoate
			ethyl butanoate
			methyl butanoate
			ethyl-2-methyl propanoate
Raspberry	sucrose	citric	H-(4-Hydroxyphenyl-butan-2-one) (raspberry ketone)
	glucose		α-ionone
	fructose		β-ionone
			geraniollinalool
			benzyl alcohol
			ethyl hexanoate
			ethyl butanoate
			methyl butanoate
			γ-decalactone
			2-heptanone
			cis-3-hexanal
			β-damascenone
Grape	glucose	tartaric	methyl anthranilate
Concord	fructose	malic	0-aminoacetophenone
(*Vitis labruscana*)			furaneol
			methyl furaneol
			β-damascenone
Muscadine	glucose	tartaric	β-phenylethanol
Vitis rotundifolia	fructose	malic	butyl alcohol
			hexyl alcohol
			hexanal
			trans-2-hexenal
			isoamyl alcohol
			acetaldehyde

Contd...

Table 56.2–*Contd...*

Fruit	*Major Sugars*	*Major Acids*	*Important Aroma Compounds*
			isobutyraldehyde ethyl acetate ethyl propionate butyl acetate propyl acetate 2-methylbutanol
Muscat varieties (*Vitis vinifera*)	glucose fructose	tartaric malic	linalool geraniol methoxyisobutylpyrazine
Banana	sucrose glucose fructose	malic citric oxalic	decan-1-ol 2-phenylethanol 3-oxy-pentanoic acid 3-methylbutanoic acid 3-methylbutyl acetate butanoate 3-methylbutanoate eugenol 5-methoxyeugenol eugenol-methylether elemicin
Sweet Orange	sucrose glucose fructose	citric	geranial neral acetaldehyde decanal octanal nonanal ethyl acetate ethyl propionate ethyl butanoate methyl butanoate ethyl-2-methyl butanoate ethyl-3-hydroxy hexanoate linalool α-terpineol limonene myrcene α-pinene valencene
Tangerine	sucrose glucose fructose	citric	acetaldehyde decanal octanal dimethyl anthranilate thymol α-sinensal γ-terpinene β-pinene
Grapefruit	sucrose glucose fructose	citric	acetaldehyde decanal ethyl acetate

Contd...

Table 56.2–*Contd...*

Fruit	*Major Sugars*	*Major Acids*	*Important Aroma Compounds*
			methyl butanoate ethyl butanoate 1-*p*-menthene-8-thiol nootkatone limonene naringin
Mango	sucrose glucose fructose	citric malic	ethyl butanoate ethyl-2-butanoate hexanal cis-3-hexanal trans-2-hexanal γ-octalactone γ-dodecalactone furaneol α-pinene β-pinene 3-carene myrcenelimonene *p*-cymene terpinolene α-Copaene caryophyllene
Melon: Cantaloupe Honeydew Watermelon	sucrose fructose	malic citric watermelon - malic only	ethylbutyrate ethyl-2-methyl butyrate ethyl butyrate ethyl hexanoate hexyl acetate 3-methylbutyl acetate benzyl acetate *cis*-6-nonenyl acetate *trans*-6-nonenol *cis*, *cis* -3,6-nonadienol *cis*-6-nonenal 4-oxononanal 2-hydroxy-5-pentyltetrahydrofuran *cis*-non-6-enyl acetate methyl acetate ethyl acetate isopropyl acetate ethyl propanoate ethyl isobutanoate propyl acetate butyl acetate methyl-2-methylbutanoate ethyl butanoate 2-methylpropanoate 2-methylbutyl acetate 2-methylpropyl acetate methyl (methylthio) acetate ethyl (methylthio) acetate ethyl (methylthio) propanoate

Contd...

Table 56.2–*Contd...*

Fruit	*Major Sugars*	*Major Acids*	*Important Aroma Compounds*
Tomato	glucose fructose	citric malic	hexanal *trans*-2-hexenal *cis*-3-hexenal *cis*-3-hexenol β-ionone β-damascenone 1-penten-3-one 3-methylbutanal 3-methylbutanol 2-isobutylthiazole1-nitro-phenylethane *trans*-2-heptenal phenylacetaldehyde 6-methyl-5-hepten-2-one methyl salicylate geranylacetone

Table 56.3: Classification of Food Flavors

Flavor Class	*Subdivision*	*Representative Example*
Fruit flavor	Citrus-type flavors (terpeny) berry-type flavors (non-terpeny)	Grapefruit, orange, apple, raspberry, banana
Vegetable flavors	-	Lettuce, celery
Spice flavors	Aromatic, lachrymo-genic, hot	Cinnamon, peppermint, onion, garlic
Beverage flavors	Unfermented flavors Fermented flavors Compounded flavors	Juices, milk, wine, beer, tea, soft drinks
Meat flavors	Mammal flavors Sea food flavors	Lean beef, fish, clams
Fat flavors	-	Olive oil, coconut fat, pork fat,butter fat
Cooked flavors	Broth vegetable and fruit	Beef bouillon, legume, potatoes, marmalade
Processed flavors	Smoky flavors, broiled, fried flavors	Ham, processed meat products, coffee, snack foods, processed cereals
Stench flavors	-	Cheese

Aroma Compounds Classified by Structure

1. Esters

Compound Name	*Fragrance*	*Natural Occurrence*	*Chemical Structure*
Methyl formate	Ethereal		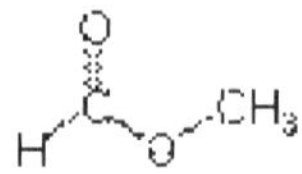
Methyl acetate	Sweet, nail polish Solvent		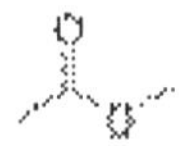
Methyl butyrate Methyl butanoate	Fruity, Apple Pineapple		
Ethyl acetate	Sweet, solvent	Wine	
Ethyl butyrate Ethyl butanoate	Fruity, Orange Pineapple		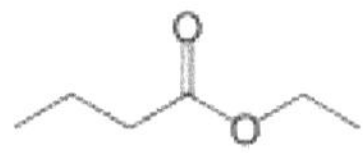
Isoamyl acetate	Fruity, Banana Pear	Banana plant	
Pentyl butyrate Pentyl butanoate	Fruity, Pear Apricot		
Pentyl pentanoate	Fruity, Apple		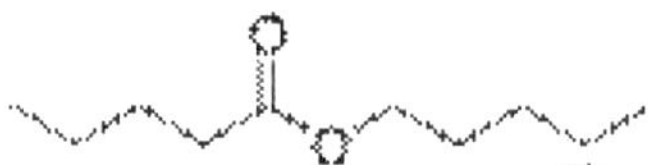
Octyl acetate	Fruity, Orange		

2. Linear Terpenes

Compound Name	*Fragrance*	*Natural Occurrence*	*Chemical Structure*
Myrcene	Woody, complex	Verbena, Bay	
Geraniol	Rose, flowery	Geranium, Lemon	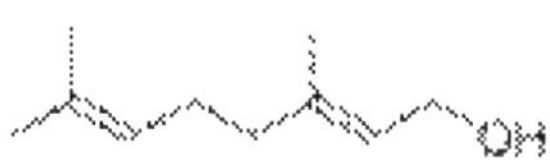
Nerol	Sweet rose, flowery	Neroli, Lemongrass	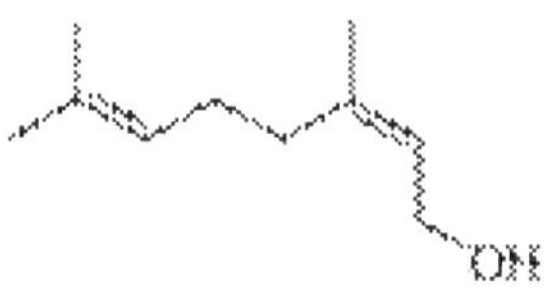

Contd...

Compound Name	*Fragrance*	*Natural Occurrence*	*Chemical Structure*
Citral, lemonal Geranial, neral	Lemon	Lemon myrtle, Lemongrass	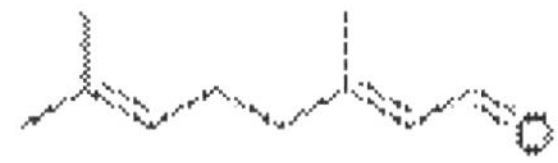
Citronellal	Lemon	Lemongrass	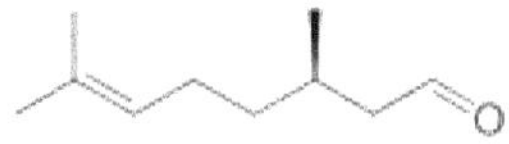
Citronellol	Lemon	Lemongrass, rose Pelargonium	
Linalool	Floral, sweet Woody, Lavender	Coriander, Sweet basil Lavender	
Nerolidol	Woody, fresh bark	Neroli, ginger Jasmine	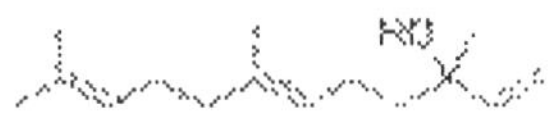

3. Cyclic Terpenes

Compound Name	*Fragrance*	*Natural Occurrence*	*Chemical Structure*
Limonene	Orange	Orange, lemon	
Camphor	Camphor	Camphor laurel	
Terpineol	Lilac	Lilac, Cajuput	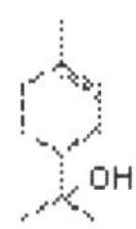
α-Ionone	Violet, woody	Violet	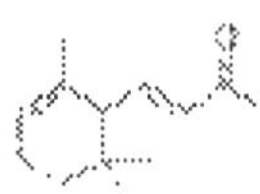 α
Thujone	Minty	Cypress, lilac Juniper	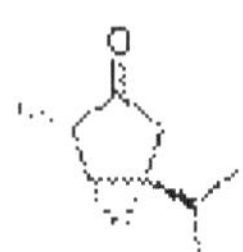

4. Aromatic

Compound Name	*Fragrance*	*Natural Occurrence*	*Chemical Structure*
Benzaldehyde	Almond	Bitter almond	
Eugenol	Clove	Clove	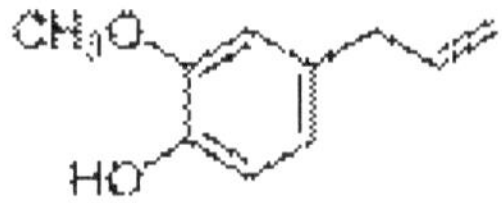
Cinnamaldehyde	Cinnamon	Cassia Cinnamon	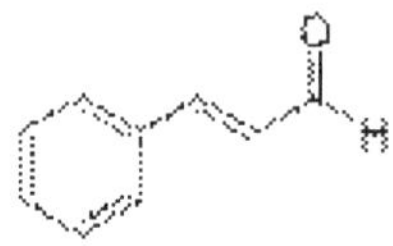
Ethyl maltol	Cooked fruit Caramelized sugar		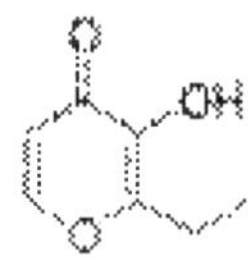
Vanillin	Vanilla	Vanilla	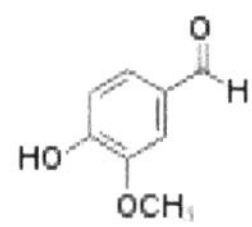
Anisole	Anise	Anise	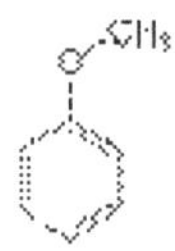
Anethole	Anise	Anise Sweet basi	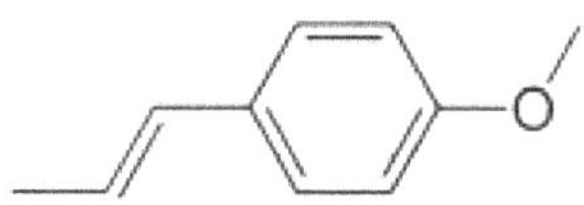
Estragole	Tarragon	Tarragon	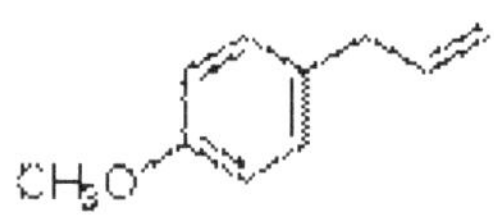
Thymol	Thyme	Thyme	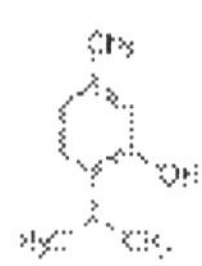

5. Amines

Compound Name	*Fragrance*	*Natural Occurrence*	*Chemical Structure*
Trimethylamine	Fishy Ammonia		
Putrescine Diaminobutane	Rotting flesh	Rotting flesh	
Cadaverine	Rotting flesh	Rotting flesh	
Pyridine	Fishy	Belladonna	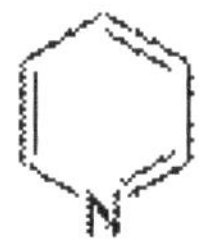
Indole	Feca lFlowery	Feces Jasmine	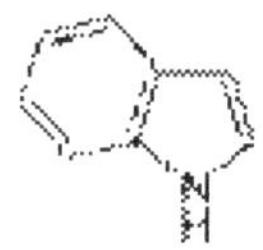
Skatole	Fecal	Feces	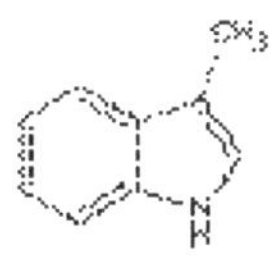

Other Aroma Compounds

6. Alcohols

- ☆ Furaneol (strawberry)
- ☆ 1-Hexanol (herbaceous, woody)
- ☆ *cis*-3-Hexen-1-ol (fresh cut grass)
- ☆ Menthol (peppermint)

7. Aldehydes

- ☆ Acetaldehyde (pungent)
- ☆ Hexanal (green, grassy)
- ☆ *cis*-3-Hexenal (green tomatoes)
- ☆ Furfural (burnt oats)
- ☆ Hexyl cinnamaldehyde

8. Esters

- ☆ Fructone (fruity, apple-like)
- ☆ Hexyl acetate (apple, floral, fruity)
- ☆ Ethyl methylphenylglycidate (strawberry)

9. Ketones

- ☆ Dihydrojasmone (fruity woody floral)
- ☆ Oct-1-en-3-one (blood, metallic, mushroom-like)
- ☆ 2-Acetyl-1-pyrroline (fresh bread, jasmine rice)
- ☆ 6-Acetyl-2,3,4,5-tetrahydropyridine (fresh bread, tortillas, popcorn)

10. Lactones

- ☆ γ-Decalactone intense peach flavor
- ☆ γ-Nonalactone coconut odor, popular in suntan lotions
- ☆ δ-Octalactone creamy note
- ☆ Jasmine lactone powerful fatty fruity peach and apricot
- ☆ Massoia lactone powerful creamy coconut
- ☆ Wine lactone sweet coconut odor
- ☆ Sotolon (maple syrup, curry, fenugreek)

11. Thiols

- ☆ Ethanethiol, commonly called Ethyl mercaptan (added to propane or other liquefied petroleum gases used as fuel gases)
- ☆ Grapefruit mercaptan (grapefruit)
- ☆ Methanethiol, commonly called Methyl mercaptan (Durian or leek)
- ☆ 2-Methyl-2-propanethiol, commonly called tertiary-butyl mercaptan is added as a blend of other components to natural gas used as fuel gas.

12. Miscellaneous Compounds

- ☆ Methylphosphine and dimethylphosphine (garlic-metallic, two of the most potent odorants known)
- ☆ Nerolin (orange flowers)
- ☆ Tetrahydrothiophene (added to natural gas)
- ☆ 2,4,6-Trichloroanisole (cork taint)
- ☆ Substituted pyrazines

Aroma Compound Receptors

Animals that are capable of smell detect aroma compounds with olfactory receptors. Olfactory receptors are cell membrane receptors on the surface of sensory neurons in the olfactory system that detects air-borne aroma compounds. In mammals, olfactory receptors are expressed on the surface of the olfactory epithelium in the nasal cavity.

Safety

'Fragrance' was voted Allergen of the Year in 2007 by the American Contact Dermatitis Society. The compositions of fragrances are usually not disclosed in the label of products, hiding the actual chemicals of the formula, which raises concerns among some consumers.

Anthocyanin

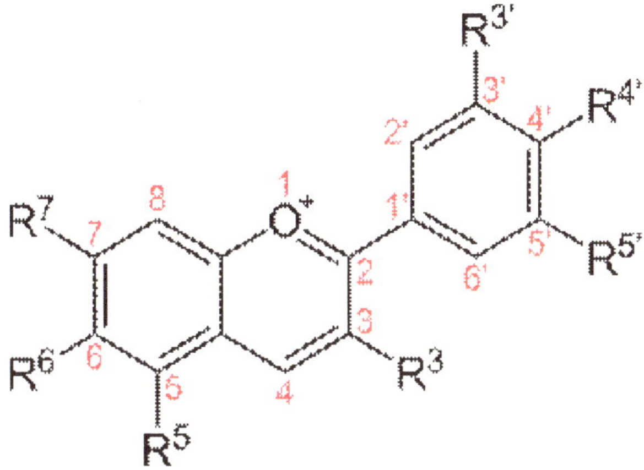

Occurrence

Anthocyanins are found in the cell vacuole, mostly in flowers and fruits but also in leaves, stems, and roots. In these parts, they are found predominantly in outer cell layers such as the epidermis and peripheral mesophyll cells.

Most frequent in nature are the glycosides of cyanidin, delphinidin, malvidin, pelargonidin, peonidin, and petunidin. Roughly 2 per cent of all hydrocarbons fixated in photosynthesis are converted into flavonoids and their derivatives such as the anthocyanins. There are no less than 10^9 tons of anthocyanins produced in nature per year. Not all land plants contain anthocyanin; in the Caryophyllales (including cactus, beets, and amaranth), they are replaced by betalains. However, anthocyanins and betalains have never been found in the same plant.

Plants rich in anthocyanins are Vaccinium species, such as blueberry, cranberry and bilberry, Rubus berries including black raspberry, red raspberry and blackberry, blackcurrant, cherry, eggplant peel, black rice, Concord grape and muscadine grape, red cabbage, and violet petals. Anthocyanins are less abundant in banana, asparagus, pea, fennel, pear, and potato, and may be totally absent in certain cultivars of green gooseberries.

The highest recorded amount appears to be specifically in the seed coat of black soybean (*Glycine max* L. Merr.) containing some 2,000 mg per 100 g and in skins and pulp of black chokeberry (*Aronia melanocarpa* L.). However, the Amazonian palmberry, açaí, having about 320 mg per 100 g of which cyanidin-3-glucoside is the most prevalent individual anthocyanin (approximately 10 mg per 100 g), is also a high-content source for which only a small fraction of total anthocyanins has been determined to date. Due to critical differences in sample origin, preparation and extraction methods determining anthocyanin content the values presented in the adjoining table are not directly comparable.

Nature, primitive agriculture, and plant breeding have produced various uncommon crops containing anthocyanins, including blue- or red-flesh potatoes and purple or red broccoli, cabbage, cauliflower, carrots, and corn. Tomatoes have been bred conventionally for high anthocyanin content by crossing wild relatives with the common tomato to transfer a gene called the anthocyanin fruit tomato ("aft") gene into a larger and more palatable fruit.

Tomatoes have also been genetically modified with transcription factors from snapdragons to produce high levels of anthocyanins in the fruits. Anthocyanins can also be found in naturally ripened olives, and are partly responsible for the red and purple colors of some olives.

Anthocyanidins: Flavylium Cation Derivatives

Selected anthocyanidins and their substitutions

Anthocyanidin	*Basic Structure*	R_1	R_2	R_3	R_4	R_5	R_6	R_7
Aurantinidin		–H	–OH	–H	–OH	–OH	–OH	–OH
Cyanidin	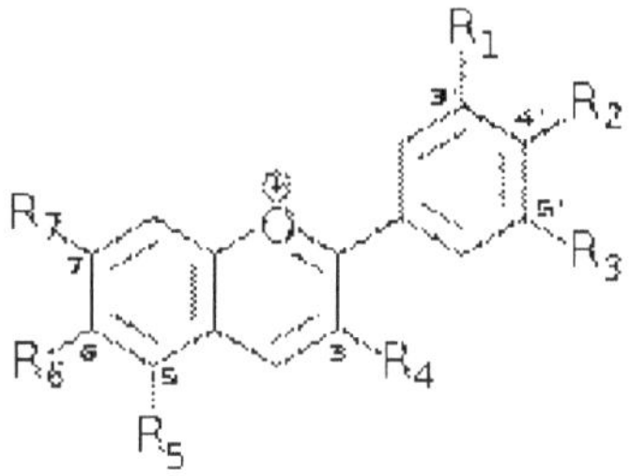	–OH	–OH	–H	–OH	–OH	–H	–OH
Delphinidin		–OH	–OH	–OH	–OH	–OH	–H	–OH
Europinidin		$–OCH_3$	–OH	–OH	–OH	$–OCH_3$	–H	–OH
Luteolinidin		–OH	–OH	–H	–H	–OH	–H	–OH
Pelargonidin		–H	–OH	–H	–OH	–OH	–H	–OH
Malvidin		$–OCH_3$	–OH	$–OCH_3$	–OH	–OH	–H	–OH
Peonidin		$–OCH_3$	–OH	–H	–OH	–OH	–H	–OH
Petunidin		–OH	–OH	$–OCH_3$	–OH	–OH	–H	–OH
Rosinidin		$–OCH_3$	–OH	–H	–OH	–OH	–H	$–OCH_3$

Anthocyanins: Glycosides of Anthocyanidins

The anthocyanins, anthocyanidins with sugar group(s), are mostly 3-glucosides of the anthocyanidins. The anthocyanins are subdivided into the sugar-free anthocyanidin aglycones and the anthocyanin glycosides. As of 2003 more than 400 anthocyanins had been reported while more recent literature (early 2006), puts the number at more than 550 different anthocyanins. The difference in chemical structure that occurs in response to changes in pH is the reason why anthocyanins are often used as pH indicator, as they change from red in acids to blue in bases.

Anthocyanins: Stability

Anthocyanins are thought to be subject to physiochemical degradation *in vivo* and *in vitro*. It has been demonstrated that B-ring hydroxylation status and pH mediate the degradation of anthocyanins to their phenolic acid and aldehyde constituents. Indeed, significant portions of ingested anthocyanins are likely to degrade to phenolic acids and aldehyde in vivo, following oral consumption. This characteristic confounds scientific isolation of specific anthocyanin mechanisms *in vivo*.

Anthocyanins and Biomedicinal Properties

Anthocyanins are members of the flavonoid group of phytochemicals, a group predominant in teas, honey, wines, fruits, vegetables, nuts, olive oil, cocoa, and cereals. The flavonoids, perhaps the most important single group of phenolics in foods, comprise a group of over 4000 C_{15} aromatic plant compounds with multiple substitution patterns. The primary players in this group include the anthocyanins (*e.g.*, cyanidin, pelargonidin, petunidin), the flavonols (quercetin, kaempferol), flavones (luteolin, apigenin), flavanones (myricetin, naringin, hesperetin, naringenin), flavan-3-ols (catechin, epicatechin, gallocatechin), and, although sometimes classified separately, the isoflavones (genistein, daidzein). Phytochemicals in this class are frequently referred to as bioflavonoids due to their multifaceted roles in human health maintenance, and anthocyanins in food are typically ingested as components of complex mixtures of flavonoid components. Daily intake is estimated from 500mg to 1g, but can be several g/d if an individual is consuming flavonoid supplements (grape seed extract, ginkgo biloba, or pycnogenol.

The colorful anthocyanins are the most recognized, visible members of the bioflavonoid phytochemicals. The free-radical scavenging and antioxidant capacities of anthocyanin pigments are the most highly publicized of the modus operandi used by these pigments to intervene with human therapeutic targets, but, in fact, research clearly suggests that other mechanisms of action are also responsible for observed health benefits. Anthocyanin isolates and anthocyanin-rich mixtures of bioflavonoids may provide protection from DNA cleavage, estrogenic activity (altering development of hormone-dependent disease symptoms), enzyme inhibition, boosting production of cytokines (thus regulating immune responses), anti-inflammatory activity, lipid peroxidation, decreasing capillary permeability and fragility, and membrane strengthening. The chemical structure (position, number, and types of substitutions) of the individual anthocyanin molecule also has a bearing on the degree to which anthocyanins exert their bioactive properties and the structure/function relationships also influence the intracellular localization of the pigments. The anthocyanin literature includes some controversy over the relative contributions of glycosylated anthocyanins versus aglycones in terms of bioavailability and bioactive potential. Originally, it was assumed that only aglycones could enter the circulation circuit, however, absorption and metabolism of anthocyanin glycosides has now been demonstrated. The nature of the sugar conjugate and the aglycone are important determinants of anthocyanin absorption and excretion in both humans and rats.

The roles of anthocyanin pigments as medicinal agents have been well-accepted dogma in folk medicine throughout the world, and, in fact, these pigments are linked to an amazingly broad-based range of health benefits. For example, anthocyanins from *Hibiscus* sp. have historically been used in remedies for liver disfunction and hypertension; and bilberry (*Vaccinium*) anthocyanins have an anecdotal history of use for vision disorders, microbial infections, diarrhea, and diverse other health disorders. But while the use of anthocyanins for therapeutic purposes has long been supported by both anecdotal and epidemiological evidence, it is only in recent years that some of the specific, measurable pharmacological properties of isolated anthocyanin pigments have been conclusively verified by rigorously controlled *in vitro, in vivo*, or clinical research trials. In many other cases, the exact roles of the anthocyanins in human health maintenance versus other phytochemicals in a complex mixture from a fruit extract or whole food have not been completely sorted out. In fact, some reports suggest that anthocyanin activity is potentiated when delivered in mixtures.

For example, visual acuity can be markedly improved through administration of anthocyanin pigments to animal and human subjects, and the role of these pigments in enhancing night vision or overall vision has been particularly well documented. Oral intake of anthocyanosides from black currants resulted in significantly improved night vision adaptation in human subjects, and similar benefits were gained after administration of anthocyanins from bilberries. Three anthocyanins from black currant stimulated regeneration of rhodopsin (a G-protein-coupled receptor localized in the retina of the eye), and formation of a regeneration intermediate was accelerated by cyanidin 3-rutinoside. These studies strongly suggest that enhancement of rhodopsin regeneration is at least one mechanism by which anthocyanins enhance visual acuity.

In both in vitro and in vivo research trials, anthocyanins have demonstrated marked ability to reduce cancer cell proliferation and to inhibit tumor formation. The capacity of anthocyanin pigments to interfere with the process of carcinogenesis seems to be linked to multiple potential mechanisms of action including inhibition of cyclooxygenase enzymes and potent antioxidant potential. Hou *et al.* revealed that anthocyanins inhibit tumorigenesis by blocking activation of a mitoses-activated protein kinase pathway. This report provided the first indication of a molecular basis for why anthocyanins demonstrate anticarcinogenic properties. In other research, fruit extracts with significant anthocyanin concentrations proved to be effective against various stages of carcinogenesis but the individual role of anthocyanins versus other components was not determined, in part because the anthocyanins were too easily degraded during bioassays if separated from stabilizing cofactors such as other phenolic constituents.

The role of anthocyanins in cardiovascular disease protection is strongly linked to oxidative stress protection. Since endothelial dysfunction is involved in initiation and development of vascular disease, four anthocyanins isolated from elderberries were incorporated into the plasma lemma and cytosol of endothelial cells to directly examine the protective roles. These tests demonstrated not only those anthocyanins could be directly incorporated into endothelial cells, but that significant oxidative stress protection was the result. Delphinidin, but not malvidin or cyanidin, provided endothelium-dependent vasorelaxation in the rat aorta, providing a pharmacological benefit comparable to red wine polyphenolics. In a rat model, little influence of feeding purified anthocyanins (cyanidin 3-*O*-glucoside) or anthocyanin-rich extracts from elderberry or blackcurrant could be detected on cholesterol levels or fatty acid patterns in liver, but the pigments were capable of sparing vitamin E. Crude anthocyanin extracts from bilberry have been administered both orally and via injection to reduce capillary permeability. In other research related to cardiovascular impairment, the roles of anthocyanin pigments versus other flavonoids delivered in the phytochemical extract have not been completely sorted out. Protection from heart attacks through administration of grape juice or wine was strongly tied to the ability of the anthocyanin-rich products to reduce inflammation and enhance capillary strength and permeability, and to inhibit platelet formation and enhance nitric oxide (NO) release. Similarly, delivery of a black currant concentrate with intense anthocyanin content caused endothelium-dependent vasorelaxation in rat aorta rings *in vitro*. The mechanism of vasorelaxation was attributed to increased levels of NO production, but the active compounds in the concentrate were not isolated. When rats were pretreated to create increased susceptibility to oxidative damage, then fed anthocyanin-rich extracts, significant reduction in indices of lipid peroxidation and DNA damage resulted. Ingestion of these extracts, which contained mixtures of delphinidin, cyanidin, petunidin, peonidin, and malvidin in the 3-glucopyranoside forms, also increased plasma antioxidant capacity.

Tsuda *et al.* recently provided evidence that anthocyanins extracted from purple corn, when provided to mice in tandem with a high-fat diet, effectively inhibited both body weight and adipose tissue increases. Typical symptoms of hyperglycemia, hyperinsulinemia, and hyperleptinemia provoked by a high-fat diet did not occur when mice also ingested isolated anthocyanins. The experiments suggest that anthocyanins, as a functional food component, can aid in the prevention of obesity and diabetes.

Anthocyanins have been credited with capacity to modulate cognitive and motor function, to enhance memory, and to have a role in preventing age-related declines in neural function. Cho *et al.* reported that administration of isolated semipurified anthocyanins from purple sweet potato enhanced cognitive performance as assessed by passive avoidance tests in ethanol-treated mice, and also effectively inhibited lipid peroxidation in rat brain tissues. By administering blueberry extracts with significant anthocyanin content (but not purified pigments), it was noted that the blueberry-supplemented diets led to effective reversal of age-related deficits in various neural and behavioral parameters (memory and motor functions). Further investigations by this laboratory team demonstrated that anthocyanins (in particular, cyanidin-3-sambubioside-5-glucoside and cyanidin-3, 5-diglucoside) were highly bioavailable in endothelial cells, which was linked to their roles in prevention of atherosclerosis and neurodegenerative disorders.

Anthocyanins exerted multiple protective effects against pleurisy in a rat model and were capable of attenuating inflammation. Anthocyanin treatment also downregulated expression of enzymes involved in inflammation in the lung. The antimicrobial activity of anthocyanins in general has been well established, including significant inhibition of aflatoxin biosynthesis. The experimental evidence demonstrating anthocyanin benefits for diabetes and pancreatic disorders has also accumulated in recent years, and again the efficacy is attributed to the multiple, simultaneous biological effects these pigments cause in the body, including prevention of generation of free radicals, decreased lipid peroxidation, reduced pancreatic swelling, and decreased blood sugar concentrations in urine and blood serum.

Conclusion

Anthocyanin pigments and associated flavonoids have demonstrated ability to protect against a myriad of human diseases, yet they have been notoriously difficult to study with regard to human health. Anthocyanins frequently interact with other phytochemicals to *potentiate* biological effects, thus contributions from individual components are difficult to decipher. The complex, multicomponent structure of compounds in a bioactive mixture and the degradation of flavonoids during harsh extraction procedures obscure the precise assignment of bioactivity to individual pigments. Extensive metabolic breakdown after ingestion complicates tracking of anthocyanins to assess absorption, bioavailability, and accumulation in various organs. Anthocyanin pigments and other flavonoids that are uniformly, predictably produced in rigorously controlled plant cell culture systems can be a great advantage for health and nutrition research because they are quickly, easily isolated, lack interferences found in whole fruits, can be elicited to provoke rapid and prolific accumulation, and are amenable to biolabeling so that metabolic fate can be investigated after ingestion.

57

Food Quality Control and Regulations

A family spends a large share of its income on food. However, the diet of many families in India is not well balanced in nutrients such as proteins, fats, carbohydrates, vitamins and minerals. All these nutrients are essential for health and need to form a regular part of the diet. Due to rising prices, many people are not in a position to buy nutritious food and that too in adequate quantity. Many times the food purchased at high cost is often found adulterated by substances which are injurious to health.

A. What is Food Adulteration?

A food is said to be adulterated if its quality is lowered or affected either by addition of substances which are injurious to health by addition of cheaper commodities resembling the food, or by removal of substances which are nutritious. Foods are adulterated by unscrupulous elements and greedy traders to become rich quickly. They take advantage of the shortage of food articles and mix fake material and sometimes material which is injurious to health. These anti-social elements have even set up factories to manufacture adulterants (materials used to adulterate) and sub-standard articles (Table 57.1). Adulteration deprives the food of its nutritional value. That is bad enough, but worse is the fact that intake of adulterated food affects health sooner or later in many ways. It may cause to name a few, blindness, paralysis or tumor. If the adulterant is contaminated or infected, it may give rise to diseases like cholera, typhoid etc. For instance, milk is adulterated with water, this lowers its quality.

But if dirty and contaminated water from a roadside pond is used for adulterating milk, it is not only diluted but becomes a serious danger to health, Butter milk or lassi can be similarly contaminated by dirty water. Many of our common food items like milk, atta, khoa, ghee, butter, edible oils, tea, coffee, dhals, turmeric powder, chilli powder, spices etc. may be adulterated with substances which cause ill effects on health.

Table 57.1: Adulterants Used for Adulteration of Food Products

Product	*Adulterant Used*
Besan and dhal from arhar and channa	Khesari dhal
Chilli powder and other spices	Saw dust
Edible oils like coconut oil, mustard oil	Mineral oil, Castor oil, Argemone oil
Sweets and aerated water	Inedible colors
Turmeric powder	Lead chromate (yellow) or Metanil yellow
Alcoholic beverages	Poisonous adulterants
Ghee or butter	Vanaspati
Milk	Water, urea, starch
Groundnut	Rock salt
Ice-cream and serbet	Metanil yellow dye and non-permitted color
Maida	Maize or tamarind seed powder
Corn flour	Tamarind seed powder
Vanilla essence	Acetic acid
Baking powder	Cereal flours
Yeast	Grain flour (white color)
Khoa	Maida
Wheat semolina	Maize semolina
Tea	Black gram hulls, and sawdust

Food quality is generally defined as the regulation by law of food manufacture, distribution and sale in order to prevent health hazards and fraud to the consumer. There are three main aspects to the application of food quality control:

1. Moral
2. Commercial, and
3. Legal

The commercial aspect requires that standard quality products are economically produced.

The legal viewpoint demands that the quality of products conforms to national and international standards.

The control of food quality by law leads to:

1. Improved quality of product.
2. Achievement of greater consumer satisfaction.
3. The promotion of quality consciousness.
4. Increased consumption and sales.

5. Employment opportunities for scientific and technical personnel.
6. Avoidance of controversy and litigation in marketing at the national and international levels.
7. Promotion of national and international trade.
8. Provision of the means for the intelligent comparison of prices in relation to quality and grade.
9. Greater confidence in the minds of consumers.

The specific responsibility of quality control is to ensure that the system used produces a standard product with acceptable quality in respect to nutritional, purity, wholesomeness and palatability.

The specific responsibilities of quality control assigned to a department or to an individual include:

1. Standardizing procedure for sampling and examining raw materials.
2. Development of test procedures.
3. Establishment and implementation of quality standards for fresh and processed products.
4. Setting up preventive quality control methods for in-plant liaison between manufacturing section and test laboratories.
5. Examination of finished products.
6. Storage controls.
7. Recording and reporting.
8. Special problems, including attendance to consumer complaints by locating their cause and eliminating them.
9. Research and development into new products and their packaging.

The sequence of operations in quality control is as follows:

1. Raw material control.
2. Process control, or the control of the manufacturing process.
3. Production inspection, including the inspection of the finished product, packaging and storage.
4. Sensory evaluation or evaluation of the acceptability of the final product.

1. Raw Material Control

1. The quality of food material is judged in terms of its nutritional value, its purity, its wholesomeness and its palatability.
2. The definition given to a "raw material" in the food industry is anything purchased by the manufacture for direct or indirect use in food processing?
3. Raw materials include: food ingredients, water and packaging materials.
4. Raw materials examinations generally include tests for genuineness and composition, freedom from contaminants and conforming to official or factory standards.

5. Examinations carried out vary with the nature and type of ingredients and its expected use.
6. Where fats and oils are included in the ingredients used, they are examined to determine their identity, purity, freshness and keeping qualities because chemical changes or rancidity taste and odour if the fat is used below the required quality.
7. The shelf-life of the end-product, signs of corrosion of equipment must be checked.
8. Approval for processing is only given after all quality specifications on the sample run have been met.

2. Process Control

1. All treatment given during the processing are standardized, ingredients are used in the correct amounts, accurate methods of preparation and mixing are employed, checks are made on the containers used to make sure that they are sound, and processing times and temperatures are standardized to make sure that the desired results are obtained.
2. There may be change in shape or structure: Cracks, cause hardening, a browning reaction, and the oxidation of unstable components due to physical, chemical or biological processes.
3. Satisfactory hygienic conditions need to be maintained.
4. For the routine bacteriological control of the plant or factory, counts on utensils, equipment, working surfaces, walls and floors are regularly carried out.
5. Periodic inspection of the plant is made by a trained inspector to make sure that adequate hygiene standards are maintained.

3. Inspection of Finished Product

1. To determine to what extent the desired quality specifications have been achieved.
2. At the end of the production line, packets are weighed individually. Manufactures are expected to supply customers with finished products of the correct weight and volume as specified.
3. Chemical analysis are carried out of the essential ingredients (to check if there are variations from the limits set.
4. To check that the composition conforms to the set legal requirements for contaminants, such as heavy metals.
5. *Physical properties*: Such as crispness, color, viscosity and texture must be checked because they can affect the palatability and acceptability of the product.
6. Microbiological examinations are carried out to check whether the finished product is safe to keep and to eat.

4. Sensory Evaluation

1. Palatability or sensory quality is of great importance to both processors and consumers.
2. Palatability attracts consumers; to the consumer, palatability satisfies his aesthetic an gustatory senses.
3. Sensory quality is a combination of different senses of perception which come into play in choosing and eating a food.

Sensory Properties

1. Color and appearance
2. Texture
3. Flavour
4. Taste

a. Expert Sensory Judgement

These tests determine the significance of variation of average scores. This small group of people works in the rigorously controlled environment of the quality control laboratory.

b. Market Testing

By selecting people and testing is very time consuming and costly so more laboratory tests are useful for this purpose.

Three major survey techniques:

1. Summarizing market data on what consumers buy.
2. Surveying consumer opinions about products of different quality.
3. Setting up experiments designed to test preferences on the spot.

5. Packaging

1. Protect the food product.
2. Preserve the flavour, aroma until it reaches to the consumer.
3. *Paper and paper board*: Paper and paper board are used in a variety of package types and forms. These include paper wrappers, sacks and labels. Others are fiberboard cases, boxes, folding cartons, paper and carrier bags etc.
4. *Metal:* Tin-plated steel cans and boxes. The cylindrical and open-top variety, general line cans aluminium boxes. Other forms in which metal is used are aluminium foil, aerosols, collapsible tubes, steel drums, boxes and crates.
5. *Plastics:* Polyethylene, polystyrene, polyvinyl chloride and polypropylene as packaging materials.
6. *Glass*: Glass is a high inert material of great cleanliness. They are, ideal for the storage of solid and liquid foods.
7. *Wood:* Boxes, crates, casks, kegs, pallets and a few other types of containers made from wood are used for food product packaging.

8. *Textiles:* Cotton bags, sacks and bales are also used in the shipping of food products.
9. *Barrier packages:* To maintain a favourable climatic environment around food products which are normally subject to deterioration from moisture, oxygen, light or heat, any of several barrier materials are employed. These include waxed paper, metal foil, plastic film or any of the flexible materials.
10. *Labeling and storage:* Labeling can reflect on the quality of a product. Clear and informative, name and address of the manufacture, list of ingredients and additives, net weight or volume, expire date.

B. Prevention of Food Adulteration Act

Food adulteration is an offence punishable under the Prevention of Food Adulteration Act, 1954. This act was amended in 1976. According to the amended act, a person found guilty of adulterating a food article with a non-injurious substance is liable to a minimum punishment of six months imprisonment. This may extend up to three years with a fine of not less than Rs. 1000. In cases where the adulterant is of an injurious nature, the punishment can rise to an imprisonment for six years with a fine of not less than Rs. 2000. If an article of food when consumed by any person is likely to cause death or harm to his/her body as would amount to grevious hurt, the imprisonment can be extended up to life with a minimum fine of Rs. 5000.

Food adulteration has become a major public health problem. This is also a social evil and can be effectively checked only if people are alert and co-operate with the Government in curbing the menace. Whenever a consumer suspects any defect in food article, either in taste or appearance, he should report the matter immediately to the local health authority or to the health officer. Besides food inspectors, even a purchaser can have an article of food analyzed by public analysts on payment of the prescribed fee. If the food is found to be adulterated, the purchaser is entitled to the refund of the fees paid.

Every state has food analysts and local health authorities. There addresses can be had either from health officer of the area or the Director of Health Services of the State or the Food and Drugs Department where such organization has been established.

Quality Control

Statutory provisions for quality control in India; Reasons for statutory provisions:

1. To maintain the quality of food produced in the country.
2. To prevent exploitation of the consumer by the sellers.
3. To safeguard the health of the consumers.
4. To establish criteria of quality of food products, since more and more foods were eaten in processed, rather than in natural forms. This has resulted in the inability of the consumer to identify the quality of the contents that could be identified easily.

1. Prevention of Food Adulteration Act 1954 and Rules 1955 (PFA Act)

Primarily intended to check adulteration of food staffs available in the country. The rules and regulations that are formulated to make available pure food materials devoid of adulterations and contaminants to the Indian population at large. Any food not conforming to these standards is said to be adulterated. The various provisions of PEA Act formulated at Central level have to be implemented at State and Local bodies' levels.

2. Fruit products order 1955 (FPO)

Regulates the manufacture, storage and sale of fruit and vegetable products. The fruit products order 1955 was issued by the Department of Food, Ministry of Food Processing Industries under the powers vested in the Government under the Essential Commodities Act to ensure the quality of fruit and vegetable products. This order controls production, distribution and quality of the fruit and vegetable products manufactured in the country as well as Registration, Licensing and Operation of manufacturing units. Under the FPO Central Government has constituted an expert committee called Central Fruit Products Advisory Committee to advise the Government on standards and policy matters pertaining to Fruit and Vegetable Preservation Industry. The various provisions of the FPO are being implemented by the Director, Fruit and Vegetables Preservation. Zonal Deputy Directors stationed at Chennai, Mumbai, New Delhi and Kolkata with technical staff. FPO mark is given to each professor after the grant of licence for manufacturing fruit or vegetable products, after the inspection of the factory for hygiene and sanitation. FPO mark and licence number is required by law to be exhibited on labels of each processed item along with other information as laid down in the FPO Rules. Continuous inspection schedule and pre shipment inspection schedule are being exercised under the FPO especially regarding export oriented Fruit and Vegetable Products. The FPO specifications cover methods of preservation, permissible colors in the preparations and also the minimum quality requirements of the final products. Fruit and vegetable products which do not conform to the FPO specifications are considered adulterated.

3. Agricultural Produce (Grading and Marketing) Act 1937 (AGMARK)

This is one of the oldest Food Laws promulgated in the country to provide for quality control of agricultural produce through grading and marketing. This Act provides compulsory standards for export and voluntary standards for vegetable oils, ghee, butter, cream, essential oil, guar, egg, groundnuts, potatoes, fruits, pulses, rice, condiments, spices etc. These standards are formulated on the physical and chemical characteristics of food, both the natural as well as those acquired during processing. This Act is administered by the Directorate of marketing and inspection with Agricultural marketing advisor to the Government of India, Faridabad, as its Chairman. This act includes all Agricultural produce and food drinks. The AGMARK seal issued by the Directorate of Marketing and inspection of the Government of India is thus a stamp of good quality. Any organization or packer who wants to use "AGMARK" symbol on the label of the sample containers can apply to the authorities. The authorities after inspection and ensuring that the necessary facilities like equipment, laboratory etc., are available allows them to use AGMARK symbol which ensures that the quality standards laid down for that product under AGMARK have been complied with. This gives a sort quality assurance to the consumers and at the same time a fair return to the farmers.

4. Sugar (control) Order 1956

Indigenous Production of raw sugar and its quality are controlled primarily under this order. The Directorate of sugar, ministry of Agriculture and Rural Development is the operating authority of this order. The quality of sugar is specified in terms of its color and size. Detailed quality standards have been specified under the provisions of PFA Act 1954 and rules thereof.

5. Vegetable Oil Products (Control) Order 1947; The Solvent Extraction Oil, Deoiled Meal and Edible Flour (Control) Order 1967; Vanaspati Control Order 1975

These orders have been promulgated under the Essential Commodities Act 1955 by the Central Government and covers commodities like vanaspati, margarine, bakery shortening, edible vegetable oils, edible flour, meal etc. The Directorate of Vanaspati, Ministry of Food and Civil Supplies is the operating authority, registration, licensing, manufacture and distribution are all controlled by this Directorate.

6. Meat Food Products (Control) Order 1975

This order covers both raw and processed meat and meat products and has been promulgated under essential commodities act 1955. This order is operated by the meat food advisory committee with agricultural marketing advisory to the Government of India as its chairman.

7. Rice Milling Industry (regulation) Act 1958 and Regulation and Licencing Rules 1976

Modernization of rice mills and safe preservation of stocks.

8. Export (Quality Control and Inspection) Act 1963 and Rules 1964

Certain notified exportable commodities are subjected to compulsory quality control and inspection before shipment. This Act is intended to provide for sound and development of export trade from India. Under this Act, the Export Inspection Council of India is exercising quality control and preshipment inspection on export oriented commodities including marine products.

9. Insecticide Act 1968

Provides for compulsory registration of all insecticides. Regulates import, manufacture, transport, storage, sale and use of all pesticides under this act.

10. Standards of Weights and Measures Act 1976

It is an act enacted by the parliament (No. 60 of 1976) to established standards of weights and measures to regulate inter-state trade or commerce in weights, measures and other goods which are sold or distributed by weight, measures or number and to provide for matters connected therewith or incidental there to. Director of legal metrology appointed under section 28 of the Act, additional, joint, Deputy or Assistant Directors and other staff appointed by the Director implements the provisions of the Act. The Central Government in consultation with the State Governments may controller of legal metrology for speedy and strict enforcement of the provisions of the Act and Rules thereof including packaged commodities.

11. State Licensing Order Governing Grain Dealers

State Governments Control grain dealers so as to protect stock against deterioration through moisture, insects, rodents etc.

12. The Consumer Protection Act 1986

This Act operated by the Ministry of Food and Civil Supplies is enacted recently where consumers through recognized consumer organizations can play the role of food samples. Rules are yet to be formulated for implementation at state levels.

Food Standardization and Regulatory Agencies in India

1. Central Committee for Food Standards

It is concerned with the prevention of Food Adulteration Act. It specifies standards for food items to check food adulteration and fraudulent practices.

Since 1941, CCFS has been functioning to advise the Central and State Government on matters arising out of the administration of the PFA Act. The PFA Act, 1954 and the PFA Rules, 1955 amended in 1976, 1980, 1981 and 1985, provide guidelines for the minimum basic requirements of food quality such as handling, storage, preparation and serving of food under sanitary conditions; freedom from extraneous matter, use of approved food additives such as preservatives, flavours, colors etc., proper packaging, branding and declaration of net weight as well as the date of manufacture and packing on the package. The guidelines are primarily intended to protect consumers from the health hazards of poisonous food. They also serve to prevent consumer exploitation by malpractices such as misbranding, adulteration, incorrect labeling, false claims, excessive and indiscriminate use of food additives etc.

There are at present four Central Food Laboratories (CFL):

1. Central Food Laboratory, Kolkata.
2. Food Research and Standardization Laboratory, Ghaziabad.
3. Public Health Laboratory, Pune.
4. Central Food Technology Research Institute (CFTRI), Mysore.

2. Central and State Food Departments

This Department come under the ministry of Food and Civil Supplies, formulates specifications for cereal food grains/pulses and selected food stuff for purchase and procurement operations.

3. State Food Laboratories/Food and Drug Administration

Generally under the department of health-there is law enforcing authorities to check food adulteration.

4. Bureau of Indian Standards

Constituted under the Act of parliament which deals with standardization of various articles including food. Agriculture Food Division Council (AFDC) deals with quality standards of various articles of food. The certification scheme under ISI is a voluntary one and any manufacturer who wants to use ISI mark on his food product may apply for the same to the authority. The authorities after inspection and other formalities permit the manufacture to use ISI Mark on their product. ISI symbol on any product ensures quality to the consumers. Recently, ISI has been renamed as Bureau of Indian Standards (BIS).

5. Food Corporation of India

Food Corporation of India has laid down specifications for several food commodities for internal purchase and procurement. To meet the essential requirements of the people in respect of rice, wheat, other cereals, pulses and gram products, the FCI procures these items in season, stores them and then distributes according to need the stored food stuffs to the public through fair price shops as well as other outlets. Different types of godowns, silos, cover and plinth are built by the FCI. Modern rice mills, maize mills, soybean processing units and solvent extraction (of oil seeds) plants are maintained by the FCI to help in food processing. The FCI has provision to make nutritious foods as fortified atta (wheat flour) and Balahar, on a very limited scale. Portable grain driers and purifies developed by the corporation serve to furnish clean and dry grains to the consumers in addition to ensuring better prices to the farmer.

6. Army supply Corps

Army supply corps has stipulated A.S.C. specifications for supply of food commodities to Army purchase organizations.

7. Central Insecticide Board

Central insecticide board under the Department of Agriculture controls the use of insecticides as applied to food crops. Under the Insecticide Act, all insecticides are required to be compulsorily registered with the registration committee established under section 5 of the Act. Registration certificates are issued after the committee is satisfied that the insecticide is effective in pest control and at the same time safe to human beings, domestic animals and wild life.

F.P.O. Specifications for Fruit and Vegetable Products

Specifications for fruit syrup, crush, squash, nectar, cordials, juice, aerated water containing fruit juice or pulp, ready-to-serve beverage, fruit juice concentrate, jam, jelly, marmalade, cheese, preserve, chutney etc.

Product (Prepared from Any Suitable Fruit and Variety)	*Specifications*	
	Minimum Per cent of Total Soluble Solids in Final Product (w/w)	*Minimum Per cent of Fruit Juice or Fruit in Final Product (w/w)*
Fruit syrup	65	25
Crush	55	25
Squash	40	25
Fruit nectar (excluding orange and pineapple nectars)	15	20
Orange and pineapple nectars	15	40
Mango nectar	15	20
Cordial	30	25
Unsweetened juice	Natural	100
Sweetened juice	10	85
Ready-to-serve fruit beverages including aerated water containing fruit juice	10	10
Barley water	30	25 (minimum of barley starch 0.25)
Fruit juice concentrate	32	100
Jam and fruit cheese	68	45
Fruit jelly and marmalade	65	45
Fruit preserve	68	55
Fruit chutney	50	40
Synthetic syrup/sharbat (prepared from herbs, flowers or essences)	65	-
Candied and crystallized or glazed fruit and peel	T.S. 70 per cent and above	R.S. 25 and above
Tomato juice	5	-
Tomato soup	7	-
Tomato puree	9	-
Tomato paste	25	-
Tomato sauce/ketchup	25	-
Sauces	15	-

Storage Life of Vegetables

Vegetable	*Temperature (ºC)*	*Relative Humidity (per cent)*	*Storage Life (Weeks)*
Asparagus	0.0	95	3-4
Beet, bunched	0.0	90	1.5
Beet, topped	0.0-1.7	90-95	8-14
Bitter gourd	0.6-1.7	85-90	4
Brinjal	10.0-11.1	92	2-3
Cabbage Early	0.0-1.7	92-95	4-6
Cabbage Late	0.0-1.7	92-95	12
Carrot topped	0.0	95	20-24
Cauliflower (snow ball)	0.0-1.7	85-95	7
Celery	0.0-0.6	92-95	8
Colocasia	11.1-12.8	85-90	21
Coriander leaves	0.0-1.7	90	5
Cucumber	10.0-11.7	92	2
Hyacinth bean pod	0.6-1.7	90	3
Garlic	0.0	65	28-36
Lettuce leaves	0.0	95	1
Ginger	7.2-10.0	75	16-24
Lima bean pod	4.4-7.2	90-95	1.5-2.0
Musk melon, Cantaloupe	1.7-3.3	85-90	1.5
Musk melon, Honeydew	7.2	85	4.5
Okra	8.9	90	2
Onion bulbs	0.0	70-75	20-24
Pea green	0.0	88-92	2-3
Pepper green	7.2	85-90	3-5
Pepper ripe	5.6-7.2	90-95	2
Potato	3.0-4.4	85	34
Pumpkin	1.7-11.6	70-75	24-36
Radish topped	0.0	88-92	3-5
Squash winter	12.8-15.6	70-75	24-36
Sweet corn	0.6-1.7	90-95	1
Sweet potato	10.0-12.8	80-90	13-20
Tapioca root	0.0-1.7	85	23
Tomato unripe	8.9-10.0	85-90	4.5
Tomato ripe	7.2	90	1
Turnip	0.0	90-95	8-16
Water melon	7.2-15.6	80-90	2
Yam	26.7	66-70	3-5

Storage Life of fruits

Sl.No.	Fruit	Temperature (°C)	Relative Humidity (Per cent)	Storage Life (Weeks)
I.	**Tropical and sub-tropical**			
1.	Banana			
	i. Green (for ripening)	12-13	80-85	1-2
	ii. Ripe	12-13	85-90	3
2.	Chiku/Sapota	3-4	85-90	6-8
3.	Citrus fruits			
a.	Mandarin			
	i. Coorg orange	6-8	85-90	10-12
	ii. Nagpur orange	5-6	85-90	10-14
b.	Sweet orange			
	i. Malta blood red	2-5	85-90	16
	ii. Malta common	5-6	85-90	16
	iii. Valencia late	5-6	85-90	20
	iv. Mosambi	7-8	85-90	16
	v. Sathgudi	7-8	85-90	16
c.	Pummelo lime	8-10	85-90	3-6
d.	Lemon	8-9	85-90	8-12
e.	Grape fruit (Marsh seedless)	8-9	85-90	12
4.	Custard apple	5-7	85-90	5-6
5.	Dates ripe	7-8	85-90	2
6.	Figs Fresh	0-2	85-90	4
	Dry	0-2	65-70	52
7.	Grapes	0-2	80-85	6-8
8.	Guava	9-10	85-90	3
9.	Jackfruit	11-13	85-90	6
10.	Litchi	0-2	85-90	10
11.	Mango	8-9	85-90	4-7
12.	Mangosteen	5-7	85-90	6-7
13.	Papaya	9-10	80-85	1-2
14.	Passion fruit	7-8	80-85	4-5
15.	Pineapple			
	Mature, green	11-13	85-90	3-4
	Ripe	8-9	85-90	4-6
16.	Pomegranate	0-2	80-85	4-6
II.	**Temperate fruits**			
17.	Apple	0-2	85-90	16-32
18.	Apricot	0-2	80-85	2
19.	Cherry	0-2	85-90	2
20.	Peach	0-2	80-85	2-4
21.	Pear	0-2	85-90	12-26
22.	Persimmon	0-2	85-90	7
23.	Plum	0-2	85-90	4-8
24.	Strawberry	0-2	80-85	2-3

Permissible Limits of Preservatives in Food Products

Sl.No.	*Food Product*	*Preservative*	*Parts per Million (ppm)*
1.	Fruit, fruit pulp or juice for conversion into jam, crystallized, glazed, cured fruit or other products		
	a. Cherries	Sulphur dioxide	3000
	b. Strawberries, raspberries	Sulphur dioxide	2000
	c. Other fruits	Sulphur dioxide	1000
2.	Fruit juice concentrates	Sulphur dioxide	1500
3.	Dried fruits		
	a. Apricots, peaches, apples, pears and other fruits	Sulphur dioxide	2000
	b. Raisins, sultanas	Sulphur dioxide	750
4.	Squashes, crushes, fruit syrups, cordials, fruit juices and barley water	Sulphur dioxide or Benzoic acid	350 or 600
5.	Jam, marmalade, preserves, canned cherry and fruit jelly	Sulphur dioxide or Benzoic acid	40 or 200
6.	Crystallized, glazed or cured fruit	Sulphur dioxide	150
7.	Fruit and fruit pulp not other-wise specified in the schedule	Sulphur dioxide	350
8.	Sweetened ready-to-serve beverage	Sulphur dioxide or Benzoic acid	70 or 120
9.	Pickles and chutneys made from fruit or vegetables	Sulphur dioxide or Benzoic acid	100 or 250
10.	Tomato and other sauces	Benzoic acid	750
11.	Dehydrated vegetables and dried ginger	Sulphur dioxide	2000
12.	Tomato puree and paste	Benzoic acid	250
13.	Syrups and sharbats	Sulphur dioxide or Benzoic acid	350 or 600
14.	Cider	Sulphur dioxide	200
15.	Wines	Sulphur dioxide	450
16.	Beer	Sulphur dioxide	70
17.	Brewed ginger beer	Benzoic acid	120
18.	Sugar, glucose, gul and Khandsari	Sulphur dioxide	70
19.	Corn flour and such like starches	Sulphur dioxide	100
20.	Corn syrup	Sulphur dioxide	450
21.	Gelatine	Sulphur dioxide	1000
22.	Coffee extract	Sulphur dioxide	450
23.	Cheese or processed cheese	Sorbic acid or its Na, K, salts or Nisin	1000
24.	Flour confectionary	Sorbic acid or its salts	1500
25.	Hard boiled sugar confectionary	Sulphur dioxide	350
26.	Sausage, cereals, condiments	Sulphur dioxide	450
27.	Cooked pickled meat including ham and bacon	Sodium or potassium nitrate	200
28.	Smoked fish (in wrappers)	Sorbic acid	Only wrappers may be impregnated with sorbic acid

Additives to be Used with Caution

Additive	*Use*	*Possible Adverse Effects*
Artificial flavourings	In soft drinks, breakfast cereals, baked goods, vegetable and fruit products, ice-cream, custards, desserts, alcoholic beverages.	Hyperactivity in some children, not adequately tested for safety.
Butylated hydroxy-anisole (BHA)	Antioxidant in cereals, chewing gum, potato chips, edible oils etc.	Appears to be safer than BHT but needs more testing.
Butylated hydroxy-toluene (BHT)	Antioxidant in cereals, chewing gum, potato chips, edible oils etc.	Cancer, allergic reactions, stored in body fat.
Caffeine	Stimulant in soft drinks	Insomnia and other adverse effects at high levels of intake. Not recommended for children and pregnant mothers.
Coal tar dyes	Colorant in vegetable and fruit products, soft drinks, candy, desserts, pastry, sausage, baked food, ice-cream, hot dogs, hamburgers, sweetmeats, snacks, confectionary, alcoholic and other beverages.	Allergic reactions, cancer and pathological lesions in vital organs.
Monosodium gluta-mate	Flavour enhancer for soup, poultry meat preservations, sauces, stews and cheese.	Damages brain cells in infants' mice, so not recommended for children, headache, tightness of head, neck and arms in sensitive adults.
Phosphoric acid and phosphates	Acidifier, chelating agent, buffer, emulsifier, nutrient, discoloration inhibitor used in baked goods, cheese, cured meat, soft drinks, dried potatoes.	Dietary imbalance that may cause bone thinking (osteoporosis) on prolonged use.
Saccharin	Non-calorie sweetener in food products also as adulterant	Bladder cancer reported in animals. Not recommended for normal people.
Sodium nitrite and nitrate	Preservative to prevent growth of bacillus. *Clostridium botulinum* and colorant for beacon, ham, meat, smoked fish, corned beef.	Formation of small amounts of cancer-producing nitrosamines.
Sulphur dioxide and bisulphites	Preservative and bleach for sliced fruit, wine, grape juice, dried potatoes, dried fruits, vegetables and fruit products.	Destroys vitamin B1, but otherwise safe at prescribed levels.
Talc and kaolin	Making dry powdery foods free flowing and as dusting agent for rice, confectionary chewing gum.	Absorbed and stored in vital organs, cancer if asbestos is present.

Food Adulteration, Food Standards and Labeling

Food is consumed to provide energy and nutrition and as such it should be wholesome and not have deleterious substances. Adulteration is a term used to designate that a product is not what it should be from biochemical, cleanliness and hygienic point of view. Consumers are rightly concerned about the safety, wholesomeness, nutritional adequacy, palatability and cost of foods, that they buy for their use. Prevention of food adulteration Act 1954 and the rules under this Act were made in 1955. The standards laid down under the PFA Act and rules are minimum standards of purity and are based on the agricultural practices followed, climatic conditions prevailing, economic conditions and nutritional status of the people in the country.

List of Adulterants Generally Used in Different Food Products

Food	*Adulterants*
Asafoetida	Resins, gum, starch
Cardamom and cloves	Extracted fruits
Chilli powder	Oil soluble dyes, added colors, brick powder, sawdust
Coffee powder	Date seed, tamarind seed powder
Coriander	Dung powder
Ghee or butter	Hydrogenated oil
Honey	Sucrose, jiggery
Milk	Water, starch, urea
Pulses	Coloring matter
Spices and condiments	Dirt, filth, sand etc.
Tea	Iron filling
Turmeric powder	Lead chromate, metanil yellow
Vegetable oil	Argemone oil, castor oil, mineral oil
Wheat	Chalk powder

Monosodium-glutamate should not be added to any food for use by the infant below 12 months. Broadly the quality enforcement under the PFA Act can be categorized under three heads: 1. Enforcement, 2. Analysis, and 3. Prosecution. Packing material should not cause any health hazard to the consumer or does not contaminate the food.

Prevention of Food Adulteration Tips to Consumer

While shopping following care should be taken:

1. Read label before purchasing
2. Purchase food articles from licenced vendors and insist on bill or cash memo.
3. Prefer foods sold in packed containers even at higher cost.

4. Prefer foods certified by Government agencies like AGMARK, ISI, certification mark and F.P.O.
5. Avoid colored foods especially sweetmeats or sharbats or ice candy.
6. Buy foods from reputed firms.
7. Do not buy cut or exposed fruits or vegetables.
8. Do not use containers or packages used for insecticide chemicals or non-edible items.

While Preparing or Serving Food Following Care Should be Taken

1. Wash hands with soap and water before the start of preparing food and after evening interruption.
2. Cover cuts in hand by bandage.
3. Cut mails short and keep them clean.
4. Cover head with hair net or band.
5. Wear cleans over cloths.
6. Keep all kitchen surfaces meticulously clean.
7. Wash food grains or vegetables, fruits, eggs, fish, meat thoroughly before cooking or eating or storing in refrigerator.
8. Avoid contact between raw foods and cooked foods.
9. Cook food thoroughly at boiling temperature.
10. Serve cooked hot food immediately.
11. Store cooked food carefully, preferably below 10 °C or above 60 °C.
12. Protect foods from flies, insects, rodents and other animals.
13. Keep the refrigerator door closed, deforest or clean refrigerator every week.
14. Do not consume stored prepared food if having off (rancid) flavour or smell or food in which froth has set in.
15. Use pure and clean water while preparing food.
16. Baby care is a difficult job and following care should be taken.
17. Mothers milk is best for baby.
18. After 4 months of age start have made weaning food.
19. Do not give left-over food to the baby.

Food Laws

The more important is that of safety; the health hazards that may arise on account of adulteration, contamination, microbiological deterioration, decays and other factors. Food is an item which could account for 50 per cent or 60 per cent of the budget of a family, and thus it is all the more important that the Government should give some protection to the hard-pressed consumer. The quality of food in India lies with organizations related to agriculture and food products, through different orders and acts like the Agriculture marketing products and grading Act 1973, prevention of food Adulteration Act 1954, vegetable oils control order 1947, Fruit products order 1955, solvent extracted de-oiled meal edible flour 1967, and so on. The meat products statutory order 1976, functioning under the directorate of marketing and inspection (Ministry of Agriculture), enables control of quality in production of raw and processed meat. There is also the export products control order.

Food Standards

A wide range of processed foods are now being manufactured in the country which include canned jams, curries, meat, fish, processed breakfast, cereals, papads, soups, noodles, potato wafers, of different varieties, protein textured foods, frozen pizzas etc.

BIS: Bureau of Indian Standards (earlier ISI).

ISO: International Organization for Standardization 1947.

ICUMSA: International Commission for Uniform Methods of Sugar Analysis.

CODEX: Codex Alimentarius Commission was set up in 1962.

Prevailing Food Laws, Acts and Implementing Agencies for Food Standards

Act Order	*Implementing Agency*	*Year of Introduction*	*Kind of Legislation*
Agricultural produce (grading x marketing) Act	Directorate of marketing inspection	1937	Voluntary: Compulsory for export only
ISI (Certification Indian market Act)	Indian standards institution (BIS)	1952	Voluntary
Prevention food adulteration Act	Ministry of health	1954	Compulsory for internal trade
Fruit products order	Department of Food, Ministry of Agriculture and Rural Reconstruction	1955	Compulsory for export/ internal trade
Export (quality control and inspection) Act	Export inspection council	1963	Compulsory for export only
Solvent extracted oil, de-oiled meal edible flour control order	Directorate of vanaspathi	1967	Compulsory for export/ internal trade
Meat products order	Directorate of marketing and inspection	1973	Compulsory for export/ internal trade

Market Standards

Economic status and quality consciousness of the consumer.

End Use Standards

Depend on the product preparation.

Health Ministry Standards

Mandatory in nature quality denotes the degree of excellence of a product.

Quality and its Down Gradation

Quality of an agricultural commodity can be divided into two categories, viz; intrinsic quality and acquired quality.

A. The *intrinsic quality* is a genetic factor grouped in to four categories.
   1. *Biochemical*: Protein, fat, fiber, vitamins, minerals, carbohydrates and enzymes as well as anti-nutritional factors.
   2. *Physical*: Color, texture, size, shape, aroma, and weight tec.
   3. *Processing*: Milling yield, cooking quality, yield of juices, pulses and other items.
   4. *Storability*: Storability of agricultural produce.

B. *Acquired quality*: Those characteristics are acquired during pre- and post-harvest conditions and practices such as discoloration, infection due to field fungi *e.g.*, ergot, smut, bunt, sprouting, admixture, contamination etc.

Quality Centers

At the international level, the WHO and FAO of United Nations are the quality centers. For ensuring safety and health of consumers, the Government of India has enacted a few Acts and detailed Rules have been framed. They are (*i*) agricultural produce (Grading and marketing) Act 1937, (*ii*) Prevention of Food Adulteration Act 1954 (PFA), (*iii*) Fruit Products Order 1955, (*iv*) Bureau of Indian Standard Act 1986 and (*v*) Export (Quality Control and Inspection) Act 1963.

Prevention of Food Adulteration Act

Broadly the control of the quality can be categorized under three heads viz: 1. Enforcement, 2. Analysis and 3. Prosecution. The Act deals with preservatives, poisonous metals, naturally occurring toxic substances, anti-oxidants, emulsifying and stabilizing agents, flavouring agents, coloring matter and other food additives, insecticides and pesticides, solvent extracted oils and edible flours, non-alcoholic beverages, starchy foods, spices and condiments and their mixtures, honey, jiggery, saccharin, coffee, tea, milk and milk products, fruit products, edible oils, cereals, backed products, sweets and confectionaries and a range of similar products.

AGMARK

The agricultural produce (Grading and Marketing) Act was enacted by the Government of India in 1937. Central Agmark Laboratory is at Nagpur in Maharashtra (21 Labs in the Country). It checks 140 commodities from agriculture produce.

The FPO covers all types of fruit products such as juices, pulp concentrates, squashes, nectar, aerated water containing fruit juices, bottled and canned fruits and vegetables, jams, jellies, fruit cheese, preserves, chutneys, soups, ketchup, paste, dehydrated vegetables and dehydrated onion. In addition to specifications for various products, limits for poisonous metals, list of permitted and harmless food colors, limits for permitted preservatives and other additives are given in the schedules.

FAQ (Fair Average Quality) Specifications for Food Grains

About 20-25 million tones of both wheat and rice procure per year for network distribution for fair price shops in the country. The procurement is done through Food Corporation of India, State Government and co-operative organizations.

Bureau of Indian Standards

Bureau of Indian Standard also prescribes quality standards for agricultural products and processed foods such as milled wheat, maize, barley, pulse products, corn flakes, macaroni, spaghetti, biscuits, bread etc. The BIS standards are quality standards and more rigid than PFA standards.

Consumer Protection Act

The consumer protection Act 1986 and the central consumer protection rules 1987 were enacted by the Government of India. Consumer councils at various levels have also been established to advise the consumers for their rights.

Export Inspection Council

Export inspection council has been set up under the export inspection act to have a check on quality of food articles exported.

Other Quality Control Legislations

Essential commodities Act of 1954 in addition to the fruit products order 1955, vegetable oils product council order 1947 and meat product control order 1973 were formulated to regulate manufacture, commerce and distribution of vegetable oils and their products and meat products respectively.

Rule: A-18.06 Food Grains

1. Foreign matter: Inorganic matter up to 1 per cent (in paddy 3 per cent); organic matter up to 3 per cent .
2. Damaged grains up to 5 per cent by weight.
3. Insect damaged grains not more than 10 per cent .
4. Uric acid content should not be more than 10 mg/100g.
5. Rodent hair and excreta should not more than 5 pieces/kg sample.

Rule: A-07.03 Honey

1. Moisture 25 per cent or less.
2. Ash 0.5 per cent
3. Sucrose 5 per cent
4. Reducing sugars 65 per cent
5. Fructose: glucose not less than 0.9 per cent

Food Labeling

Part-I: Covers general guidance on labeling.

Part-II: Covers claims and lays down.

Part-III: Covers guidelines on labeling with respect to nutritional information. To increase the sales, attractive design, name and colors etc. Mention FPO, ISI, AGMARK makes sure of standard quality.

Agro-Chemical Residues in Indian Items

To meet the needs of growing population, we require massive food production programme and this in turn is impossible without the use of pesticides. Even small quantities of these residues consumed over long periods can build up to high levels in the body fat.

Safe Levels

In order to regulate contamination of food with pesticide residues to safe levels, government has laid down principles for arriving at maximum residue limits in food commodities.

Tolerance Limit of Common Pesticides

Pesticide	*Food*	*Tolerance limit (mg/Kg)*
Aldrin, dieldrin	Fruits and vegetables	0.4
Carbaryl	Okra, leafy vegetables, potato	10.4
Chlordance	Vegetables, fruits	0.2, 0.1
Diazinon	Vegetables	0.5
Dichlorvos	Vegetables, Fruits	0.15, 0.1
Dicofol	Fruits and vegetables	5.0
Dimethoate	Fruits and vegetables	2.0
Endosulfan	Fruits and vegetables	2.0
Feniltrothion	Fruits and vegetables	0.5, 0.3
Heptachlor	Vegetables	0.05
Inorganic bromide	Fruits	30.0
Lindane	Fruits and vegetables	3.0
Parathion	Fruits and vegetables	0.5
Parathion methyl	Fruits and vegetables	0.2
Phosphamidon	Fruits and vegetables	0.2
Pyrethrine	Fruits and vegetables	1.0

Safety Periods

Every pesticide has some safety period or waiting period. Safety period is the number of days to lapse before the pesticide reaches the tolerance limit.

Effects of Processing

The extent of removal depends on the nature of the chemical, type of food, length of contact with food, and environmental conditions. No residues are found in cauliflower when harvested 9 days after spray. Washing residues 50-75 per cent residues in cauliflower and cooking reduces almost completely. Washing, soaking in 2 per cent salt solution, cooking in slightly acid medium like lime juice or tamarind extract removes more than 75 per cent residue. In green leafy vegetables it is difficult to remove the residues with one washing.

Removal of Insecticides by Common Processing Produce in Vegetables

Vegetable/Insecticides	*Washing with 2 Per cent Salt Water and Cooking for 15 min.*	*Washing and Steam Cooking*
Beans		
Malathion	60	69
Monocrotophos	42	47
Carbaryl	58	69
Chillies		
Monocrotophos	28	30
Quinalphos	22	29
Tomatoes		
Carbaryl	67	75
Monocrotophos	31	32
Quinalphos	29	30

Biological Effects

Pesticide residues causes' acute and long term toxic effects in humans, animals, fish and birds. DDT accumulates in the fat and causes more toxicity in malnourished population. Due to sensitization of body, allergic reaction can occur and in such causes, each additional exposure to matter how small will cause serious health problems. Immediate effects may be dizziness, burning eyes and skin rash. Continuous exposure for longer periods can cause liver or kidney damage or damage to nervous system; it can also cause mutations resulting in birth defects. WHO has indicated that maximum safe limit for DDT to be 0.01 mg/kg body weight/day, new born consuming breast milk from mothers exposed to pesticide sprays are consuming 100 mg DDT/day.

Points for Action

1. Random checking of market samples.
2. Educate farmers, labourers and public etc.
3. Enforcement of regulation on pesticides utilization.
4. Create facilities for analysis of food samples.
5. To see that food produced is safe for human consumption.

C. Fruit Products Order (FPO) 1955

Every manufacturer shall manufacture fruit products In conformity with the sanitary requirements and the appropriate standard of quality and composition specified in the second schedule to this order or according to the standard of quality and composition laid down by the licensing officer.

S.R.O. 1952. The powers conferred by section 3 of the (Essential Commodities Act, 1955), the Central Government makes the order.

1. a) This order is called the fruit products order, 1955,
   b) It extends to the whole of India except the state of Jammu and Kashmir
2. In this order unless the context otherwise requires
   a) The act means the Essential Commodities Act, 1955 (10 of 1955)
   b) "Committee" means the central fruit products Advisory Committee constituted by the Central Government under clause-3
   c) 'Form' means a form set forth in the first schedule
   d) "Fruit product" means any article prepared from fruit and vegetable and named according to their specification.

The First Schedule

FORM 'A'

Application for license under the Fruit Products Order, 1955

1. Name and address of the applicant
2. Address of the Factory
3. Description of the fruit products which the applicant wishes to manufacture
4. Period for which the licence is required
5. Plan of the factory and list of equipments
6. Whether any power is used in the manufacture of fruit products. If so, state the exact horse power used
7. Licence fee paid during the previous year
8. Total value of fruit products manufactured during the previous year
9. I/we hereby undertake to comply with all the provisions of the fruit products order, 1955
10. I/we have forwarded a sum of rupees in respect of the licence fee due, according to the provisions of fruit products order, 1955

Signature (s) of the applicant (s)

FORM 'B'

Government of India,
Ministry of Food Processing,

Government of India
Emblem

Licence under Fruit Products Order, 1955
Licence No. FPO ——————————

1. Name and address of licensee

2. Address of authorized premises for manufacture

3. Change of premises if any

This licence is granted under and is subject to the provisions of F.P.O., 1955 all of which must be complied with by the licensee.

Place:

Date:

Licensing Officer
to the Government of India

Validation and Renewal

Period of validity	Categories fee of fruit products authorized to manufacture	Rate of Licence fee per category	Licence paid

Signature of licensing officer

FORM 'C'

1. Name and address of licensee

2. Address of the authorized premises for the manufacture of fruit products

3. F.P.O. Licence No.

4. Statement showing quantities of fruit and vegetable products manufactured in kg. with their sale value during the term

Sl.No.	*Name of the Fruit Product*	*Size of Can or Bottle*	*Quantity in Kg*	*Sale Price per kg or per Unit of Packing*	*Value*

Quantity Exported (kg)	*Name of the Country or Part of Export*	*Rate per kg or per Unit of Packing (C.I.F./F.O.B.)*	*Value*	*Remark*

Signature of the licensee

A register in the form of the above table shall be maintained by each licensee for inspection.

D. Preservatives

Preservatives means a substance which when added to food is capable of inhibiting, retarding or arresting the process of fermentation, acidification or other decomposition of food.

Class-I Preservative

1. Common salt
2. Sugar
3. Dextrose
4. Glucose (syrup)
5. Spices
6. Vinegar or acetic acid
7. Honey
8. Edible vegetable oils

Addition of class-l preservatives in any food is not restricted.

Class-II Preservative

1. Benzoic acid and their salts
2. Sulphurous acid and their salts
3. Nitrites of sodium or potassium
4. Sorbic acid including its sodium, potassium and calcium salts
5. Nisin
6. Sodium and calcium propionate
7. Methyl or propyl parahydroxy-benzoate
8. Propionic acid and its esters or salts
9. Sodium diacetate
10. Sodium, potassium and calcium salts of lactic acid.

The use of more than one class-II preservative is prohibited by Prevention of Food Adulteration Act, 1954. The use of class-II preservatives shall be restricted to the following group of foods in concentration not exceeding the proportions given in Table 57.2.

E. Food Colors

These are the food additives added to the product to improve the sensory properties of the processed products. Food colors used are mostly the permitted coal-tar-dyes which are classified as synthetic color. The natural colors are extracted from plant products and are used in foods. In India, the Fruit Product Order (F.P.O.) and Prevention of Food Adulteration (PFA) Act have laid down stringent regulation regarding the use of artificial colors in foods. The Indian Standard Institution has examined the possibilities of certifying food colors used in India. The list of important colors used in fruit and vegetable products is given in Table 57.3.

Table 57.2: Permissible Level of SO_2 in Processed Products of Fruits and Vegetables

Product	*Preservative*	*Permissible Level (ppm)*
Fruit, fruit pulp or juice, jam, cherries, strawberries and raspberries	Sulphurdioxide	3000
Fruit juice concentrate	Sulphurdioxide	1500
Dried fruits	Sulphurdioxide	2000
Raisins and sultanas	Sulphurdioxide	750
Non-alcoholic beverages, wines, squashes, crushes, fruit syrups, cordials, fruit juices	Sulphurdioxide	350
Jam, marmalade, preserve, canned cherry and fruit jelly	40 or 200	Sulphurdioxide or Benzoic acid
Cured fruit	Sulphurdioxide	150
White sugar, dextrose and jaggery	Sulphurdioxide	70
Cider	Sulphurdioxide	200
Alcoholic wines	Sulphurdioxide	450
Pickles and chutneys	Sulphurdioxide or Benzoic acid	100 or 250
Tomato and other sauces	Benzoic acid	750
Dehydrated vegetables	Sulphurdioxide	2000
Tomato puree and paste	Benzoic acid	250
Syrups and sherbets	Sulphurdioxide or Benzoic acid	350
Fruits and vegetable flakes, powder, figs	Sulphurdioxide	600

Table 57.3: Food Colors and their Uses in Fruits and Vegetables Products

Colors	*Use in Fruit/Vegetable Product*
Tartrazine NSS	Lemon jelly, lime juice cordial
Sunset Yellow FCF	Apricot jam
Orange AG	Orange jelly, Orange squash
Ponceau RNS	Canned straw berries, Strawberry jam and jelly
Carmoisine WNS	Raspberry jelly
Caramel W	Vinegar, Sauces
Black Currant A	Black current jam and jelly
Erythrosin AS	Canned Cherries, Victoria plums
Pea Green B	Canned Fresh and processed peas
Red BR	Canned beet root

The PFA act defines the quantitative limits of these colors in the products. Recently, some colors have been banned for use in foods. Acid Magnets II and Blue VRS have been deleted from the list of permitted colors. The naturally occurring colors are freely allowed to be added to foods. According to FPO (1955), the natural colors allowed to be used in food products are, carmani, carotene and carotenoids, chlorophyll, lactoflavin, caramel, annatto, saffron (safflower) and curcumin (turmeric).

Bibliography

Anderson, R. A. (1962). A note on wet-milling of high amylose corn containing 75 per cent-amylose starch, *Cereal Chem.,* 39, 406-408.

Anderson, R. A. (1963). Wet-milling Properties of grains: Bench- Scale Study, *Cereal Sci. Today,* 8, No.6, 190-92, 195, 221.

Anon. (1964). Joint United States-Canadian tables of feed composition, *Natl. Acad. Sci.,* Nath. Res. Council Publication 1232.

Anonymous, (1982). Prevention of Food Adulteration Act, Government of India Publication, New Delhi.

Arumughan, C., Gopalkrishnan, N., Thomas, P.P., Narayanan, C. S. and Mathew, A.G. (1985). Refining and bleaching of indigenous palm oil at pilot plant scale. J. Food Sci. Technol. 22: 330-333.

Arya, P.R. and Rastogi, P.P. (1984). Fruit-Vegetable Preservation at Home (In Hindi). Directorate of Translation and Publication, G.B. Pant University of Agriculture and Technology, Pantnagar, Nainital.

Athalye (1992). Plastics in Packaging. Tata McGraw-Hill Publishing Company Limited, New Delhi.

Atkins, T. A., and Geddes, W. F. (1939). The relationship between protein content and strength of flours in glutenenriched flours, *Cereal Chem.,* 16, 223-31.

Au, P. M. and Fields, M. L. (1981). Nutritive quality of fermented sorghum. J. Food Sci. 46: 652-654.

Autrey, H. S. (1953). Effect of variables upon milling yields, *Rice J. Annual.*

Autrey, H. S., Grigorieff, W. W., Attschul, A. M., and Freeman, E. E. (1955). Effect of milling conditions on breakage of rice grains, *J. Agr. Food Chem.,* 3(7), 593.

Badi, S. M. and Hoseney, R. C. (1976). Use of sorghum and pearl millet flours in cookies. Cereal Chem. 53: 733-738.

Bailey, A. E. (1948). Cotton Seed and Cotton Products; Inter-science Publishers, Inc., New York.

Bailey, A. E. (1951). Oil and Fat Products. Inter-science Publishers, Inc., New York.

Baird, P. D., Mac Masters, M. M., and Rist, C. E. (1950). Studies on a rapid test for the viability of corn for industrial use, *Cereal Chem.;* 27, 508-13.

Banerjee, S. N. (1975). Report on the evaluation of the modern dal mill, Department of Food, Govt. of India.

Bankar, J. R., Chavan, J. K. and Kadam, S. S. (1986). Effect of grain conditioning on physico-chemical properties of sorghum bhakari. J. Maharashtra Agric. Univ., 11: 221-223.

Banwart, G.J. (1979). Basic Food Microbiology. The AVI Publishing Company, Inc., Westport, Connecticut.

Bau, H. M., Mohatadi-Nia, D. J., Mejean, L. and Debry, G. (1983). Preparation of colorless sunflower protein products: Effect of processing on physiochemical and nutritional properties. J. Am. Oil Chem. Soc. 60:1141-1148.

Bender, A. E. (1973). Composition of infant foods, In Nutritional and Dietetic Foods, International Textbook Co. Ltd., London, U. K., p. 64.

Bhat, C.M., Sharma, R.N. and Sehgal, S. (1982). A Manual on Food Preservation at Home. Directorate of Publication, Haryana Agricultural University, Hissar.

Bishop, W., and Dustin, J. (1951). Topnotch wet Corn mill house, *Food Ind.,* 23, No.3, 121-24, 211-12.

Blession, C. W., Inglett, G. E., Garcia, W. J. and Deatherage, W. L. (1974). An edible defatted germ flour from a commercial dry-milled corn fraction. Cereal Sci. Today. 19:224-255.

Board, P.W. (1988). Quality Control in Fruit and Vegetable Processing. FAO Food and Nutrition Paper 39. Food and Agriculture Organisation of the United Nations, Rome.

Borasio, L., and Gariboldi, P. (1957). Illustrated glossary of rice processing machines, Food. and Agricultural Organization of the United Nations, Rome.

Bose, T K. and M. K. Mitra, Eds. (1986). Fruits of India Tropical and Subtropical. Naya Prokash, Culcutta.

Bose, T K. and M.G. Som., Eds. (1986). Vegetable Crops in India, Naya Prokash, Culcutta.

Boundy, J. A., Woychik, J. H., Dimler, R. J., and Wall, J. S. (1967). Protein composition of dent, waxy. and high-amylose corns, *Cereal Chem.,* 44, 160-169.

Bradbury, D., Wolf, M. J., and Dimler, R. J. (1962). The hilar layer of white com, *Cereal Chem.,* 39, 72-78.

Brautlecht, C.A. (1953). *Starch, Its sources, Production and Uses,* Reinhold Publishing Corp., New York.

Brekke, O. L., and Weinecke, L. A. (1964). Corn dry-milling, A comparative evaluation of commercial degermer samples, *Cereal Chem.,* 41, 321-28.

Brekke, O. L., Weinecke, L. A., Boyd, J. N., and Griffin, E. L., Jr. (1963). Corn dry milling: Effects of first-temper moisture, Screen perforations, and rotor speed on Beall degerminator throughput and products, *Cereal Chem., 40,* 423-29.

Brekke, O.L, (1967). Corn dry milling: Pretempering low-moisture corn, *Cereal Chem.,* 44, 521-31.

Brekke, O.L. (1968). Corn dry-milling: Stress crack formation in tempering low-moisture corn, and effect on degerminator performance, *Cereal Chem.,* 45, 291-303.

Brekke, O.L. (1970). Dry milling artificially dried corn: Roller milling of degerminator stock at various moistures, *Cereal Science Today,* 15, 37-42.

Brekke, O.L. and Kwolek, W.F. (1969). Corn dry milling: Cold tempering and degermination of Corn of various initial moisture contents, *Cereal Chem.,* 46, 545-49.

Burton, W.G. (1989). Specific gravity as a guide to the content of dry matter and of starch in potato tubers. In 'The Potato'. 3rd Edn. Longman Scientific and Technical. Essex, England.

C.F.T.R.I. (1981). Home-Scale Processing and Preservation of Fruits and Vegetables. Central Food Technological Research Institute, Mysore.

Canella, M. and Bernardi, A. (1983). Changes in phenolics and oligosaccharides during sunflower seed germination. Riv. Ital. Sost. Grasse. 60: 761-762.

Chadha, K. L. and G. Kalloo, Eds. (1993). Advances in Horticulture. Vol. 6-7, Malhotra Pub. Co. New Delhi.

Chadha, K. L. and O. P. Parekh, Eds. (1993). Advances in Horticulture. Vol. 1-5, Malhotra Pub. Co. New Delhi.

Chakraverty, A. (2007). Postharvest technology of cereals, pulses and oilseeds. Oxford & IBH Publishing Co. Pvt. Ltd. New Delhi.

Chandra, P. K. (1972). Methodology and Procedures for evaluating performance of commercial rice mills, Unpublished M. Tech. Thesis, L I. T., Kharagpur.

Chavan, J. K. and Kadam S. S. (1989). Nutritional improvement of cereals by fermentation. CRC in Food Sci. and Nut. 28: 349-399.

Chavan, J. K. and Salunkhe, D. K. 1984. Structure of sorghum grain. In Nutritional and processing quality of sorghum. Pp. 21-31.

Chavan, J. K., Chavan, U. D. and Nagarkar, V. D. (1989). Effect of malting and fermentation on nutritional quality of sorghum. J. Maharashtra Agric. Univ., 14: 246-247.

Chavan, U. D. and Chavan, J. K. (1991). Utilization of malted sorghum, mung bean and black gram in preparation of bhakari. J. Maharashtra Agric. Univ., 16: 141-142.

Chavan, U.D. (1987). "Effect of Fermentation on the Nutritional Quality of Grain Sorghum (*Sorghum bicolor* L. Moench). "M. Sc. (Agric.) Thesis, Mahatma Phule Agriculture University Rahuri, Maharashtra, India. pp.i-95.

Chavan, U.D. and Chavan, J.K. (1992). Utilization of groundnut flour in preparation of sorghum bhakari. J. Maharashtra Agric. Univ., 17 (20): 346-347.

Chavan, U.D., Chavan, J.K. and Kadam, S.S. (1988). Effect of fermentation on soluble proteins and *in-vitro* protein digestibility of sorghum, green gram and sorghum + green gram blends. J. Food Sci. 53: 1574-1575.

Chevassua-Agnes, S., Favier, J. C. and Joseph, A. (1976). Traditional technology and nutritive value of sorghum beer from Cameroon (in French) Cah. Nutr. Diet. 11: 89-104.

CSIR, (1980). The Wealth of India: Raw Materials, Council of Scientific and Industrial Research, New Delhi.

Dandekar Machinery Works, Catalogue for rice and pulse milling machinery, *Mis.* G. G. Dandekar Machine Works Ltd., Bhiwandi, Thana, Maharashtra, India.

Dauthy, M.E. (1995). Fruit and Vegetable Processing. FAO Agricultural Services Bulletin 119. Food and Agriculture Organisation of the United Nations, Rome.

Desikachar, H. S. R. (1977). Processing of sorghum and millets for versatile food uses in India. In Sorghum and millets for human food. Pp. 41-45. Symp. Proc. IACC, Vienna, 1976. Tropical Products Institute, London.

Desrosier, N.W. and Desrosier, J.N. (1987). The Technology of Food Preservation. CBS Publishers and Distributors, New Delhi.

Desrosier, N.W. and Tressler, D.K. (1977). Fundamentals of Food Freezing. The AVI Publishing Company, Inc., Westport, Connecticut.

Earle, F.R., Curtis, J. J., and Hubbard, J. E. (1946). Composition of the component parts U1 the Corn kernel, *Cereal Chem.*, 23, 504-11.

Elias, D. G., and Scott, R. A. (1957). British flour milling technology, *Cereal Sci. Today,* 7, 180-84.

Ezaki, H. (1973). Paddy husker, Group training course-Fiscal, Institute of Agricultural Machinery, Japan.

Fennema, O.R. Ed. (1976). Principles of Food Science, Part-l. Food Chemistry, Marcel Dekker Inc, New York.

Finney, K. F., and Barmore, M. A. (1948). Loaf volume and protein content of hard winter and spring wheats, *Cereal Chem.*, 25, 291-311.

Frazier, W.E. and Westhoff, D.C. (1983). Food Microbiology. Tata McGraw-Hill Publishing Company Limited, New Delhi.

Garibaldi, F. (1974). Rice Milling Equipment Operation and Maintenance, Agricultural Services Bulletin No. 22, FAO, Rome.

Gauri Shanker (1984). Practical Manual in Horticulture. Kitabistan, Allahabad.

Girdhari Lal, Siddappa, G.S. and Tondon, G.L. (1986). Preservation of Fruits and Vegetables. Publications and Information Division, Indian Council of Agricultural Research, New Delhi.

Godin, V. J. and Spensley, P. C. (1971). Oils and oilseeds crop and product Digest No. 1. London: Tropical Products Institute.

Goel, A. K., Kumar, R., and Mann, S. S. (2007). Postharvest Management and Value Addition. Daya Publishing House, Delhi.

Gopalan, C, Rama Sastri, B.V. and Bala Subramanian, S.C. (1982). Nutritive Value of Indian Foods. National Institute of Nutrition, R.CM.R., Hyderabad.

Grist, D. H. (1959). Rice (3rd edn.), Longmans Green and Co. Ltd., London.

Haard, N.F. and D.K. Salunkhe, Eds. (1975). Postharvest Biology and Handling of Fruits and Vegetables, AVI, Westport, CT.

Hagenmaier, R. D. (1979). Experimental coconut protein products. J. Am. Oil Chem. Soc. 56:448-449.

Hagenmaier, R. D., Quinitio, P. H. and Clark, S. P. (1975). Coconut flour: technology and cost of manufacture. J. Am. Oil Chem. Soc. 52:439-443.

Hahn, R. R. (1969). Dry milling of grain sorghum. *Cereal Sci. Today*. 14: 234-237.

Hardenburg, R. E., A.E. Watada and C. Y. Wang, (1986). The Commercial Storage of Fruits, Vegetables and Florist and Nursery Stocks, Agricultural Handbook 66, US Dept. Agric. Washington D.C.

Harris, RS. and Karmas, E. (1975). Nutritional Evaluation of Food Processing. 2nd Edition. AVI Publishing Co. Westport, Connecticut.

Hartley, C. W. S. (1967). The oil palm, pp. 1-70; 608-692. London; Longman.

Haung, M. T, O. Toshihiko, C. T Ho and R.T. Rosen, (1994). Food Phytochemicals for Cancer Prevention, I. Fruits and Vegetables ACS Symposium Series 546, American Chemical Soc. Washington, D. C.

Hofman, V., *et al.* (1979). Sunflower Oil as a Fuel Alternative. North Dakota Ext. Bul. 13 AENG-735. 4 pp.

Hofman, V., *et al.* (1981). Sunflower for Power: North Dakota Ext. Circ. AE-735. 12 pp

Hogan, J. T. (1967). Rice bran oil and wax. FAO Agr. Eng. Informal Working Bull. 30:1.

Hogan, J.T. and Deobald, H. J. (1995). Measurement of the degree of milling of rice, *Rice* 1, 68 (10), 10.

Houston, D. F. (1973). Western Marketing and Nutrition Research Division, Agricultural Research Service, U. S. Department of Agriculture, Berkeley, California.

Hulme A.C. (1978). Biochemistry of Fruits and Their Products, Vol. I &. II. Academic Press, London.

ICAR (1970). Pulse Crops of India, I.CA.R. Publication, New Delhi.

Ilyas, S. M. (2003). Importance of postharvest processing and value- addition for additional income and employment generation. Presented at National Symposium on Resource Management for Eco-friendly Crop Production, held at CSAUA&T, Kanpur, February, 26-28.

Ilyas, S. M., Wanjari, O. D. and Goyal, R. K. (2002). Prospects of income and employment generation through agro-processing and value-addition. Presented at Conference on Emerging Opportunities of Agro and Food Processing Industries in Jharkhand held at Indian Institute of Coal Management, Ranchi during October, 8-9.

Ilyas, S.M. and Coyal, R.K. (2002). Quality assurance in food processing-impact of WTO, In: Proc. National Seminar on Development of Food Processing Industries in U.P., held at Lucknow during January 16-17.

Jacob, T. (1987). Poisons in our Food. Publications Division, Ministry of Information and Broadcasting, Government of India, New Delhi.

Jasperson, H. and Pritchard, J. L. R. (1965). Factors influencing the refining and bleaching of palm oil. Paper presented at the TPI oil palm conference. P.96. London: Ministry of overseas Development.

Javaraman, J. (1981). Laboratory Manual in Biochemistry. Wiley Eastern Limited, New Delhi.

Jones, CR. (1940). The production of mechanically damaged starch in milling as a governing factor in the diastatic activity of flour, *Cereal Chem.,* 17, 133-69.

Kader A.A. Ed.(1992). Postharvest Technology of Horticulture Crops. University of California, Publication No. 3311.

Karel, M., O.R. Fennema and D.S. Lund, Eds. (1975). Principles of Food Science Part II. Physical Principles of Food Preservation. Marcel Dekker, Inc. New York.

Karim, A. and Rooney, L. W. (1972). Characterization of pentosans in sorghum grain. J. Food Sci. 37: 369-371.

Kazanas, N. and Fields, M. L. (1981). Nutritional improvement of sorghum by fermentation. J. Food Sci. 46: 819-821.

Kent-jones, D. W., and Amos, A. J. (1957). Modern Cereal Chemistry, 5th Edn., Northern Publishing Co. Ltd., Liverpool.

Kerr, R. W. (1950). Chemistry and Industry of starch, 2nd Edition, Academic Press, New York.

Kester, E. B., and Matz, S. A. (1970). Rice Processing, In: Cereal Technology, (Ed.) S.A. Matz. AVI Pub. Co., Westport, Conn.

Kewdar, A. A. (1986). Biochemical and physiological basis for effects of controlled and modified atmospheres on fruits and vegetables. Food Technol. 40: 99-104.

Klein, K. and Crauer, L. S. (1974). Further developments in crude oil processing. J. Am. Oil Chem. Soc. 51: 382A-385A.

Koga, Y. (1969). Drying, husking and milling in Japan IV and V., *Farming,* Japan.

Kordylas, J.M. (1991). Processing and Preservation of Tropical and Subtropical Foods. Macmillan Education Ltd., Houndmills, Basingstoke, Hampshire.

Krishnamurthy, K., Girish, G. K, Rarnasivan, T., Bose, S.K., Singh, Karan and Tomar, R. P. S. (1972). A new process of removal of husk of red gram using sirka, Bulletin of Grain Technology, 10: 18l.

Kuprits, Y., ed. (1967). Technology of grain processing and provender milling, Tekhnologiya perer atbotki zernai Kombikormovoe, Izd. 'Kolos': Moscow (1 per cent 5). (Translation by Israel Program for Scientific Translation, Jerusalem, Israel)

Kurien, P. P. (1977). Grain legume milling technology. International Proceedings of FAO Expert Consultation on Grain Legume Proceeding. Rome, Italy, AGS/GLP/77/11.

Kurien, P. P. (1979). Pulses Milling in Food Industries. C.F.T.R.I., Mysore, pp. 3.1-3.20.

Kurien, P. P., and Parpia, H. A. B. (1968). Pulse milling in India, I, Processing and milling of tur and arhar *(Caianas cajan), Food Sc. and Tech.,* 5(6): 203-207.

Kurien, P.P. and Parpia, H. A. B. (1968). Pulse milling in India 1- Processing and milling of tur, arhar *(Caianus caian* Linn). CFTRI, Mysore, 11 June.

Lal, G., G.S. Siddappa and G. L. Tandon, (1986). Preservation of Fruits and Vegetables, Indian Council of Agricultural Research, New Delhi.

Lockwood, J. F. (1952). *Flour Milling,* 3rd edn, Northern Publishing Co. Ltd., Liverpool.

Lockwood, J. F. (1960). *Flour Milling,* The Northern Publishing Co., Liverpool.

Macrae, R., R. K. Robinson and M. J. Sadler, Eds, (1993). Encyclopaedia of Food Science, Food Technology and Nutrition, Academic Press, London.

Markley, K. S. (1950). Soybean and Soybean Products. Inter-science Publishers, Inc., New York.

Matanhelia, K. G. (1980). Commercial *dal* milling in India. Proceedings of the International Workshop on Pigeon pea, Vol-I. ICRISAT, Patanchery 15-19 December.

Matz, S. A. (1959). The Chemistry and Technology of Cereals as Food and Feed, Avi Publishing Co., Westport, Conn.

Mayande, V. M. (1987). Pigeonpea (*Caianus cajan* L.) milling Technology: Biochemical and Engineering Studies, unpublished M. Tech. Thesis, G. B. Pant University of Agric. and Technology, Pantnagar.

Meena, RK and Yadav, J.S. (2001). Horticulture Marketing and Postharvest Management Pointer Publishers Jaipur (Raj.).

Miller, O. H. and Burns, E. E. (1970). Starch characteristics of selected grain sorghum as related to human food. J. Food Sci. 35: 666-668.

Misra, J.B. (1983). A simple arrangement for the accurate determination of the specific gravity of potato tubers. J. *Indian Potato Assoc.* 10: 121-128.

Modi B. S. (1972). Factors affecting the performance of paddy separator, Unpublished M. Tech. Thesis, 1. 1. T. Kharagpur.

Mudambi, S.R and Rao, S. (1986). Food Science. Wiley Eastern Limited, New Delhi.

Murty, D. S. and Subramanian, V. (1982). *Sorghum roti*. I. Traditional methods of consumption and standard procedures for evaluation. ICRISAT, 1982, Proc, Intl. Symp. On Sorghum grain quality. 28-31 October, 1981, Patancheru, A.P. India. Pp. 73-78.

Murty, D. S., Patil, H. D., Prasad Rao, K. E. and House, L. R. (1982). A note on screening the Indian sorghum collection for popping quality. J. Food Sci. Technol. 19: 79-80.

Nagy, S. and P.E. Shaw, Eds. (1980). Tropical and Subtropical Fruits AVI, Westport, CT.

Narsirnha, H. V, Ramakrishnan. N., Pratape, V M., Sashikala, V B. and Narsirnhan, K. S. (2000). Status of pulse milling industries in India. Presented in Workshop on Recent Development in Pulse Milling, CFTRI, Mysore on 14 February.

Negi, J.P., Singh, B. and Dagar, KS. (2000). Indian Horticulture Database-2000. National Horticulture Board, Ministry of Agriculture, Government of India, Gurgaon (Haryana).

Nissen, M. (1967). The weight of potatoes in water. Further studies on the relation between the dry matter and starch content. *Eur. Potato* J. 10: 85-99.

Novellie, L. (1977). Beverages from sorghum and millets in "Proc. of Symp. on Sorghum and Millets for Human Food". International Association for Cereal Chemistry. Symposium. Vienna. 1976. Tropical Products Institute, London. Pp. 73-77.

Odunfa, S.A., Nordsrom, J. and Adeniran, S.A. (1994). Development of starter cultures for nutrient enrichment of ogia West African fermented cereal gruel. Report submitted to HBVC research grants program. USAID, Washington, U.S.A.

Pandey, P.H. (1997). Postharvest Technology of fruits and Vegetables (Principles and Practices). Saroj Prakashan, 646-47, Katra, Allahabad.

Pantastico, Er. B. (1975). Postharvest Physiology, Handling and Utilization of Tropical and Subtropical Fruits and Vegetables. The AVI Publishing Company, Inc., Westport, Connecticut.

Patil, U. S, Dalvi, U. S. and Chavan, J. K.(2003). Studies on the production of starch from mould infected discolored sorghum. J. Food Sci. Technol. 40: 115-117.

Paulus, K. O. (1984). Modelling in Industrial Cooking. In: *Thermal Processing and Quality of Foods.* Zeuthen *et al.* (eds.), Elsevier Applied Science Publishers, London.

Peplinski, A. J, and Pfeifer, V. F. (1970). Gelatinization of corn and sorghum grits by steam-cooking, *Cereal Science Today,* 15, 144, 149-51.

Potter, N. N. (1984). Food Science. 3rd Edition. The AVI Publishing Company, Inc., Westport, Connecticut.

Potter, N.N. (1973). Food Science. The AVI Publishing Company, Inc., Westport Connecticut.

Potter, N.N. and Hotchkiss, J.H. (1996). Food Science. 5th Edition. CBS Publishers and

Potty, VH. (1988). Horticultural Industry in India. UNIDO Consultation Meeting, Beijing, Nov. 22-24.

Powar, C.B. and Daginawala, H.F. (1986). General Microbiology Vol II Himalaya Publishing Company, Bombay.

Primo, E., Barber, S., Tortosa, E., Camacho, J., Ulldemolins, J., Jimenez, A, and Vega, R, (1970). Chemical composition of the byproducts obtained in the different steps of the rice milling process (In Spanish), Rev. Agroquim Teenol, Alimentos 10': 244.

Pryde, E. H. (1981). Vegetable Oil vs. Diesel Fuel: Chemistry and Availability of Vegetable Oils. Alcohol and Vegetable Oil as Alternative Fuels. Proceedings of Regional Workshop: USDA Peoria, IL.

Quick, G. (1980). Developments in use of Vegetable Oils as fuel for Diesel Engines. ASAE Paper 80-1525r 15 pp.

Quick, Graeme R. (1980). An In-depth Look at Farm Fuel Alternatives. Power Farming Magazine, Feb. 10-17 pp.

Raghavandra Rao, S. N. and Desikachar, H. S. R. (1964). Pearling as a method of refining jower and wheat and its effect on their chemical composition. J. Food Sci. Technol. 1: 40-42.

Raghavandra Rao, S. N., Malleshi, N. G., Sreedharamurty, S., Viraktamath, C. S. and Desikachar, H. S. R. (1979). Characteristics of roti, dosa and vermicelli from maize, sorghum and bajra. J. Food Sci. Technol. 16: 21-24.

Rai, Mangla and Mauria S. (1999). Coarse cereal: Not for the poor alone. The Hindu Survey of Indian Agriculture. pp. 55-58.

Rai, Mangla. 2002. Amazing nutrient composition. The Hindu Survey of Indian Agriculture. pp. 59-62.

Ranganna, S. (1979). Manual of Analysis of Fruit and Vegetable Products. Tata McGraw-Hill Publishing Company Limited, New Delhi.

Raymond, W. D. (1961). The palm oil industry. Trop. Sci. 3:69-89.

Reddy, V. P. R (1981). Dehusking of arhar grain (*Cajanus cajan*). Unpublished M. Tech. thesis, G. B. Pant University of Agric. and Technology, Pantnagar.

Reichert, R. D. and Young, C. G. (1976). Dehulling cereal grains and grain legumes for developing countries. I. Quantities comparison between attrition and abrasive type mills. Cereal Chem. 53: 829-839.

Ryall A. L. and W. T. Pentzer, (1982). Handling Transportation and Storage of Fruits and vegetables, Vol. 1. Vegetables and Melons, AVI, Westport.

Ryall, A.L. and W.T. Pentzer, (1982). Handling, Transportation and Storage of Fruits and vegetables, Vo1. 2. Fruits and Tree Nuts AVI, Westport CT.

Saeed, M. and Cheryan, M. (1988). Sunflower protein concentrates and isolates low in polyphenols and phytates. J. Food Sci. 53:1127-1131.

Salunkhe D. K. and S.S. Kadam, Eds. (1995). Handbook of Vegetable Science and Technology: Production, Composition, Storage and Processing, Marcel Dekker lnc., New York.

Salunkhe, D. K. and B. B. Desai, (1984). Postharvest Biotechnology of Fruits (Vol. I & II), CRC Press, Boca Raton, Florida.

Salunkhe, D. K. and B. B. Desai, (1984). Postharvest Biotechnology of Vegetables (Vol. I & II), CRC Press Boca Raton, Florida.

Salunkhe, D. K. and S.S. Kadam, Eds, (1995). Handbook of Fruit Science and Technology: Production, Composition Storage and Processing, Marcel Dekker, Inc. New York.

Salunkhe, D. K., Chavan, J. K., Adsule, R. N. and Kadam, S. S. (1992). World Oilseeds. Chemistry, Technology and Utilization. AnaviBook, Van Nostrand Reinhold, New York.

Salunkhe, D. K., Kadam, S. S. and Chavan, J. K. (1977). Nutritional quality of proteins in grain sorghum. Qual. Plant. Pl. Foods Human Nutr. 27: 187-205.

Salunkhe, D.K., S.S. Kadam and S. J. Jadhav, (1991). Potato: Production, Processing and Products, CRC Press, Soca Raton.

Salunkhe, DK, H. R. Bolin and N. R. Reddy. (1991). Storage, Processing and Nutritional Quality of Fruits and Vegetables 2nd Edn. Vol. II. Processed Fruits and Vegetables, CRC Press, Boca Raton, Florida.

Sastry, M. C. S., Subramanian, N. and Parpia, H. A. B. (1974). Effects of dehulling and heat processing on nutritional value of sesame proteins. J. Am. Oil Chem. Soc. 51: 115-118.

Satake Engineering Co. Ltd. (1973). Rice Milling Machinery, Technical Note No. 601, Extension and Training Institute, Satake Engineering Co. Ltd., Tokyo, Japan.

Satake Owner Manual, Type-03, 91973). Satake Enzz, Co., Ltd., Japan.

Saxena, R P., Singh, B. P. N.; Singh, A. K. and Singh, J. K. (1981). Effect of chemical treatment on husk removal of arhar *(Cajanus Cajan)* grain. Paper presented at the Annual Convention of the Indian Society of Agricultural Engineering (ISAE), New Delhi, India.

Schoch, T. J. (1941), Physical aspects of starch behavior, *Cereal Chem.,* 18, 121-28.

Schrimshaw, N. S. (1974). Nutritional requirements for high protein foods, Proc. Int. Symp. Protein Foods and Concentrates, Central Food Technol. Res. Institute, Mysore, India, pp. 13-18.

Scott, J. H. (1951), *Flour Milling Processes,* Second ed., Chapman and Hall, London.

Seetharam A. (1997). Coarse cereals: Waiting for incentives. The Hindu Survey of Indian Agriculture. pp. 55-59.

Seetharam A., Kadalli, G G and Halaswamy B H. (2001). Results of front line demonstrations and Technology for increasing Prod uction of finger millet and small millets in India. All India Coordinated Small Millets Improvement Project, University of Agril. Sciences, GKVK, Bangalore 560 065.

Seetharama, N. and Rao, B.D. (2004) Sustaining nutritional security. The Hindu Survey of Indian Agriculture. pp. 37-38.

Shakuntala Manay, N. and Shadaksharaswamy, M. (1987). Foods: Facts and Principles. Wiley Eastern Limited, New Delhi.

Shepherd, A. D. (1974). How grain structure influences sorghum quality, presented to 5th E. African Cereals Res. Conf. Malawi.

Shibano, M. (1973). Construction and Function of Abrasive Roll Type Rice Whitening Machine, Technical note, JRMA, Japan.

Shreve, N. R. (1956). The Chemical Process Industries, McGraw-Hill Book Co. Inc., Tokyo.

Shrivastava, H. C. (1985). Oilseeds Production Constraints and Opportunities. Oxford & IBH Pub. Co., New Delhi.

Siew, W. L. and Mohammad, Y. (1989). Effects of refining on chemical and physical properties of palm oil products. J. Am. Oil Chem. Soc. 66:1116-1119.

Singh, A. (1986). Fruit Physiology and Production. Kalyani Publishers, New Delhi.

Singh, G. (1997). Data book on Mechanization and Agro-Processing Since Independence, CIAE, Bhopal-462038.

Smith, L. (1944). *Flour Milling Technology*, 3rd edn. Northern Publishing Co., Liverpool.

Sosulski, F. W., Sabir, M. A. and Fleming, S. E. (1973). Continuous diffusion of chlorogenic acid from sunflower kernels. J. Food Sci. 38: 468-470.

Sprague, G. F. and Dudley, J. W. (1988). Corn and corn improvement, 3rd Edn., Madison, WI: American Society of Agronomy.

Srivastasva, M (1985). Essentials of Food and Nutrition Vol. II (Applied Aspects). The Bangalore Printing and Publishing Company Limited, Bangalore.

Srivastava, R.K and Singh, S. (1987). Utilization of Mushrooms for Preparation of Different Types of Preserved Products. Government Fruit Preservation and Canning Institute, U.P., Lucknow.

Srivastava, R.P. (1992). Preservation of Fruit and Vegetable Products. Bishen Singh Mahendra Pal Singh, New Connaught Place, Dehra Dun.

Srivastava, V.; Mishra, D.P.; Laxmi Chand, Gupta, RK. and Singh, B.P.N. (1988). Influence of soaking on various biochemical changes and dehusking efficiency in pigeon pea *(Cajanus caian L.,)* seeds. Journal of Food Science and Technology 25,: 267-271.

Stermer, R A, (1968). Environmental conditions and stress cracks in milled rice, *Cereal Chem.,* 45, 365.

Stivers, T. *E.,* Jr. (1955), American corn milling systems for degermed products. *Assoc. Operators Millers Bull.,* 2168-79.

Sugden, G. H. (1956). Various aspects of wheat conditioning, *Cereal Science Today,* I, 136-42.

Sukumaran, N.P. and Ramdass, C. (1980). A simple variable load potato hydrometer. *J. Indian Potato Assoc.* 7: 32-37.

Sullins, R. D. and Rooney, L. W. (1974). Microscopic evaluation of the digestibility of sorghum lines that differ in endosperm characteristics. Cereal Chem. 51: 134–142.

Sulllins, R. D., Rooney, L. W. and Rosenow, D. T. (1975). The endosperm structure of high lysine sorghum. Crop Sci. 15: 599-600.

Tauro, P., Kapoor, KK and Yadav, KS. (1986). An Introduction to Microbiology. Wiley Eastern Limited, New Delhi.

Thompson, R. A., and Roster, G. H. (1963). Stress cracks and breakage in artificially dried corn, U. S. Dept. Agr., AMS, 631.

Tressler, D.K. and Joslyn, M.A. (1971). Fruit and Vegetable Juice Processing technology. The AVI Publishing Company, Inc., Westport, Connecticut.

Tuite, J., and Foster, G. H. (1993). Effect of artificial drying on the hygroscopic properties of corn, *Cereal Chem., 40,* 630-37.

Verma, P.; Saxena, R. P.; Sarkar, B. C. and More, P. K. (1993). Enzymatic pre-treatment of pigeon pea (*Cajanus cajan*) grain and its interaction with milling. Journal of Food Science and Technology. 30: 368-370.

Verma, S.c. (1991). Potato Processing in India-An Appraisal. Central Potato Research Institute, Shimla.

Wang, Y. D. and Fields, M. L. (1987). Germination of corn and sorghum in the home to improve nutritive value. J. Food Sci. 43: 1113-1115.

Watson, H. and Helmer, J. D. (1964). Cottonseed quality as affected by the ginning process-a progress report, ARS, USDA, Publ. 42, 107.

Watson, S. A. (1970). Wet milling process and products. In Sorghum production and utilization. J. S. Wall and Ross, W. M. Eds. The AVI Publishing Co. Inc, Westport, Connecticut USA. Pp. 602-626.

Watson, S. A. (1976a). Manufacture of corn and milo starches, In: *Starch: Chemistry and Technology,* Vol. II, R. L. Whistler, and E. F. Paschall (Editors), Academic Press, New York.

Watson, S. A., and Yahl, K. R. (1976b). Comparison of wet-milling properties of opaque-2 high-lysine corn and normal corn, *Cereal Chem.,* 44, 488-98.

Watson, S. A., Hirata, Y. (1962). Some wet-milling properties of artificially dried corn, *Cereal Chem.,* 39, 35-43.

Watson, S. A., Sanders, F. W., Wakely, R. D., and Williams, C. B. (1955). Peripheral cells of the endosperms of grain sorghum and corn and their influence on starch purification, *Cereal Chem.,* 32, 165-82.

Weichmann, J. (1987). Postharvest Physiology of Vegetables. Marcel Dekker Inc. New York.

Weinecke, L. A., Brekke, O. L., and Griffin, E. L., jr. (1963). Corn dry milling: Effect of Beall degerminator tailgate configuration on nroduct streams, *Cereal Chem., 40,* 575-81.

Weiss, E. A. (1983). Oilseed crops. New York: London.

Wills, R.HH, T.H. Lee, D. Graham, W.B. Megalasson and E. G. Hall, (1981). Postharvest: An Introduction to Physiology, Handling of Fruits and Vegetables, New South Wales University Press, Kensington, Australia.

Wolf, M. J., Buzan, C. L., Mac Masters, M. M., and Rist, C. E. (1952a). Structure of the mature Corn Kernel, Gross anatomy and structural relationship, *Cereal Chem.,* 29, 321-33.

Wolf, M. J., Buzan, C. L., Mac Masters, M. M., and Rist, C. E. (1952b). Structure of the mature Corn Kernel, II, Microscopic structure of pericarp, seed coat, and hilar layer of dent corn, *Cereal Chem., 29,* 334-48.

Woodroof, J. G. (1983). Peanut: Production, Processing, Products. Westport, CT: AVI.

Woodroof, J.G. and Luh, B.S. (1975). Commercial Fruit Processing. The AVI Publishing Company, Inc., Westport, Connecticut.

Woods, A.E. and Aurand, L.W. (1977). Laboratory Manual in Food Chemistry. The AVI Publishing Company, Inc., Westport, Connecticut.

Yamaguchi, M. (1983). World Vegetables: Principles, Production and Nutritive Values, AVI Westport CT.

Appendices

APPENDIX–I
Discoveries and Definitions of Food Science and Technology

The definitions given below are mainly drawn from the book Preservation of Fruits and Vegetables by G. Lal, G.S. Siddappa and G. L. Tondon, Indian Council of Agricultural Research, New Delhi, 1986 and other Food Science and Technology books.

Discoveries in Food Science

Scientist	*Year*	*Discovery*
Greek Physician Hippocrates	300 BC	Father of medicine had many ideas about nutrition and diets
Drummond	1400 BC	Noted night blindness
Carl Wilhelm & Scheele	1742-1786	Isolated & studied properties of lactic acid, Discovered chlorine, glycerol and oxygen 3 years before priestly but unpublished
Antoine Laurent & Lavoisier	1743-1794	Principles of Modern Chemistry
Marggraf	1747	Milk sugar
Needham	1749	First time he explained the cause of spoilage of stored food
James Lind	1753	Scurvy can be cured by fresh fruits and vegetables among sailors
Spallanzani	1765	Microorganisms are responsible for spoilage of fruits and vegetables during storage
Nicolas	1767-1845	Principles of Agricultural and Food Chemistry
Lavoisier	1770-1794	Nature of respiration, oxidation of organic compounds into CO_2, H_2O and heat (energy)
Josephlouis, Gay-Lussac & Louis-Jacques, Thenard	1778-1850	Percentage of carbon, hydrogen and nitrogen in dry vegetable substances
Jons Jacob Berzelius & Thomas Thomson	1779-1848 1773-1852	Beginnings of organic formulas
William Beaumont	1785-1853	Gastric digestion

Scientist	*Year*	*Discovery*
Michel Eugene & Chevreul	1786-1889	Listed O, Cl, I, N, S, P, C, Si, H, Al, Mg, Ca, Na, K, Mn, Fe, exist in organic substances
Fr.de. Fourcroy	1789	Identified proteins and fats in plants and animal tissues
Arthur Hill Hassall	1800	Differentiated pure and adulterated food stuffs
Jean Baptiste Dumas	1800-1884	Only protein, carbohydrate and fat are inadequate for support of life
Beaccornot	1802	First hydrolyzed protein with acid
Justus Von Liebig	1803-1873	Vinegar fermentation, First book on Food Chemistry (1847), Research on the chemistry of Food. He also discovered nitrogenous and non-nitrogenous compounds in food
M. Nicholas	1804	First to report the successful preservation of food in glass containers
Prout	1806	Glucose and malt sugar & protein analysis
Sir Humphry Davy	1807-1808	Isolated K, Na, Ba, Sr, Ca and Mg
M. Nicholas	1809	Awarded the prize; Book on "The art of Preserving Animal and Vegetable Substances for Many Years"
Wollaston	1810	Discovered cystine (amino acid)
Courtois	1811	Presence of iodine in burnt sponge (for the treatment of goiter)
Chevereul	1814	Fats are composed of fatty acids and glycerol
Fastier	1824	"Hole-and-cap" cans development; patent granted in 1839
Angilbert	1833	Full aperture containers
Prout	1834	Put forward the theory that there are three nutrients in food: CHO, Fats and Proteins
Mulder	1838	The word protein was coined
Leibig	1842	No organic compounds but carbohydrates, fats and proteins oxidized
Winslow and Raymond Chevalien Appert	1843	Canned foods could be processed by steam and water under pressure
-	1852	Development of pressure cookers

Scientist	*Year*	*Discovery*
Takaki	1857	Observed addition of meat, vegetables, condensed milk and barley in the diet of sailors prevent beriberi disease
-	1857	Fruit and Vegetable processing first started in an organized manner
In Germany	1860	Established first Agricultural experiment station, Hanneberg, W. – Director & Stohmann, F. – Chemist
Papin	1861	First time cooking of foods by means of pressure
Isaac Newton	1862	Established United States Department of Agriculture (USDA), First Commissioner
Harvey Washington Wiley	1863	First Chief Chemist of the USDA
Louis Pasteur	1864	Role of microorganisms in food spoilage
Peltier and Paillard	1868	Used varnish for internal coating of cans
Shriver	1874	Autoclave provided with inlet for steam from external source
Howe	1876	Soldering of cans
Parry and Cobley	1882	Used sodium, potassium or calcium silicate and a serum made of proteinous materials
Baumann	1896	Presence of iodine in thyroid gland
Government Policy	1906	Pure Food and Drug Act (PFDA)
Hopkins, Osborne and Mendel	1906 1912	Showed zein a maize protein did not support for growth, deficient in lysine and tryptophan
Funk	1912	Coined the term Vitamin for the growth factors, Vitamin theory
McCollum & Davis	1913	Vitamin "A"
McCollum & Coworkers	1913	Water soluble Vitamin "B"
Cold berger	1915	Found that the addition of milk and eggs to poor maize diet to prevent occurrence of pellagra
-	1927	Canning of fruits and vegetables started for export

Scientist	*Year*	*Discovery*
Filby, F. A.	1934	A history of food adulteration and analysis
Cicely Williams	1935	Demonstrated Kwashiorkor disease in children due to protein deficiency and can be cured by feeding milk
Farrow and Green	1941	Classified lacquers into five groups
Government of India	1935	First fruit and vegetable processing factory was established at Bombay
Government of Uttar Pradesh	1949	Fruit Preservation and Canning Institute was established at Lucknow
Government of India	1950	Established Central Food Technological Research Institute, Mysore
Government of India	1955	Passed Fruit Products Order
Government of India	1973	For licensing, a Food and Nutrition Board was established
Darby	1977	Noted several deficiencies in human health in ancient Egypt
Van Leeuwenhock	-	Developed microscope
Harvey	-	Blood circulation in body
Rutherford	-	Discovered nitrogen
Priestly	-	Discovered oxygen
Black	-	Carbondioxide
Lavoisier	-	Showed that respiration is the life process
Von Liebig	-	Founder of the sciences of Agricultural chemistry and with Pasteur of Biochemistry
Lunin	-	Vitamins in milk
Eijkman	-	Vitamins in rice
Szent-Gyorgyi, King & Waugh	-	Isolation and identification of Vitamin "C"
Lavoisier	-	Established the law of conservation of mass. Father of modern chemistry and Nutrition
Scottish Physician	-	Lime and lemons are good source to prevent scurvy disease

Definitions/Terminologies

Terminology	*Definition/Explanation/Principles*
Abnormal behavior	Behavior that is characterized by the convergence of the four criteria of relative infrequency, social deviance, impaired social functioning and personal distress.
Absolute threshold	The minimum amount of stimulation necessary to produce a sensation.
Accommodation	The process by which schemata are revised completely or developed further and made more complex to fit new incoming information.
Acetyl value	The number of milligrams of potassium hydroxide required to neutralize the acetic acid liberated by the hydrolysis of 1 g of the acetylated fat or oil.
Achievement	Performance of children in mid-year and inspire examinations
Acid value	The number of milligrams of potassium hydroxide required to neutralize the free fatty acids in 1 g of fat or oil.
Acquisition	During conditioning, the process by which an organism learns a new response.
Active absorption	Absorption in which a carrier is used and ATP energy is expended.
Active listening	In humanistic psychotherapy, the process by which the therapist tries to grasp both the content of what the client is saying and the feeling behind it.
Active oxygen method (AOM)	It involves maintenance of the sample at 97.8 °C while air is continuously bubbled through it at a constant rate. The time required to obtain a specific peroxide value is determined.
Adaptation	The process of changing behavior to fit changing circumstances.
Additives	Chemical substance added to food during processing or packaging to improve their sensory properties, textural properties and extend their shelf life.
Adenosine triphosphate (ATP)	The main energy currency for cells. ATP energy is used to promote ion pumping, enzyme activity, and muscular contraction.
ADF (acid detergent fiber)	A fraction of crude fiber consisting mostly of lignocellulose.

Terminology	*Definition/Explanation/Principles*
Adjuvant	Substance which increase immunity response to an antigen.
Adolescence	The period that begins with the onset of puberty and ends somewhere around age 18 or 19 years.
Adsorption	The tendency of a molecule to attach itself to a finely.
Adult	An individual who is capable of assuming the responsibility of daily work and committed love.
Aerobic fermentation	The fermentation carried out in the presence of air/oxygen is known as aerobic fermentation.
Affinity chromatography	Separation of biological mixture based on highly specific biological interaction like receptor and ligand.
Affinity chromatography	Separation of biological mixture based on highly specific biological interaction like receptor and ligand.
Aflatoxin	Toxin produced in groundnut due to mold, *Aspergillus flvus*
Aftertaste	The experience that, under certain conditions, follows removal of the taste stimulus; it may be continuous with the primary experience or may follow as a different after a period during which swallowing, saliva, dilution, and other influences may have affected the stimulus substance. The result of the persistence of flavor note, particularly after swallowing.
Ageusia	Lack or impairment of sensitivity to taste stimuli.
Agnosia	Inability to recognize sensations; may be primarily in one sense, *e.g.* olfactory agnosia.
Agricultural marketing	Human activity directed at satisfying the needs and wants through exchange process. It can be also depended as comprising of all activities involved in supply of farm inputs to the farmers and movement of agricultural product from the farms to the consumers.
Alcohol yield	With corn containing 60 per cent starch, distillers traditionally obtain 19-19.7 L (5.0-5.2 proof gallons)/0.03 m (bushel). Theoretical yields as liters of absolute alcohol/100 kg of starch are, stoichiometric 72.0 L (100 per cent), maximum level 68.4 L (95 per cent), industrial standards 62-65 L (86-90 per cent). For the maximum level, 5 per cent or more of carbon substrate is consumed by yeast growth and by-product formation.
Alcoholism	Alcohol abuse characterized by psychological dependence, physical dependence and impaired social functioning.

Terminology	*Definition/Explanation/Principles*
Alkali refining or neutralization	A process in which the oil is heated with a calculated amount of sodium/potassium hydroxide to saponify and remove the free fatty acids from the crude oil.
Alpha (α) helix	A spiral shape constituting one form of the secondary structure of proteins, arising from a specific hydrogen-bonding structure.
Alveolus	Smallest drogans involved in biosynthesis of milk, having lumen and provided with myoepithelial secretary cells.
Amino acid	An organic molecule possessing both carboxyl and amino groups. Amino acids serve as the monomers of proteins.
Amnesia	A psychological disorder in which memories for certain events are unavailable to recall.
Amylase	Starch digesting enzymes from the salivary glands or pancreas.
Amylopectin	A digestible branched-chain polysaccharide made of glucose units.
Amylose	A digestible straight-chain polysaccharide made of glucose units.
Analgesic	Any agent that produces insensitivity to pain without loss of consciousness.
Anemia	A condition in which the haemoglobin content of the blood is lower than normal
Anesthesia	Loss of sensation with or without loss of consciousness.
Angle of Nip	Angle formed by the tangents to the roll faces at the point of contact between particle and the rolls.
Angle of repose	When a grain mass is arranged in a heap, the angle between its horizontal base and the slope of cone formed is called angle of repose.
Animal fats	It mostly contains large amounts of C_{16} and C_{18} fatty acids (oleic and linoleic acids).
Anisidine value	A measure of oxidation of fats/oils beyond peroxide stage. With aldehydes, anisidine forms compounds (Schiff's base), which show a strong absorbance at 350 nm. In the presence of acetic acid, p-anisidine reacts with aldehydes producing a yellowish color. The molar absorbance at 350 nm increases if the aldehyde contains a double bond conjugated to the carbonyl double bond. Thus the anisidine value is mainly a measure of 2-akenals. An expression termed the Totox or oxidation value (OV), which is equivalent to 2X peroxide value + anisidine value, has been suggested for the assessment of oxidation in oils.

Terminology	*Definition/Explanation/Principles*
Annealing	It is the process in which glass is heated up to 1000 F for 15 minutes and slowly cooled to impart stress and crack resistance is called annealing. Term annealing generally refers to removal of stress annealing temperature or point being defined as temperature at which stresses in glass are relived in few minutes.
Anorexia	It is the term used to describe the lack of appetite for food
Anorexia nervosa	An eating disorder involving a psychological loss of appetite and self-starvation, resulting in part from a distorted body image and various social pressures associated with puberty.
Anosmia	Inability to smell, either totally or a particular substance or group of substances.
Antagonist	It is a chemical substance that counteracts the effects of another chemical substance
Anterograde amnesia	A memory disorder in which humans lose the ability to store information.
Antetaste	A prior taste, or foretaste, usually of short duration, preceding the main taste or flavor characteristic.
Antherosclerosis	It is a disease in which the walls of the arteries become narrowed as a result of the deposition of lipid containing material
Anthropometry	Measurement of physical and gross composition
Antibiotic	A chemical agents produced by one organism that inhibits or harmful to other organism.
Anticaking agents	Chemical substance used to avoid formation of cakes during storage in powdered food.
Antifoaming agents	An agent that reduces foaming often caused by the presence of dissolved proteins and other stabilizers.
Antimicrobial Agent	Substance used to preserve food by preventing growth of microorganisms and subsequent spoilage, including fungistats, mold and rope inhibitors. Also includes antimicrobial agents, antimyotic agents, pre-servatives and mold preventing agents (indirect additives).
Antioxidants	Substances that retard the oxidative rancidity and thereby improve the shelf life of fats *e.g.* propyl gallate (PG), octyl gallate (OG), butylated hydroxyanisole (BHA), butylated hydroxy toluene (BHT), tertiary butyl hydroquinone (TBHQ) and tocopherol.

Terminology	*Definition/Explanation/Principles*
Antioxidants	The substances, which are responsible for the control of free radical-mediated lipid oxidation, are known as antioxidants.
	The substances which can prevent the occurrence of oxidation process in food and avoid rancidity as well as free radical formation are known as antioxidants.
Anxiety hierarchy	A rank ordering of anxiety-provoking situations used in systematic desensitization.
AOM (active oxygen method) time	The time required to obtain a predetermined peroxide value washed air is bubbled through a fat sample held at 97.8 °C.
Aphagia	It means the loss of the power of swallowing
	Impairment in the ability to speak or to understand spoken language.
Aroma	The fragrance or odor of food, perceived by the nose by sniffing. In wines, the aroma refers to odors derived from the variety of grape, *e.g.,* Muscat aroma. It is the overall odor impression as perceived by the nasal cavity.
Arousal	An internal state of excitement or tension. Drive theorists believe that a state of arousal demands the reduction or elimination of tension.
Aseptic packaging	The technique in which food is sterilize the package and fill in the previously sterilized container and packed in it under sterile environment is known as aseptic packaging.
Aspartame	A sweetener made of two amino acids and methanol; it is 200 times sweeter than sucrose.
Aspiration	A process of cleaning grains by using a large volume of air through grain layer to separate lighter particles.
	It is the process of separation of husk or hull by use of air currents.
Assignment problem	Scheduling jobs to person at the least cost so that one person gets one and only one job is called as assignment problem.
Astringency	Astringency is a taste-related phenomenon, perceived as a dry feeling in the mouth along with a coarse puckering of the oral tissue. Astringency usually involves the association of tannins or polyphenols with proteins in the saliva to form precipitates or aggregates.

Terminology	*Definition/Explanation/Principles*
Attention	The process that determines which sensations will be perceived; *i.e.*, which information will be transferred from sensory to short-term memory.
Auditory nerve	Axons from neurons within the cochlea that conduct neural impulses to the auditory areas of the brain.
Autosmia	Disorder of the sense of smell in which odors are perceived when none are present.
Autotrophic nutrition	A mode of obtaining organic food molecules without eating other organisms. Autotrophus use energy from the sun or from the oxidation of inorganic substance to make organic molecules from inorganic ones.
Autoxidation	The breakdown of fatty acids into peroxides and hydroperoxides upon addition of molecular oxygen across the double bonds during the storage of oil.
Auxotrophs	Bacterial mutant that require one or more growth factors that wild type or phototropic can synthesis.
Axon	The portion of a neuron that transmits neural impulses away from the soma toward the synapse.
Backset	Backset is the screened aqueous by-product from distillation. It is recycled and added to the cooked grain mash prior to fermentation.
Balling	Balling is a measure of the sugar concentration in a grain mash, expressed in degrees. It approximates percent by weight of the sugar in solution.
Basal metabolic rate (BMR)	The minimal number of kilocalories a resting animal requires to fuel itself for a given time.
Beer	Beer is the alcoholic product arising from the yeast fermentation of saccharified grain mash. It may or may not include stillage from a previous fermentation/distillation.
Beer Lambert's law	It states that the absorbance of a solution is directly proportional to the solutions concentration.
Behavior	Any activity that can be observed recorded and measured.
Beriberi	The thiamin deficiency disorder characterized by muscle weakness, loss of appetite, nerve degeneration and sometimes edema.
Beta (β) pleated sheet	A zigzag shape constituting one form of the secondary structure of proteins formed of hydrogen bonds between polypeptide segments running in opposite directions.

Terminology	*Definition/Explanation/Principles*
Beverage	It is a fruit juice considerably altered in composition before consumption. It may be diluted before it is served.
Biennial	These are defined as plant which requires two years or at least two growing seasons.
Bio functional membrane	Used in principle of bio sensors for molecules recognition in the process.
Bioavailability	The degree to which the amount of an ingested nutrient actually gets absorbed and so is available to the body.
Biochemical engineering	Biochemical engineering is concerned with conducting biological processes on an industrial scale, providing the link between biology and chemical engineering.
Biological value	Biological value of protein is the body's ability to retain protein absorbed from a food.
Bitter pit	Small brown dry area disfigures the flesh location below the skin.
Bitter taste	Some have speculated that bitter taste serves as a detergent from poisonous foods; however, we enjoy many bitter foods, *e.g.*, coffee. Many different types of molecules produce a bitter taste including divalent cations, alkaloids, and some amino acids. With >30 receptor systems for bitterness, it is the least discriminating of the taste modalities. Bitterness could arguably be broken down into several additional taste classifications. Due to the broad range of chemical structures, multiple analytical approaches may be necessary to analyzing bitter components, through HPLC is often applicable.
Blanching	Blanching is an important heating then holding and cooling process to inactivate enzymes.
Bland	Having no distinctive taste or odor property.
Bleaching	A process for the removal of coloring matter from the oil.
BV (biological value)	The percent of the absorbed nitrogen actually retained in the body.
Blind spot	A portion of the retina through which the optic nerve exits from the eye and travels to the brain. This area has no rods or cones and is not responsive to light.
Blood	Blood is the fluid which circulates through the body bringing nourishment too and removing waste products from the cells
Blood brain barrier	A specialized capillary arrangement in the brain that restricts the passage of most substances into the brain, thereby preventing dramatic fluctuations in the brain's environment.

Terminology	*Definition/Explanation/Principles*
Blood doping	A technique by which an athlete's red blood cell count is increased.
Blood pressure	The hydrostatic force that blood exerts against the wall of a vessel.
Body mass index	Weight (kg) divided by height squared (meter). A value of 30 or greater shows obesity related health risks.
Boiler Water Additive	Substance used in a steam or boiler water system as an anticorrosion agent, to prevent scale or to effect steam purity.
Boiling point	A liquid boils when its vapour pressure is equal to the external pressure. The boiling point is thus constant for any external pressure. The normal boiling point refers to an external pressure, which is equal to the atmospheric pressure (760 mm Hg), which for water is 100 °C.
Bomb calorimeter	An instrument used to determine the Calories content of a food.
Bond	A sharing of electrons, charges or attractions. This links two atoms.
Bonded whiskey	Bonded whiskey is whiskey stored at least four years in wooden containers where the spirits have been in contact with the wood surface. It is unaltered from the original character by the addition or subtraction of any substance other than by filtration or chill proofing, is reduced in proof by the addition of water to 100 proof (5 vol. per cent) and bottled at 100 proof, and is produced at the same distillery in the same season (January through June or July through December).
Bound water	Water that is attached to organic substances. This is not available for microbial use. In certain foods some of the water may be bound very closely and very different to separate. Often it is extremely different to separate the water without decomposing other molecules present in the food sample.
Brainstem	The bottommost protein of the brain, an enlarged extension of the spinal cord consisting of the medulla, pons, midbrain and thalamus.
Bran	Corresponding fraction from the last break of grind in milling of wheat or any cereal grains or the brownish covering of cereal grains below the husk, that consists of pericarp, testa and aleurone layer is called bran..
Bread staling	Change in crust and crumb of bread.

Terminology	*Definition/Explanation/Principles*
Break-even point	The point at which the level of sales just equal to total expenses.
Browning	Certain enzymatic and non-enzymatic changes occurring in food mostly fruits and vegetable due to oxidation reaction cause change in color of the food product is known as browning.

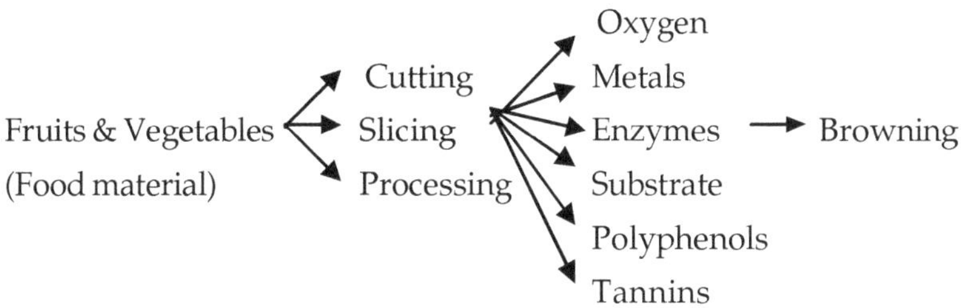

Browning of foods is due to oxidative or non-oxidative reactions. Oxidative or enzymic browning is a reaction between oxygen and a phenolic substrate catalyzed by polyphenol oxidase. Nonoxidative or nonenzymic browning involves the phenomena of cartelization and or the interaction of proteins or amines with carbohydrates.

Browning reaction contributes to the aroma, flavour, and color of many foods such as ready-to-serve cereals, toffees, roasted coffee, malted barley and baked goods.

Bubble gum	It is confection which can produce bubbles in mouth. It is prepared by using gum, resins and plasticizer with sugar and glucose syrup.
Buffer	Compounds that can take up or release hydrogen ions to maintain a certain pH value in a solution.
Bulk density	Mass of particles occupied by unit volume of bed.
Butter	It is a fat or cream separated from other milk constituents.
Cacogeusia	Persistent or intermittent unpleasant taste in the mouth.
Cacosmia	Perception of persistent or intermittent unpleasant odor.
Cake	A product obtained by baking a leavened and shortened batter containing flour, sugar, egg, milk, flavor and leavening agents.
Calories	A measurement of energy provided by foods or The amount of heat energy required to raise the temperature of 1 g of water 1 °C; the amount of heat energy that 1 g of water releases when it cools by 1 °C. The Calorie (with a capital C), usually used to indicate the energy content of food, is a kilocalorie.

Terminology	*Definition/Explanation/Principles*
Calorific value	The amount of energy (kcal) produced when one gram of substance is completely oxidized.
Cancer	A condition characterized by uncontrolled growth of abnormal body cells.
Canning of peas	Heat processing in hermetically sealed container.
Caramel	Sugar heated beyond their melting point decompose and form a brown mass known as caramel.
Caramelization	Direct heating of carbohydrates, particularly sugars and sugar syrups, produces a complex group of reactions termed Caramelization.
Carbohydrate	A compound containing carbon, hydrogen and oxygen atoms; most are known as sugars, starches and dietary fibers.
Carbon dioxide injury	In CA storages were sometimes CO_2 concentration goes higher then, the periphery of the internal tissue turns slightly brown initially then deep brown.
Cariogenic	A substance often carbohydrate-rich that promotes dental caries.
Carotenes	Pigment substances in plants that can often form vitamin A. β-carotene are the most active form.
Casein	Proteins in milk that form hard curds. These are difficult for infants to digest. It is principle milk proteins, having heterogeneous mixture of various proteins.
Catabolism	Breaking down compounds.
Catalyst	A compound that speeds reaction rates but is not altered by the reaction.
	A compound that speeds reaction rates but is not altered by the reaction.
Cavitations	If the pressure in the suction line is less than the vapour pressure, some of the liquid flashes into vapour or if the liquid contains gases, they may come out of the solution resulting into gas pockets. This phenomenon is known as cavitations.
Celiac disease	Also known as gluten-induced enteropathy. It is caused by an allergy to protein found in wheat, rye, oats and barley. If untreated, it causes a severe flattening of the villi in the intestine, leading to severe malabsorption of nutrients.
Cellulose	A straight-chain polysaccharide of glucose molecules that is undigestible because of the presence of beta carbohydrate bonds.
Cereals	Seeds of grass belonging to graminiae family

Terminology	Definition/Explanation/Principles
CGRE	Combine glass reference electrode used to measure the difference in pH in bioreactor.
Chalky grains	The grains which appear milky/white/opaque called chalky grains.
Cheilosis	Cheilosis is the condition characterized by lesions of the lips and at the angles of the mouth, which is caused by a deficiency of the vitamin riboflavin
Chemesthesis	Chemesthesis in the mouth is the chemical irritation (e.g., pain, heat, cooling) due to stimulation of the trigeminal nerve. Chemesthesis also occurs in other parts of the body including the eyes, nose, and throat. Some examples of chemesthesis are burning from jalapenos, cooling from mint, and pain from carbonation.
Chemical score	The percentage of the ratio of the most limiting amino acid in the test protein to that in the standard (whole egg) protein.
Chilling injury	Commodities (mainly tropical and subtropical) held at temperature above their freezing point and below 5 to 15 °C depending on the commodity.
	Exposure of tissues of commodities to temperature below critical level of storage temperature leads to imbalance of metabolism and causes injury to tissue.
Cholecystokinin (CCK)	A hormone that stimulates enzyme release from the pancreas and bile release from the gallbladder.
Cholesterol	A waxy lipid. It has a structure containing multiple chemical rings.
Chromatography	Chromatography can be defined as a primarily as a process separation which is used for the separation of molecular mixture.
Chronic	Long-standing, developing over time; slow to develop or resolve. When referring to disease this indicates that the disease progress, one developed, is slow and tends to remain; a good example is coronary heart disease.
Chronic toxicity	Refers to an effect that requires some time to develop toxicity.
Chylomicrons	Dietary fat surrounded by a shell of cholesterol, phospholipids and protein. These are made in the intestine after fat absorption and travel through the lymphatic system to the bloodstream.

Terminology	*Definition/Explanation/Principles*
Cis isomer	An isomer form seen in compounds with double bonds, such as fat where the hydrogen on both ends of the double bond lie on the same side of that bond.
Cleaning of wheat	Process of removal of impurities from wheat.
Climacteric fruits	Fruits which ripens after harvesting.
Clinching	Loosely covering of lids before exhausting in canning technology.
Cling film wrapping	It is the pretreatment in which the fresh fruits are wrapped in very thin film of polyethylene having a glossy edible adhesive, reduced water vapour transpiration rate, high gas permeability and do not require vents which helps to reduce the physiological loss in weight and increases shelf life.
Clostridium botulinum	A bacterium that can cause a fatal type of food poisoning.
Cloud point	The temperature at which the first permanent cloud, due to the crystallization of high-melting materials, appears in the body of the oil when cooled under standardized conditions.
Cloud test	The time (h) required for the appearance of visible cloud in the salad oil held at 0 °C.
Cloying	A taste sensation that stimulates beyond the point of satiation; frequently used to describe overly sweet products.
Coagulation	Random aggregation reactions with denaturation and aggregation reactions where protein-protein interactions predominate over protein-solvent interactions are defined as coagulation.
Coarse dispersion	Dispersion having particle size greater than 0.5 μm is coarse dispersion.
Coccus	Comma shape bacteria.
Cold chain	It is the provision of lowest safe temperature to the commodities immediately after its harvest till it reaches the consumer.
Cold test	The temperature at which the oil becomes solid.
Colostrum	The first fluid secreted by the breast during late pregnancy and the first day after birth. This thick fluid is rich in immune factors and protein.
	It is first milk secreted by mammary/milking glands of all living females after birth of young one. It is rich in immunoglobulins, anti-microbial peptides (lactoferrin and lactoperoxidase) and other bioactive molecules, including growth factors. Human colostrums have much higher concentrations of epidermal growth factor.

Terminology	*Definition/Explanation/Principles*
Color blindness	A condition in which individuals lack the ability to discriminate among different wavelengths of light.
Color constancy	The tendency for a familiar object to be perceived as a constant color even though the light that it reflects changes the sensation.
Coma	A state of deep unconsciousness below normal sleep.
Compatibility	In flavor terminology the ability of one substance to enhance the flavor characteristic of another.
	It is nothing but package neither should change its color, flavor, foreign test to food material nor should it absorb it from material.
Compensation	The result of interaction of the components in a mixture of stimuli, each component of which is perceived as less intense than it would be alone.
Competence	Uptake of free DNA fragments and incorporation into its genome by a prokaryotic cell.
	Ability of recipient cell to take up DNA through changes in cell wall.
Complete proteins	Proteins that contain ample amounts of all ten essential amino acids.
	Proteins that contain ample amounts of all ten essential amino acids (PVT TEAM HALL).
Compressible fluid	When the density of fluid is sensitive to change in temperature and pressure, the fluid is said to be compressible fluid.
Concentrate	This is a fruit juice, which has been concentrated by removal of water either by heat or by vacuum drying.
Conditioning	Process of adjustment and throughout distribution of moisture in grain
	Process of soaking grains in water to dehydrate them and for even distribution of moisture in grain.
Conduction	It is the transfer of heat between adjacent molecules without appreciable displacement of the particles of the body.
Conformity	Adopting the attitudes and behaviors of other people as a result of real or imagined pressure from others.
Congeners	Congeners are the flavor constituents in beverage spirits that are responsible for its flavor and aroma and that results from the fermentation, distillation, and maturation processes.
Constraints	Physical limitations on the set of decision variable.

Terminology	*Definition/Explanation/Principles*
Contribution	The difference between sales revenue and variable cost is known as contribution.
Convection	Convection is the transfer of heat from one point to another within a fluid, gas or liquid by the mixing of one portion of the fluid with another. The motion of the fluid may be entirely the result of differences in density resulting from the temperature differences. This is known as natural convection. The motion may also be produced by mechanical means as in forced convection.
Convergence	The tendency of a test sample, regardless of quality, to be perceived as similar to prior sample(s); sometimes called the halo effect.
Conversion	Conversion describes the enzymatic starch hydrolysis processes, liquification, and saccharification.
Convolutions	The irregular "hills and valleys" that make up the surface of the cortex in humans and other complex animals.
Convulsion	It is a condition marked by involuntary contraction of the muscles
Cooking	Cooking is the gelatinization by heat treatment and alpha-amylase liquification, and raw material starch.
Cooling	Cooling sensations occur when certain chemicals contact the nasal or oral tissues and stimulate a specific saporous receptor.
Cooperation	Form of association of people to work together in order to active a particular end.
Cordial	It is a sparking, clear sweetened fruit juice from which all pulp and other suspended materials have been completely eliminated. Limejuice cordial is a typical example of cordial.
Cornea	The transparent, domelike, outer covering of the lens and iris that allows light to enter the eye.
Cortical bone	Dense, compact bone that comprises the outer surface and shafts of bone.
Cortisol	A hormone made by the adrenal gland that, among other functions, stimulates the production of glucose from amino acids.
Covalent bond	A union of two atoms formed by the sharing of electrons.
Critical control point	Identify critical control points in the process at which the potential hazard can be controlled or eliminated.

Terminology	*Definition/Explanation/Principles*
Critical point in grain drying	It is the point on the drying curve at which the constant rate period ends and falling rate period starts.
Critical speed of mill	Speed at which the small sphere inside the mill just begins to centrifuge.
Crop insurance	Crop insurance scheme was launched to protect the farmers from heavy loses of crop due to rain fall, floods, drought etc.
Crude fiber	What remains of dietary fiber after acid and alkaline treatment? This consists of primary cellulose and lignin.
Crush	Type of fruit beverage which contains at least 25 per cent fruit juice or pulp and 55 per cent TSS.
Crusting	Formation of dry crust on surface of dough due to evaporation of water.
Cryptosmia	Impairment of olfaction by obstruction of the nasal passages.
Crystallized fruits/vegetables	Candied fruits/vegetables covered or coated with crystals of sugar or finely powdered sugar are called crystallized fruits/vegetables.
Curing or Pickling Agent	Substance imparting a unique flavor and/or color to food, usually producing an increase in self-life stability.
Custard	Sweet mixture of egg and milk baked or cooked over hot water
Daily Reference Values (DRV)	Standards of intake for certain parts of a diet, such as carbohydrate, fat, saturated fat, cholesterol, sodium, potassium and dietary fiber, set by FDA for which no U. S. RDA exist. These values are intended to be used for comparing intake of these factors to desirable levels of intake.
Dampness	The paper boards observed moisture during rainy season when humid atmosphere is there the absorption of moisture package become loose its strength is a dampness.
Data	Records of the observations or measurements in an empirical study.
Deaeration	Freshly extracted and screened juices contain appreciable quantity of oxygen which should be removed before packing this process is known as deaeration.
Debittering	The process of removing bitter substances from oils or fats.
Decimal reduction time	Time over which the spore concentration is reduced by tenfold.
Defatted meal	Meal obtained after extraction of oil, utilization, defatted meal.
Degerming	The process of separating germ or embryo from the seed or kernel.

Terminology	*Definition/Explanation/Principles*
Degree of milling	The extent to which undesirable parts like husk, bran etc. are separated from paddy.
Degreening	It is the process in which treatment with ethylene under controlled conditions hastens the loss of chlorophyll.
Delinting of cotton seed	Cottonseed contains residual fibers after ginning called as lint. The process of removal of these lints in battery machines is known as delinting of cottonseed.
Denaturation	A process in which a protein unravels and loses its native conformation, thereby becoming biologically inactive. Denaturation occurs under extreme conditions of pH, salt concentration, and temperature.
Dental caries	Erosions in the surface of a tooth caused by acids made by bacteria as they metabolize sugars.
Deodorization	A process in which the heated oil is placed in a vaccumized tower and allowed to cascade over steam moving in countercurrent direction, to remove steam-volatile substances contributing to undesirable flavor and odor to the oil.
Depression	An extreme mood of despondency, hopelessness, and lowered self-esteem.
Desolventizers	Used for removal of traces of solvent from oil and miscella.
Dextrin	Partial breakdown products of starch that contain few too many glucose molecules. These appear while starch is being digested into many units of maltose.
Dextrose equivalent	It is total reducing power of glucose syrup expressed on a dry solids basis as dextrose. It is measure of degree of hydrolysis.
Diabetes mellitus	A disease characterized by high blood sugar levels (hyperglycemia), resulting from either an insufficient insulin released by the pancreas or a general inability for insulin to act on certain body cells, such as adipose cells.
Dietary fiber	Substances in food that are not digested by the processes present in the stomach and small intestine.
Dietetics	It is a science that deals with the adequacy of diet during normal life cycle and modification required during diseased condition.
Digestion	The process of breaking down food into molecules small enough for the body to absorb.
Direct calorimetry	A method to determine energy use by the body by measuring heat that emanates from the body, usually using an insulated chamber.

Terminology	*Definition/Explanation/Principles*
Disaccharides	Class of sugars formed by the chemical bonding of two monosaccharides.
Discrimination	In conditioning, a process whereby the organism learns to respond differently to stimuli that are differentially reinforced. In social psychology, the expression of prejudice in hostile behavior toward members of a particular group.
Distillation	The separation of components of a solution on the basis of their volatilities is known as distillation.
Distortion	The process by which information in memory is altered, making it inaccurate.
Dockage	The impurities found in grains such as chaff, stalks, grain dust, other seeds, immature grains are collectively called as dockage.
Docosahexaenoic acid (DHA)	An omega 3 fatty acid with 22 carbons and six carbon–carbon double bonds. DHA is also present in fish oils and also may be synthesized from alpha-linolenic acid [22:6].
Documentation	Establish effective record-Keeping procedures that document and provide a historical record of the facility's food safety performance.
Dominant gene	A member of a gene pair that controls the expression of a physical trait regardless of the nature of the other member of the gene pair.
Doubler	A doubler is a pot still used to redistill whiskey and low wines from a beer still. The low wines are fed into the doubler where they are redistilled by way of steam enclosed in a scroll at the bottom of the still. The bottoms, the organic components remaining at the bottom of the still, are returned to the still to extract the alcohol.
Dough improvers	The chemical substances used to improve viscosity, loaf volume, cellular texture, freshness, extensibility etc. of the bread.
Dough Strengthener	Substance used to modify starch and gluten, thereby producing more stable dough.
Down syndrome	A form of organic retardation resulting from an abnormality in which the child has one more chromosome than normal.
Drying Agent	Substance with moisture-absorbing ability, used to maintain an environment of low moisture.

Terminology	*Definition/Explanation/Principles*
Drying oil	Oil which when applied as a thin film readily absorbs oxygen from the air and dries to form a relatively tough elastic substance.
Dysomia	Difficulty in ability to smell.
Edema	Edema is the presence of an abnormally large amount of fluid in the tissue spaces of the body
Edible films and coating	It can be defined as thin layers of edible material applied on foods by wrapping immersing, brushing or spraying in order to offer selective baser against transmission of gases vapours and solutes while also offering mechanical protection.
Eicosanoids	Hormone like compounds synthesized from polyunsaturated fatty acids. Within this class of compounds are prostaglandins, thromboxanes and leukotrienes.
Eicosapentaenoic acid (EPA)	An omega-3 fatty acid with 20 carbon atoms and five double bonds; present I fish oils and may be synthesized from alpha-linolenic acid [20:5].
Electrocardiogram	A plot of electrical activity of the heart over the cardiac cycle; measured via multiple skin electrodes.
Electrophoresis	It is the migration of colloidal particles through a solution under the influence of an electric field.
Emotion	A state made up of a characteristic subjective experience, a characteristic pattern of physiological arousal, and a characteristic pattern of overt expression.
Emulsification/ Emulsion	Operation in which two normally immiscible liquids are intimately mixed or when an immiscible liquid is dispersed as small droplets in another immiscible liquid by mechanical agitation, an emulsion is formed.
Emulsifier or Emulsifier Salt	Substance, which modifies surface tension in the component phase of an emulsion to establish a uniform dispersion or emulsion.
Emulsion	Emulsion is a mixture of two immiscible liquids. Milk cream, mayonnaise, salad dressings, ice cream mix, and cake batters are all O/W emulsions. Butter and margarine are W/O emulsions.
Emulsion capacity (EC)	The volume (ml) of liquid emulsified by 1 g of flour or protein. It is the volume of oil (milliliters) that can be emulsified per gram of protein before phase inversion occurs.
Emulsion stability (ES)	It is ratio between volumes of final emulsion x 100 divided by volume of initial emulsion.

Terminology

	Definition/Explanation/Principles
Enrichment	Addition of specific amounts of selected nutrients in accordance with a standard identity as defined by the U.S. Food and Drug Administration.
Enzymes	Proteins that catalyze specific biological reactions with extraordinary catalytic power.
Equilibrium moisture content	It is moisture content of solid in equilibrium with surrounding conditions when there vapour pressure are equal.
Erythroprotein	A protein secreted by the kidneys that enhances red blood cell synthesis and stimulates red blood cell release from bone marrow.
Essence recovery	An essence recovery system which employs a "stripping column" to separate the volatile essence from the less volatile water.
Essential amino acids	Amino acids not efficiently synthesized by humans that must therefore be included in the diet. There are ten essential amino acids [PVT TIM HALL].
Essential fatty acids	The fatty acids, which are not synthesized in the human body that must therefore be included in the diet. They are linoleic and alpha-linolenic acid.
Essential mineral	It is an inorganic substance which is needed by at least two species of animals for good health
Essential nutrient	A chemical element required for a plant to grow from a seed and complete the life cycle. A nutrient substance that an animal cannot make itself from raw materials but that must be obtained in food in prefabricated form.
Essential oil	The volatile material, derived by a physical process, usually distillation, from odorous plant material of a single botanical form and spices with which it agrees in name and odor.
Ester value	The difference between the saponification value and the acid value.
Esterification	With regards to fats, the process of attaching fatty acids to a glycerol molecule, creating an ester bond. Removing a fatty acid is called deesterification; reattaching a fatty acid is called reesterification.
Ethylene	The only gaseous plant hormone, responsible for fruit ripening, growth inhibition, leaf abscission and aging.

Terminology	*Definition/Explanation/Principles*
Expectancy	One aspect of the cognitive view of conditioning, which holds that the organism comes to expect an enforces.
Expel out	Separation of liquids from solids by the application of compressive forces.
Experiment	A study in which a scientist treats an object of study in a specific way and then observes the effects of the treatment.
Extra lean	Meat, poultry, seafood, and game meats containing < 5g of fat per serving.
Falling rate period	Period of drying in which the role of drying decreases with time due to slow removal of internal moisture.
False fruit	Fruit is said to be false fruit when it develops collectively along with progressive development of ovary and associated parts.
Famine	A time of massive starvation often associated with crop failures, war and political strife.
Fat	A biological compound consisting of three fatty acids linked to one glycerol molecule.
Fat analog	A compound that provides food with many of the characteristics of fat, but altered digestibility and altered nutritional value.
Fat barrier	An ingredient system that provide a barrier for products that use fat as a heat exchange medium.
Fat extender	A fat replacement system, containing a proportion of fat combined with other ingredients, designed to optimize the functionality of fat, thus allowing a decrease in the usual amount of fat in the product.
Fat-free	A product containing 0.5g or less fat per serving.
Fat mimetics	A fat replacer that can mimic one or more of the organoleptic and physical functions of fat, and usually requires a high water content to achieve its functionality.
Fat replacer	An ingredient that can be used to replace fat, yielding fewer calories than fat and may or may not provide nutritional value
Fat substitute	A synthetic compound, usually having a similar chemical structure to conventional fats and oils, designed to replace some or all of the functions of fat without any energy contribution, and is usually indigestible or unabsorbable.

Terminology	*Definition/Explanation/Principles*
Fatty acid	A long carbon chain carboxylic acid. Fatty acids vary in length and in the number and location of double bonds; three fatty acids linked to a glycerol molecule form fat.
Feints	Feints are the third fraction of the distillation cycle derived from the distillation of low wines in a pot still. This scotch term is also used to describe the undesirable constituents of the wash that are removed during the distillation of grain whiskey in a continuous patent still (Coffey). These are mostly aldehydes and fuel oils.
Fermentable sugars	Fermentable sugars like glucose [50-99-7], maltose [69-79-4], and maltotriose [1109-28-0], can be fermented by distiller's yeast.
Fermentation	A catabolic process that makes a limited amount of ATP from glucose without an electron transport chain and that produces a characteristic end product, such as ethyl alcohol or lactic acid.
Fermentation kinetics	Fermentation kinetics is concerned with the rate of cell synthesis and/or fermentation product formation and the effect of environment on these rates.
Fermented fruit beverage	It is a fruit juice, which has undergone alcoholic fermentation by yeast. The product contains varying amounts of alcohol.
Firming Agent	Substance added to precipitate residual pectin, thus strengthening the supporting tissue and preventing its collapse during processing.
Ferritin	A protein compounds that serves as the storage form of iron in the blood and tissues.
Fertility	It refers to ability of plant not able to set and mature fruits but develop a viable seed.
Fetal alcohol syndrome	A group of physical and mental abnormalities in the infant that result from the mother consuming alcohol during pregnancy.
Fiber	A lignified cell type that reinforces the xylem of angiosperms and functions in mechanical support; a slender, tapered sclerenchyma cell that usually occurs in bundles.
Fire point	The temperature at which a substance, which ignited burns freely when the ignition agent is withdrawn.
First class fruits	Fresh fruits which do not satisfy the minimum quality requirement of the extra class and are superior to lower class.

Terminology	Definition/Explanation/Principles
Flaking of soybean	Heat treatment prior to extraction of oil, increases efficiency of oil extraction.
Flash point	The temperature under standardized conditions at which a liquid begins to evolve inflammable vapours.
Flavour	The sensation produced by a material taken into the mouth, perceived principally by the sense of taste and smell, but also by the common chemical sense produced by pain, tactile, and temperature receptors in the mouth.
Flavour adjunct	A substance used in or with a flavor but not essentially a part of it. These include solvents, antioxidants, enzymes, adjusting agents, emulsifiers, and acidulants.
Flavour enhancer	A substance added to supplement, modify, or enhance the original taste and/or aroma of a food without imparting a characteristic taste or odor of its own.
Flavour extract	The dilute extract of flavor prepared from spices and aromatic plant parts usually in alcohol, fortified with WONF, if needed.
Flavour or Flavorant	A substance, added to food, whose significant function is to affect odor, imparting a characteristic flavor to that food.
Flavour reversion	This problem is unique to soybean oil and other linolenate containing oils. The off flavor has been described as beany and grassy and usually develops at low peroxide values.
Flavouring ingredients	Any single chemical entity or natural mixture added to food, drugs or other products taken in the mouth, the clearly predominant purpose and effect of which is to provide all or part of the particular flavor of the final product.
Flavours	Those mixtures of ingredients whose exact composition is usually known only to their suppliers, sold in bulk to food and beverages manufactures. They are to be labeled as flavours per CFR 21 part 101 and may contain adjuncts that are nonflavour ingredients.
Flipper	Can appears normal, but when struck against a tabletop one or both ends become convex and springs or flips out, but can be pushed back to normal condition by little pressure such can is termed as flipper.
Flocculation	It refers to random aggregation reactions in the absence of denaturation.
Flour bleaching agent	The chemical substances used to oxidize carotenoid pigments to make maida creamy white.

Terminology	*Definition/Explanation/Principles*
Flour Treating Agent	Substances added' to milled flour, at the mill to improve its color and/or baking qualities, inducing bleaching and maturing agents.
Flow point	The temperature at which upon heating the solidified fat becomes soft and flows downward through an orifice.
Fluorescence	Fluorescent compounds may develop from the interaction of carbonyl compounds (produced by lipid oxidation) and certain cellular constituents possessing free amino groups.
Foam	Foam is dispersion of gas bubbles in liquid phase or semisolid phase.
Foam spray drying	In foam spray drying a compressed gas is injected into the fluid prior to the spray nozzle. The injected air causes the solids associated with a droplet to agglomerate in a manner, which increases its buoyancy as compared with conventional spray drying.
Foam stability	The ability of foaming agents (flour or protein) to retain the foam volume upon standing.
Foam value	Number of minutes to specific temperature required to destroy a specific number of organisms having a specific 'Z' value.
Foaming capacity	The volume (ml) of foam formed when a known quantity of flour or protein is mixed with distilled water and stirred for 5 min.
Fold	Strength of concentrated flavouring materials. The concentration is expressed as a multiple of a standard, *e.g.* citrus oil is compared to cold pressed oil. In the case of vanilla, folded flavours are compared to a standard extract with minimum bean content.
Folding carton	Folding carton made from sheets of paperboard which have been cut and scored for bending into desired shapes, they are delivered in a collapsed state for erection at the packaging point.
Food	Any edible material, which is consumed by the individual and his/her family. Whatever we eat and that nourishes the body is called food.
Food Chemistry	Food chemistry is major aspect of food science, which deals with the composition and properties of food and chemical changes it undergoes during handling, processing and storage.
Food intolerance	An adverse reaction to food that does not involve an immune response.

Terminology	*Definition/Explanation/Principles*
Food retailing	Food retailing is a sale of food items in small lots to the consumers.
Food Science	The Science deals with the food, which is edible by living matter
Food Science and Technology	It is a science deals with food and their technology.
Food sensitivity	A mild reaction to a substance in a food that might be expressed as slight itching or redness of the skin.
Food web	The elaborate, interconnected feeding relationship in an ecosystem.
Food-borne illness	Sickness caused by ingestion of foods containing toxic substances produced by microorganisms.
Foods for body building	Foods rich in protein, minerals, vitamins and water.
Foods for energy	Foods rich in carbohydrates, fats and protein.
Foods for regulating body process	Foods rich in minerals, vitamins, water and fiber.
Foots	The saponification fatty acids in crude oil which settle down to bottom on agitation of oil with alkali.
	The saponified fatty acids in crude oil which settle down to known to bottom on agitation of oil with alkali.
Fore milk	The first breast milk delivered in the nursing session.
Foreshots	Foreshots is the first fraction of the scotch distillation cycle derived from the distillation of low wines in a pot still.
Forgetting	The inability to recall a particular piece of information accurately.
Formulation Aid	Substance used to promote or to produce a desired physical state or texture in food. Including carriers, binders, fillers, plasticizers, film-formers, and tableting aids, etc.
Fortification	Addition of nutrients in amounts significant enough to render the food a good to superior source of the added nutrients. This may include addition of nutrients not normally associated with the food or addition to levels above that present in the unprocessed food.
Fortified	A term generally meaning that vitamins, minerals or both have been added to a food product in excess of what was originally found in the product.

Terminology	*Definition/Explanation/Principles*
Free fatty acids	The fatty acids in the oil that are not esterified with glycerol and those liberated from triglycerides when subjected to hydrolytic rancidity.
Free radical	Short-lived form of compounds that exist with an unpaired electron in their outer electron shell. This causes it to have an electron-seeking nature, which can be very destructive to electron-dense areas of a cell, such as DNA and cell membranes.
Free water	The water not bound to the components in a food. This is available for microbial use.
Freeze drying	In this process food is first frozen at –18°C on trays in the lower chamber of freeze drier and the frozen material dried under high vacuum in the upper chamber. Direct sublimation of ice taken place without passing through the intermediate liquid stage.
Freezing or Cooling Agents	Substance that reduces the temperature of food materials through direct contact.
Freezing point	The freezing point of a material is the temperature at which it changes from a liquid to a solid. A liquid freezes when its vapour pressure is equal to the vapour pressure of its solid. The freezing point of water is 0°C.
Fructose	A monosaccharide with six carbons that form a five-membered ring with oxygen in the ring; found in fruits and honey.
Fruit	A mature ovary of a flower that protects dormant seeds and aids in their dispersal.
	The development of ovary and jaunt tissues following blossoming.
Fruit cordial	This is fruit squash from which all suspended material is completely eliminated and is perfectly clear, *e.g.*, limejuice cordial.
Fruit drink	Liquefying the whole fruit makes this and at least 10 per cent of the volume of undiluted drink must be whole fruit. It may be diluted before being served.
Fruit juice	This is a natural juice pressed out of a fruit, and is unaltered in its composition during preparation and preservation.
Fruit juice concentrate	This is fruit juice, which has been concentrated by the removal of water either by heat or by freezing.

Terminology	*Definition/Explanation/Principles*
Fruit punches	Mixing the desired fruit juices at the time when it is served makes these.
Fruit squash	This consists essentially of strained juice containing moderate quantities of fruit pulp to which sugar is added for sweetening *e.g.*, orange squash, lemon squash, mango squash, etc.
Functional Foods	Foods with a specific additional benefit that goes beyond the nutritional benefits of the nutrients they contain. Or Food that encompasses potentially healthful products, including any modified food or food ingredient that may provide a health benefit beyond that of the nutritional nutrients it contains.
Fuel oil	Fusel oil is an inclusive term for heavier, pungent tasting alcohols produced during fermentation. Fusel oils are composed of mixture of n-propyl, isobutyl, and isoamyl alcohols.
Fumigant	Volatile substance used for controlling insects or pests.
Fusel Oil	Fusel oil is an inclusive term for heavier, pungent tast-alcohols produced during fermentation. Fusel oils are composed of a mixture; propyl, isobutyl, and isoamyl alcohols.
Galactosemia	A disease characterized by the buildup of the monosaccharide galactose in the bloodstream resulting from the inability of the liver to metabolize it. If present at birth and left untreated, this disease causes severe growth and mental retardation in the infant.
Gelatinization	The tendency of starch molecules to swell and form gel when starch solution is heated.
Gelation	When denaturation molecules aggregate to form an ordered protein network the process is referred to as gelation. Gelation is the irreversible change in viscosity of egg yolk after thawing.
Gene cloning	Transfer of sequence of DNA (gene) from one organism to other organism.
Gene stability	Major of resistance to change with time of the sequence of gene within DNA molecule or of nucleotide sequence with gene.
Generally recognized as safe (GRAS)	A group of food additives that in 1958 were considered safe, therefore allowing manufactures to use them thereafter when needed in food products. The FDA bears responsibility for providing they are not safe and can be remove unsafe products.

Terminology	*Definition/Explanation/Principles*
Generation time	The time required by micro-organism to double its population.
Genetics	The branch of biology concerned with study of heredity and variation.
Germination	It is the process in which soaked grains are allowed to sprout and develop vegetative growth or roots.
Gills	Sexual spore *i.e.* basidiospores are formed, borne on the underside of fruiting bodies, mostly seen in *Agaricus* sp. of mushroom.
Glazed fruits/ vegetables	Covering of candied fruits/vegetables with a thin transparent coating of sugar, which imparts them glossy appearance, is known as glazing fruits/vegetables.
Glucagon	A hormone made by the pancreas that stimulates the breakdown of glycogen in the liver into glucose; this raises the blood glucose level. Glucagon also performs other functions.
Gluconeogesis	The production of new glucose molecules by metabolic pathways in the cell. The source of the carbon atoms for these new glucose molecules is usually amino acids.
Glucose	A six-carbon atom carbohydrate found in blood and in table sugar bound to fructose; also known as dextrose, it is one of the simple sugars.
Glycemic index	A ratio used to measure the relative ability of a carbohydrate to raise blood glucose levels as opposed to the ability of white bread (or glucose) to raise blood glucose levels.
Glycerol	An alcohol containing three hydroxyl groups (-OH); used to help from triglyceride molecules.
Glycogen	An extensively branched glucose storage polysaccharide found in the liver and muscle of animals; the animal equivalent of starch.
Glycolysis	The splitting of glucose into pyruvate. Glycolysis is the one metabolic pathway that occurs in all living cells, serving as the starting point for fermentation or aerobic respiration.
Goiter	An enlargement of the thyroid gland often caused by a lack of iodine in the diet.
Goitregens	Substances in food that interferes with thyroid hormone metabolism and so may cause goiter if consumed in large amounts.

Terminology	*Definition/Explanation/Principles*
Grading	Sorting on the basis of quality criteria.
Grain whiskey	Grain whiskey s an alcoholic distillate from fermented wort derived from malted and unmalted barley and corn, in varying proportions, and distilled in a continuous patent still (Coffey).
Growth	The progressive development or increase in size of a living thing, such as the growth of a child
Gustation	A taste sense, the receptors of which lie in the mucous membrane covering the tongue, and the stimuli for which consist of certain soluble chemicals, *e.g.* salts, acid, and sugar.
Haemoglobin	The iron-containing protein in the red blood cell that carries oxygen to the cells and carbon dioxide away from the cells. It is also responsible for the red color of blood.
Haemolysis	It is the process whereby the red cell bursts, releasing the haemoglobin from it
Haemorrhage	It is a condition in which blood escapes from the vessels. Some time it also called as bleeding
HANES	Health and Examination Surveys: A survey conducted between 1971-1972 on the United States Population aged 1 to 74 years to evaluate their nutritional status. The sample size of the population was about 20,000.
Hard to cook legume	Rate of water absorption and cooking is low, retards cooking.
Hard water	Water containing chlorides and sulphates of calcium and magnesium.
Harris-Benedict equation	An equation that predicts resting metabolic rate based on a person's weight, height, and age.
Haugh unit	Measurement of height of thick white in relation to the weight of egg.
Heads	Heads is distillate containing a high percentage of low boiling components such as aldehydes.
Health	A state of complete physical, mental and social well-being and not merely the absence of disease or infirmity (WHO, 1948)

Terminology	*Definition/Explanation/Principles*
Health	It is defined as state of complete physical, mental and social well being and not merely the absence of disease and infirmity.
Heartburn	A pain emanating from the esophagus, caused by stomach acid backing up into the esophagus and irritating the esophageal tissue.
Heat	Heat is energy transferred from one system to another solely by reason of a temperature.
Heat transfer	The transfer of heat is the principal unit operation in many food processes.
Heat transfer coefficient	Heat transfer coefficients for drop wise condensation of vapor: The heat transfer coefficients for drop wise condensation are very high in the range of 10,000 to 70,000 Btu/hr. (sq.ft.) (°F).
Hehner number (value)	The percentage of water and fatty acids in 1 g of fat.
Herd immunity	Resistance of a group to a pathogen as a result of immunity of a large proportion of the group to that pathogen.
Heterotrophic nutrition	A mode of obtaining organic food molecules by eating other organisms or their by-products.
High wines	High wines are an all-inclusive term for beverage spirit distillates that have undergone complete distillation.
High-density-lipoprotein (HDL)	The lipoprotein synthesized by the liver and intestine that picks up cholesterol from drying cells and other sources and transfers it to the other lipoproteins in the bloodstream. A low HDL level increases the risk for heart disease.
Hind milk	The milk secreted at the end of a nursing session; it is higher in fat than fore milk.
HIV (Human immunodeficiency virus)	The infectious agent that causes AIDS; HIV is an RNA retrovirus.
HLB value	Hydrophile-lipophile balance is ratio of weight of percentage of hydrophilic and hydrophobic groups in an emulsifier.
HMG-CoA reductase	An enzyme in the cytosol that catalyzes the conversion of hydroxymethylglutaryl-CoA (HMG-CoA) to form mevalonate. The action of HMG-CoA reductase is the committed step in cholesterol biosynthesis.
Holding cost	The cost associated with holding inventories in a stock is known as holding cost.
Hops	Hops are dried female flowers of hope plants which are grow in Ciregon and Washington.

Terminology	*Definition/Explanation/Principles*
Hormone	A compound secreted into the bloodstream that acts to control the function of target organ cells. Hormones can be either protein like or fatlike, such as insulin or estrogen.
Hue	Hue is the aspect of color that we describe by words such as green, blue, yellow and red.
Humectant	Hygroscopic substances incorporated in food to promote retention of moisture. Includes moisture retention agents and antidusting agents.
Hydrogenation	The chemical addition of hydrogen to a material or The addition of hydrogen atoms to the double bonds of polyunsaturated and monounsaturated fatty acids to reduce the extent of unsaturation. This process turns liquid vegetable oils into solid fats.
Hydrogenation of oil	It is the process in which liquid fat or oil is converted into solid or semi-solid fat by addition of hydrogen.
Hydrophilic	Attracts water (water loving).
Hydrophilic-Lipophilic balance (HLB)	It is a ratio between hydrophilic emulsifier and lipophic emulsifier. As rule emulsifiers with HLB values in the range 3-6 promote W/O emulsions; values between 8 and 18 promote O/W emulsions.
Hydrophobic	Repels water (water fearing).
Hyperglycemia	High blood glucose levels, above 140 milligrams per 100 milliliters of blood.
Hyperosmia	Unusually keen olfactory sensitivity.
Hypertension	A condition in which blood pressure remains persistently elevated, especially when the heart is between beats.
Hypobaric storage	The storage in which pressure in the storage atmosphere is less than atmospheric pressure.
Hypogeusia	Diminished sense of taste.
Hypoglycemia	Low blood glucose levels, below 4 to 50 milligrams per 100 milliliters of blood.
Hyposmia	Diminished sense of smell.
Hypothesis	A proposition or assertion about the possible relationship between variables.
Imbibed water	May not be different from water held as hydrate some substances pickup water and swell, when they come in contact with water.

Terminology	*Definition/Explanation/Principles*
Immobilized enzymes	An immobilized enzyme or cell is chemically or physically restricted in movement so that it can be physically reclaimed from the reaction medium.
Immune system	A complex of organs and cells that function to protect the body against disease agents such as viruses, bacteria and pollutants.
Imperfect flower	Species that have only stamens or stigma within a flower.
Incidental food additives	Additives that gain access to food products indirectly from environmental contamination of food ingredients or during the manufacturing of food process.
Incomplete proteins	Food protein that lacks ample amount of one or more of the essential amino acids needed to support human protein needs.
Index of nutrition quality	It is percentage of nutrient need provides by food divided by percentage of calorie need provided by from food.
Infusion mashing	Infusion mashing is the process of simultaneously cooking and converting small grains (rye, barley and wheat).
Inherent toxicants	These are metabolites produced via biosynthesis of food under normal growth conditions and by organisms.
Inland fishery	Process of raising and harvesting fresh water fish, rivers, lakes are the places of habitat of fresh water fish. *e.g.* rohu, katla and mrigal etc.
Insipid	Tasteless, fat, vapid.
Insoluble fiber	Fibers that, for the most part, do not dissolve in water nor are digested by bacteria in the large intestine. These include cellulose, some hemicelluloses and lignin.
Instant dhal	Less time required for cooking as compared to ordinary dhal/ ready to eat dhal.
Insulin	A hormone produced by the beta cells of the pancreas. Insulin increases the synthesis of glycogen in the liver and the movement of glucose from the bloodstream into muscle and adipose cells.
Integral protein	A protein of biological membranes that penetrates into or spans the membrane.
Intelligence	The capacity to understand the world and the resourcefulness to cope with its problems.
Intelligence quotient (IQ)	The relationship between mental age and chronological age, expressed as $MA/CA \times 100$.

Terminology	*Definition/Explanation/Principles*
Intermediate density lipoprotein (ILD)	The product formed after a very low-density lipoprotein (VLDL) has most of its triacylglyceride removed.
International unit (IU)	A crude measure of vitamin activity, often based on the growth rate of animals. Today these units have been replaced by more precise microgram quantities.
Intoxication	The state in which an excess of a sedative poisons the body, leading to uncoordinated actions, slowed reaction times and slurred speech.
Invert sugar	The hydrolyzed mixture of sucrose, containing glucose and fructose is known as invert sugar.
Iodine value	The number of iodine absorbed by 100 g of oil. It is a measure of the unsaturation of fat.
	This test is a measure of the unsaturated linkages in a fat and is expressed in terms of percentage of iodine absorbed.
Iron	A metallic element occurring in the haemoglobin of red blood cells, stored in tissues in the form of ferritin and is an essential part of important respiratory enzymes
Isoelectric point pH	It is defined as the pH at which a solute (compound) has no net electric charge or carry equal positive and negative charge (*i.e.* zwitterions).
Isolate	A relatively pure chemical produced from natural raw materials by physical means, *e.g.* distillation, extraction, crystallization, etc. and therefore natural; or by chemical means, *i.e.* via hydrolysis, bisulfate addition products, and regeneration, etc. and therefore artificial by 1993 U.S. labeling regulations.
Isomer	Different chemical structures for compounds that share the same chemical formula.
Isotope	An alternate form of a chemical element. It differs from other atoms of the same element in the number of neutrons in its nucleus.
Isozymes/ Isoenzymes	Multiple forms of the same enzyme arising from genetically determined differences in primary amino acid sequences.
Jam	It is a product prepared by boiling fruit pulp with sufficient quantity of sugar to reasonably thick consistency, firm with enough to hold fruit tissues in position. It should contain not less than 68.5 per cent T.S.S.

Terminology	Definition/Explanation/Principles
Jaundice	A yellow staining of the skin and sclera resulting from a buildup of bile pigments in the bloodstream. Liver or gallbladder disease is often the cause.
Jelly	It is a product prepared by mixing strained extract of fruit with sugar and boiling the mixture to a stage at which it will set a clear gel.
Juice	Juice is the liquid that is naturally contained in fruit and vegetables. It can also refer to liquids that are flavored with these or other biological food sources such as meat and seafood. It is commonly consumed as a beverage or used as an ingredient or flavoring in foods. Juice is prepared by mechanically squeezing or macerating fruit or vegetable flesh without the application of heat or solvents. For example, orange juice is the liquid extract of the fruit of the orange tree, and tomato juice is the liquid that results from pressing the fruit of the tomato plant. Juice may be prepared in the home from fresh fruit and vegetables using a variety of hand or electric juicers. Juice is one of the most popular drinks to go with breakfast in the morning.
JunK DNA	The genetic material of animal and plants consists of mainly two types coding and non coding. The non coding is also known as no functional or junk DNA.
Kernel Hardness	Kernel hardness measurements are mainly useful in differentiating between soft or hard grains in plant breeding programme. Extensively hard reflects in to increase in power for grinding while extensively softness reflects into bolting and increases the requirements for sieving space.
Ketone bodies	Products of acetyl-CoA (Fat) metabolism containing three to four carbon atoms; acetoacetic acid, beta-hydroxybutric acid, and acetone. These contain a ketone group, hence, the name.
Ketosis	The condition of having high levels of ketones in the bloodstream.
Key column	The optimum column, the one with largest negative index number.
Kreis test	This is one of the first tests used commercially to evaluate oxidation of fats. It measure a red color believed to result from reaction of epihydrin aldehyde or other oxidation products with phloroglucinol.
	A test in which the oxidized fats react with phloroglucinol in acid solution to give a red color.

Terminology	*Definition/Explanation/Principles*
Kwashiorkor	A disease seen primarily in young children who have an existing disease and who consume a marginal amount of calories and considerably insufficient protein in the face of high needs. The child will suffer from infections and exhibit edema, poor growth, weakness, and an increased susceptibility to further illness. Means a sickness in a child develops when another baby is born.
Lamination	Combining two or more films into single film by means of solvent or heat to compliment all the properties in a single film is called lamination.
Lard	It is an animal fat from hogs.
LDL (low density lipoprotein) cholesterol	The fraction of cholesterol carried in the blood as a part of low-density lipoproteins.
Lean body mass	The part of the human body that is free of all but essential body fat. About 2 per cent of body weight as fat is essential to retain. The rest of the fat in the body represents storage and so is not part of lean body mass. Lean body mass includes muscle, bone, organs, connective tissue, skin and other body parts.
Lean fat	Meat, poultry, seafood, and game meats containing <10g of fat per serving.
Leathery crust	Crust of bread is crisp and breaks easily when pulled but crust becomes tough and exhibits the quality of leatheriness when pulled.
Leavening Agent	Substance used to produce or stimulate production of carbon dioxide in baked goods in order to impart a light texture, including yeast, yeast foods, and calcium salts.
Lecithin	A group of phospholipids containing two fatty acids, a phosphate group, and a choline molecule.
Lectins	Antinutritional factor present in plant foods some time it is called hemaglutinins.
Liarrage	The place where animals are brought 24 hrs before slaughtering.
Life span of spore	The life span of spore is defined as the length of time for which spores exposed to given temperature, remain viable.
Light fat	An altered product containing one-third fewer calories or 50 per cent of the fat in the reference food.

Terminology	*Definition/Explanation/Principles*
Lignin	An insoluble fiber made up of a multiringed alcohol (no carbohydrate) structure.
Limit dextrin	Limit dextrin are oligosaccharides containing one or more 1, 6-α-linkages.
Limiting amino acid	The essential amino acid in the lowest concentration in a food in comparison with the body's need.
Lipogenesis	The building of fatty acids using derivatives of acetyl-CoA molecules.
Lipogenic	Means creating lipid. The liver is the major lipogenic organ in the human body.
Lipolysis	The breakdown of lipids. Hydrolysis of lipids to produce glycerol and free fatty acids. Hydrolysis of ester bonds in lipids may occur by enzyme action or by heat and moisture, resulting in the liberation of free fatty acids. The release of short chain fatty acids by hydrolysis is responsible for the development of a rancid flavour (hydrolytic rancidity) in raw milk. Lipolysis is a major reaction occurring during deep fat frying due to the large amounts of water introduced from the food and the relatively high temperature at which the oil is maintained.
Lipoprotein	A compound found in the bloodstream containing a core of lipids with a shell of protein, phospholipid and cholesterol.
Low calorie	A product with <40 cal per serving.
Low fat	A serving containing no >40 cal, and 3g or less fat.
Low wines	Low wines is the term for the initial product obtained by separating (in a pot or Coffey still) the beverage spirits and congeners from the wash. Low wines are subjected to at least one more pot still distillation to attain a greater degree of refinement in the malt whiskey.
Lozenges	These are the sugar dough which has been flavoured, cut to shape and subsequently dried to remove the most of the added water.
Lubricant or Release Agent	Substance added to food contact surfaces to prevent ingredients and finished products from sticking to them (direct ad-ditives). Includes release agents, lubricants, surface lubricants, waxes and anti-blocking agents (indirect additives).
Macrosmatic	Abnormally keen olfactory sense.

Terminology	*Definition/Explanation/Principles*
Malnutrition	Impairment of health resulting from a deficiency or imbalance of nutrients. It can be under nutrition or over nutrition.
Malnutrition	The chronic deficiency of one or more nutrient in daily diet.
Malt whiskey	Malt whiskey is an alcoholic distillate made from fermented wort derived from malted barley only and distilled in pot stills. It is the second fraction (heart of the run) of the distillation process.
Malting	Malting is a controlled germination process, which activates the enzymes of the resting grain, resulting in conversion of cereal starch to fermentable sugars, partial hydrolysis of cereal proteins and other macromolecules.
	It is processing of grain where grains are steeped, germinated and kilned to increase maltose content of grain.
Marasmus	Results from a person consuming insufficient protein and kcalories; usually seen in infancy. It is the equivalent of protein-energy malnutrition in adults. The person will have or no fat stores and show muscle wasting.
Margarines	Mixtures of fats blended to provide desired properties for use as table spreads or for bakery use.
Marketing channels	It is the chain of intermediaries through whom the various food grains pass from producer to consumers.
Marmalade	It is a fruit jelly in which slices of fruit or of the peel are suspended. The term marmalade is generally associated with product made from citrus fruit like oranges. Lime/Lemon shredded peel is included as the suspended materials.
Masticatory Substance	Substance that is responsible for the long lasting and pliable property of chewing gum.
Mayonnaise	An emulsion of oil-in-water that readily breaks if crystals begin to form in oil.
Meat	Meat may be defined as contractive tissue from all animals used for food.
Megaloblast	A large, immature red blood cell that results from an inability for cell division during red blood cell development.
Melanoids	The brown to black amorphous, unsaturated heterogeneous polymers are called melanoids.
Melting point	The temperature at which a fat becomes completely clears.
Memory	The ability to store information so that it can be used at a later time.

Terminology	*Definition/Explanation/Principles*
Mental age	A measure of an individual's performance on an intelligence test expressed in terms of months and years.
Merosmia	A condition analogous to color blindness, in which certain odors are not perceived.
Micro-aerophillic	Microbe requiring small amount of air.
Microsmatic	Having a poorly developed sense of smell.
Milk	Milk is normal secretion of mammary gland is defined as the lacteal secretion. Practically free from colostrums, obtained by complete milking of one or more than one cows which contains not less than 8 per cent MSNF and not less than 3–35 per cent milk fat.
Milk fats	Milk fats mostly contain palmitic, oleic and stearic fatty acids and its fat is unique among animal fats in that it contains C_4 to C_{12} chain fatty acids.
Mill feed	Mixture of bran, germ and shorts.
Millard reaction	The browning reactions occur in the presence of an amino bearing compound, usually a protein, a reducing sugars and some water. It changes color, flavor, odour and texture of the product. These types of reaction occur by different four ways. 1. Sugars + nitrogenous compounds + heat 2. Sugar + organic acids + heat 3. Nitrogenous compounds + organic acids + heat 4. Organic acids themselves + heat
Milling	Process of separation of bran and germ from endosperm and reduction of endosperm into fine flour.
Milling loss	Loss of moisture by evaporation during milling.
Minimum support price	This is the price fixed by the Government to protect the producer-Farmers against excessive fall in the price during bumper production year.
Miso	Fermented food product prepared from soybean and rice using microorganisms for fermentation process.
Mixing	Two or more components are interspersed in space with one another by means of flow.
Molar extinction coefficient	In the specific absorption coefficient for a concentration of one mole and path length 1 cm.
Monod equation	$\mu = \mu\ max - S/ks + S$

Terminology	*Definition/Explanation/Principles*
Motivation	The desires, needs and interests that arouse an organism and direct it toward a specific goal.
Moulds	Moulds are multicellular, filamentous fungi belonging to the division Thallophyta but are devoid of chlorophyll. *Penicillium* sp. Blue moulds *Aspergillus* sp. Black moulds *Mucor* sp. Gray moulds
Mulching	It is system of orchard soil management in which orchard soil is covered with curry material to reduce moisture loss due to evaporation.
Multiple personality	A form of dissociative disorder in which large segments of the personality are split off from conscious awareness, so that the person seems to fluctuate between two or more distinct personalities.
Mutarotation	The change in specific optical rotation representing inters conversions of α and β forms of D glucose to an equilibrium mixture.
Mycotoxin	Toxin produced by molds *e.g.* Aflatoxin.
Nature identical flavor material	A flavor ingredient obtained by synthesis, or isolated from natural products through chemical processes, chemically identical to the substance present in a natural product and intended for human consumption either processed or not; *e.g.* citral obtained by chemical synthesis or from oil of lemongrass through a bisulfate addition compound.
NCHS	National Centre for Health Statistics, United States-NCHS collected and analyzed statistics on health status, human needs and resources. NCHS also published data in the form of tables on weight and height for children and adolescents upto 18 years old (sex separated)
NDF (neutral detergent fiber)	The representative of the fibrous cell wall constitutes, containing lignin, cellulose, hemicelluloses and some fiber-bound proteins.
NDP (net dietary protein) cal, per cent	The percentage of total calories proved by the protein.
Nectar	Nectar is obtained by blending the thin pulp of the fruit with sugar and citric acid. The finished product has 15-20 °Brix and a mild acid taste.

Terminology	*Definition/Explanation/Principles*
Nectar	Type of fruit beverage which contains at least 20 per cent fruit juice/pulp 15 per cent TSS and 0.3 per cent acid.
Neurotoxin	Toxin affects central nervous system.
Neurotransmitter	A chemical substance involved in the transmission of neural impulses from one neuron to another. Neurotransmitters are released when an action potential reaches the end of the axon (axon terminal). Upon release, neurotransmitters cross the synapse and attach to receptors on either the dendrite or cell body of the adjacent neuron.
Neutralization of oil	Process for removal of free fatty acids from oil.
Newtonian fluid	Fluid obeys Newton's viscosity law.
	For the fluids the ratio of shear stress to shear rate is constant and equal to the viscosity of the fluid. Such fluids are called Newtonian fluids.
Nonbreak oil	Edible oil, which does not deposit solid material when heated to at least 300 °C.
Non-climacteric fruits	Fruits those are showing respiratory pattern slow drift down wards after detachment from parent plant.
Nonnutritive Sweetener	Substance having < 2 per cent of the caloric value of su-crose per equivalent unit of sweetener capacity.
Nosocomical infection	Hospital acquired infection.
Nougats	These are the confections prepared by using sugar, glucose syrup and whipping agents which increase the desired volume of air.
NPR	Weight gain in test protein group (g) + weight loss in non-protein group (g) ÷ Protein consumed (g).
NPSH	The NPSH is the amount by which the pressure at the suction point of pump (*i.e.* sum of velocity and pressure heads) is in excess of the vapour pressure of the liquid.
NPU	N intake – fecal N – urine N × 100 ÷ N intake + (metabolic N + endogenous N) × 100 ÷ N intake.
Nuclear magnetic resonance (NMR)	Brain-imaging technique in which the head is exposed to magnetic fields of varying strengths and a computer-assisted three-dimensional image of the brain is produced.
Nutraceuticals	Nutraceuticals are products and their derivatives that occur in Nature and are constituents of plants and animal, including humans. These constituents confer a health benefit above and beyond basic nutrition or basic fortification.

Terminology	*Definition/Explanation/Principles*
Nutrient	The combination of processes by which a subject receives and utilizes materials (foods) necessary for the maintenance of body components.
Nutrient Supplement	Substance necessary for the body's nutritional and metabolic process.
Nutrients	Nutrients are the constituents in food that has specific function in body.
Nutrition	It is a science of food and its interactions with an organism to promote and maintain health.
Nutrition assessment	Includes the measurement and description of the nutritional status of an individual or population in relation to economic, socio-demographic and physiological variables
Nutritional status	The condition of the body of individual (s) resulting from intake, absorption and utilization of food or nutrients over a period of time.
Nutritionist	A person who advise about nutrition and/or works in the field of food and nutrition. In many states in the United States a person does not need formal training to use this title. Some states reserve this title for Registered Dietitians.
Nutritive Sweetener	Substance having > 2 per cent of sucrose per equivalent unit of sweetening capacity.
Nyctalopia	It is night blindness, failure or imperfection of vision at night or in a dim light
Obesity	A condition characterized by excess body fat, usually defined as 20 per cent above desirable weight.
	When the energy intake exceeds over expenditure, the excess is deposited as a fat over a period of time obesity occurs.
Objective evaluation	Objective evaluation involving instruments. These instruments may be categorized into two types, viz., imitative measures and non-imitative measurements.
Odor and odorant	That which is smelled. Odor may refer to the odorant or to the sensation resulting from the stimulation of olfactory receptors in the nasal cavity by gaseous material.
Odoriphore or Osmophore	Odor-producing group.
Olericulture	It deals with cultivation of vegetable crops.
Oligosaccharides	Carbohydrates containing three to ten monosaccharide units.

Terminology	*Definition/Explanation/Principles*
Omega-3 (w-3) fatty acid	A fatty acid with its first double bond first appearing at the third carbon atom from the methyl end ($-CH_3$).
Omega-6 (w-6) fatty acid	A fatty acid with its first double bond first appearing at the sixth carbon atom from the methyl end ($-CH_3$).
Opsonization	Promotion of phagocytosis by a specific antibody in combination with complement.
Optimum feasible solution	Any basic feasible solution which optimizes objective function of general linear programming problem is called as optimum feasible solution.
Osmics	The science of smell.
Osmosis	The movements of micro-organism in response to salt or sugar solution.
Osmyl	An odorant.
Osteoporosis	A bone disease that develops primarily after menopause in women and is characterized by a decrease in bone density.
Oven spring	It is phenomenon of increases in loaf volume of dough during first 15 minutes of baking.
Overweight	Weight in excess of the average for given sex, age and height in relation to the NCHS reference tables
Oxidation	The loss of electron from a substance involved in a redox reaction.
Oxidizing or Reducing Agent	Substance, which chemically oxidizes or re-duces another food ingredient, thereby producing a more sable product.
Oxirane test	This method is based on the addition of hydrogen halides to the oxirane group. Epoxide content is determined by titrating the sample with HBr in acetic acid, in the presence of crystal violet to a bluish green end point.
Oxygen absorption	The amount of oxygen absorbed by the sample as determined by the time to produce a specific pressure decline in a closed chamber or the time to absorb a pre-established quantity of oxygen under specific oxidizing conditions is taken as a measure of stability. This test has been particularly useful in studies of antioxidant activity.
Oxygen bomb method	The samples are dispersed in filter pulp in the glass liner of a sealed bomb at 50 psig oxygen pressures. The bomb is placed in boiling water and the time required to reduce the pressure to 2 psig is noted.
Packaging	It is an external means of preserving the food during storage, transportation, and marketing.

Terminology	*Definition/Explanation/Principles*
Parageusia	Gustatory disturbance resulting in erroneous identification of taste stimuli.
Parasite	A microbe which depends on living matter for its growth and survival.
Parboiling	It is a hydrothermal treatment given to paddy followed by drying before milling.
Parboiling of legumes	Partial boiling of legumes to reduces leaching losses.
Parosmia	A disturbance to the sense of smell resulting in smelling the wrong odors, usually perceived as repulsive.
Partition chromatography	If the moving phase is liquid and the stationary phase is a liquid film on same support, the resulting chromatography is called partition chromatography.
Passive absorption	Absorption that uses no energy. It requires permeability for the substance through the wall of the intestine and a concentration gradient higher in the lumen of the intestine than in the absorptive cell.
Pasteurization	Pasteurization means making free the food from human pathogens and most vegetative microorganisms by heat treatment upto 100 °C for shorter period.
Pathogencity	The ability of a parasite to infect or to cause the disease in host.
Pelagic fish	Fishes that found on ocean surface.
Pellagra	A disease characterized by inflammation of the skin, diarrhoea and eventual mental incapacity resulting from the lack of the vitamin niacin in the diet.
	Disease associated deficiency of nicotinic acid, characterized by dermatitis, diarrhoea, and dementia, usually seen in area where sorghum is staple food.
PEM	Protein energy malnutrition is defined as protein energy malnutrition occurring due to deficiency of protein and energy in the supplied food.
Pepsin	A protein-digesting enzyme produced by the stomach.
PER	The gain in body weight per gram of protein or Mean gain in body weight of rat (g) ÷ Protein consumed (g).
Per cent fat free	A product must be low fat or fat-free, and the percentage must accurately reflect the amount of fat in 100g of the food.

Terminology	*Definition/Explanation/Principles*
Percentile	A value on a scale of one hundred that indicate the percent of a distribution that is equal to or below it.
Perception	The process, by which the organism selects, organizes and interprets sensations.
Pernicious anemia	The anemia that results from a lack of vitamin B-12 absorption. It is pernicious (deadly) because of the associated nerve degeneration that can result in eventual paralysis and death.
Peroxide value	It is expressed as milliequivalents of oxygen present in one kilogram of fat. Peroxides are the main initial products of autoxidation. They can be measured by techniques based on their ability to liberate iodine from potassium iodide or to oxidize ferrous to ferric ions. Their content is usually expressed in terms of milliequivalents of oxygen per kilogram of fat.
PGI_2 (prostaglandin type I_2)	A type of prostaglandin (hormone-like substance) derived from arachidonic acid.
pH Control Agent	Substance added to change or to maintain active acidity or basicity. Includes buffers, acids, alkalis and neutralizing agents.
Phosphatides	A group of lipids which on hydrolysis give fatty acids, phosphoric acid and nitrogenous base.
Photo-oxidation	The oxidation of fatty acids by attack of singlet oxygen molecules produced from the usual triplet from by the action of light energy and sanitizers.
Photophos-phorelation	The process of generating ATP from ADP and phosphate by means of a proton-motive force generated by the thylakoid membrane of the chloroplast during the light reactions of photosynthesis.
Photorespiration	The metabolic pathway that consumes oxygen, evolves carbon dioxide, generates no ATP, and decreases photosynthetic output; generally occurs on hot, dry, bright days, when stomata close and the oxygen concentration in the leaf exceeds that of carbon dioxide.
Phytate phosphorus	Phosphorus binds with phytic acid, no free phosphorus is available.
Phytic acid	Inositol hexaphosphoric acid.
Pickles	The process of preservation of food in common salt or in vinegar is called pickling and the product is called pickles. Spices and edible oil are added for taste.

Terminology	*Definition/Explanation/Principles*
Pitting of legumes	Scratching of seed coat for oil penetration in to seed and it is given as a premilling treatment to the legumes.
Pk_a	Pka is defined as the pH at which given ionic compound is half dissociated.
Plant growth regulators	These are enzymes or chemicals which stimulate the vegetative growth of plant.
Plant layout	The plan or act of planning an optimum arrangement of industrial facilities is called as plant layout.
Polenske value	The number of milliliters of 0.1 N alkalis required neutralizing the volatile water-insoluble fatty acids present in 5 g of fat.
Polymerization	Polymerization or aggregation reactions generally involve the formation of large complexes.
Polymorphism	Certain triglycerides exist in several different crystal systems each of which has a characteristics melting point, X-ray diffraction pattern and infrared spectrum. This phenomenon is known as polymorphism. Polymorphic forms are solid phases of the same chemical composition that differ among themselves in crystalline structure but yield identical liquid phases upon melting (a, b′ and b polymorphic structure).
Pork	Uncured meat of pig or swine.
Postharvest technology	It is defined as techno economical activity applicable to all agricultural commodities or produced originated from farm, forest livestock and aqua culture for their conservation, handling, storage, marketing and value addition to make them useful as food, fed, fiber, fuel and industrial raw material.
Pour point	The lowest temperature at which a liquid will flow when a test container is inverted.
Poverty	A situation in which the level of living of an individual family or group is below the standard of living of the community either in terms of subsistence or in contrast to normal standards of income required for at least modest participation in community life.
Powder	When juice is converted into free flowing highly hydroscopic powder to which natural flavour in powder from is incorporated to compensate for any loss of flavour in concentration.

Terminology	*Definition/Explanation/Principles*
Power number	The ratio of external to internal forces per unit volume of liquid is defined as the power number.
Pre cooling	Removal of field heat from the freshly harvested commodities (Fruits and vegetables) using cooling media.
Pre packaging	Operation of packaging of fresh produce in thin plastic films.
Prebiotics	Prebiotics are a collective term for non-digestible but fermentable dietary carbohydrates that may selectively stimulate growth of certain bacterial group resident in the colon, such as biofidobacteria, lactobacilli and eubacteria, considered to be beneficial for human host.
Precipitation	It includes all aggregation reactions leading to a total or partial loss of solubility.
Preservatives	Compounds that extend the shelf life of foods by inhibiting microbial growth or minimizing the destructive effect of oxygen and metals.
Preserves	It is a product made from properly matured fruit by cooking it whole or in the form of large pieces in heavy sugar syrup until it becomes tender and transparent. In this case cooking is continued until a concentration of T.S.S. is reached to 68.5 °Brix
Primary metabolite	A metabolite produced by organism during the growth phase.
Primary packaging	It is direct contact with the contained product.
Principle of canning	Destruction of spoilage organisms within the sealed container by means of heat.
Prions	Smallest infective forms of life.
Probes	Different instruments used for instrumentation and their control (sensitive)
Probiotics	Probiotics are living microorganisms that following ingestion from part of the colonic flora at least temporarily and are used with a view to improving the health and well being of the host, *e.g.* Lactobacillus strain GG.
Processing Aid	Substances used as a manufacturing aid to enhance the appeal or utility of a food or component. Includes clarifies, clouding agents, catalysts, flocculants, filter aids, crystallization inhibitors, etc.

Terminology	*Definition/Explanation/Principles*
Proof	The alcoholic concentration of beverage spirits is expressed in terms of proof in Canada, the United Kingdom, and the United States. U.S. regulations defines this standards as follows: proof spirit shall be held to be that alcoholic liquor which contains one-half its volume of alcohol of a specific gravity of 0.7939 at 15.6 °C *i.e.* the figure for proof is always twice the percent alcohol content by volume. For example, 100° proof means 50 per cent alcohol by volume. In the United Kingdom as well as Canada, proof spirit is such that at 10.6°C alcohol weighs exactly twelve-thirteenths of the weight of an equal bulk of distilled water. A proof of 87.7° indicates an alcohol concentration of 50 per cent. A conversion factor of 1.142 can be used to change British proof to U.S. proof.
Proof gallon	Proof gallon is a U.S. gallon of proof spirits or the alcoholic equivalent thereof, *i.e.*, a U.S. gallon 3785 cm^3 (231 cubic in.) containing 50 per cent of ethyl alcohol by volume. Thus a gallon of liquor at 120° proof is 1.2 proof gallons; a gallon at 86° proofs is 0.86 proof gallons. A British and Canadian proof gallon is an empirical gallon of 4546 cm^3 (277.4 cubic in.) at 100° proof (57.1 per cent of ethyl alcohol by volume). An empirical gallon is equivalent to 1.2 U.S. gallons. To convert British proof gallons to U.S. proof gallons, multiply by 1.37. Since excise taxes are paid on the basis of proof gallons, this term is synonymous with tax gallons.
Proofing	Process of exposing dough to 31–32°C with 88 per cent relative humidity in closed proofing chamber to increase loaf volume of dough by stimulating process of fermentation.
Pro-oxidants	Chemical substances and basic natural conditions or environmental states, which actively promote the oxidation of fats.
Propellant	Gas used to supply force to expel a product or used to reduce the amount of oxygen in contact with the food in packaging.
Protein	A three-dimensional biological polymer constructed from a set of 20 different monomers called amino acids.
Protein association	Protein association reactions generally refer to changes occurring at the subunits or molecular level.
Protein calorie malnutrition (PCM)	Clinical and biochemical disorders caused by various degrees of deficiency and additional physical insults and stress. Inadequate dietary intake of good quality protein and calories

Terminology	*Definition/Explanation/Principles*
Protein efficiency ratio (PER)	A measure of protein quality determined by the ability of a protein to support the growth of a young animal.
Protein isolate	Percent protein content is more than 90 per cent.
Protein-energy malnutrition (PEM)	This results when a person regularly consumes insufficient amounts of kcalories and protein. The deficiency eventually results in body wasting and an increased susceptibility to infections.
Prothrombin	A blood protein needed for blood clotting that requires vitamin K for its synthesis.
Proton pump	An active transport mechanism in cell membranes that consumes ATP to force hydrogen ions out of a cell and in the process generates a membrane potential.
Psychometric chart	A psychometric chart is a graph of the properties of air-water vapour mixtures as a function of temperature. Lines of constant wet bulb in psychometric chart are a plot of equation. If the 'wet bulb' temperature Tw and the dry bulb temperature Ta are known it is possible to establish the humidity Ha.
PUFA	Poly unsaturated fatty acids, more in vegetable oil.
Pungency	Certain compounds found in several spices and vegetables cause characteristic hot, sharp, and stinging sensations that are known collectively as pungency.
Radiation	A hot body gives off heat in the form of radiant energy, which is emitted in all directions. When this energy strikes another body, part is reflected and part may be transmitted unchanged through the body, depending on its degree of opacity. The remainder is absorbed and quantitatively transformed into heat except where photochemical or special reactions are induced.
Radiation dose	The quantity of radiation energy absorbed by a food as it passes through in processing.
Radioactivity	It is property of an atom whose nucleus or center is physically unstable and spontaneously releases radiation energy.
Rancidity	The development of off-flavour in stored oil/fat due to lipolysis or autoxidation.
	Development of off flavor in food items due to break down of unsaturated fatty acids to saturated one.
Reactor	Reactor is a well designed vessel used for different bioconversion reaction.

Terminology	*Definition/Explanation/Principles*
Recommended dietary intake (RDI)	Recommendations from the original tenth edition RDA Committee that were published in 1987 in the American journal of Clinical Nutrition after the National Academy of Sciences refused to publish the original tenth edition of the RDA.
Recommended Nutrient Intake (RNI)	The Canadian version of RDA.
Reduced fat	25 per cent or less fat than a full-fat product per serving.
Reduced or fewer calories	25 per cent or fewer calories than regular product per serving.
Reducing agent	In one sense a compound capable of donating electrons (also hydrogen ions) to another compound.
Reduction	To gain an electron or hydrogen atom.
Reference Daily Intake (RDI)	Standards of expressing nutrient content on nutrition labels. RDI figures are based on average 1989 RDA values set for a nutrient that span a particular age range, such as children over 4 years through adults. RDI soon should replace U.S. RDA.
Refractive index	A measure of the bending or refraction of a beam of light on entering a dense medium. It is the ratio of the sine of the angle of incidence of the ray of light to the site of the angle of refraction.
Refrigeration	If the primary purpose is to discharge heat to a certain high temperature region, the system is called a heat pump. If the purpose is to absorb heat from a certain low temperature region, the system is called a refrigerator.
Regulated market	In these markets business is done accordance with rules and regulations.
Reichert-Meissel value	The R-M value of a fat is the number of milliliters of 0.1 N potassium hydroxide required to neutralize the stem-volatile fatty acids in 5 g of fat.
REM sleep	A periodic state during sleep identified by rapid eye movements and correlated with vivid dreaming.
R-enamel cans	Lacquered cans used for acid are known as R-enamel cans.
Rendering	Steam heating to oil seeds in vacuum container.
Rennin	An enzyme formed in the kidney in response to low blood pressure; it acts on a blood protein to produce angiotensin I.

Terminology	*Definition/Explanation/Principles*
Respiration	Oxidative breakdown of the complex materials of cell into simples' molecules with the concurrent production of energy required by the cell for the completion of chemical reactions.
	Fundamental process of conversion of potential energy to kinetic energy.
Respiration rate	It is postharvest phenomenon where grains take in O_2 and give out CO_2.
Respiratory quotient	It is an arithmetic ratio of mole of CO_2 evolved to the moles of O_2 absorbed.
Restoration	Addition to restore the original nutrient content.
Retention time	Time required coming out of eluent.
Retina	The innermost layer of the vertebrate eye containing photoreceptor cells (rods and cones) and neurons; transmits images formed by the lens to the brain via the optic nerve.
Retinal	The light-absorbing pigment in rods and cones of the vertebrate eye.
Retrogradation	The insolubilizating effect, initiated when the long and somewhat unwieldy molecules begin to crystallize is retrogradation when it occurs in starch.
Reversion	The development of a characteristic beany, buttery, fishy, grassy, hay like, or painty flavor in oil before the onset of autoxidation.
Reynolds number	The ratio inertial force to viscous force.
Rheology	Rheology is concerned with stress-strain relationships of materials that show, flavor, intermediate between those of solids and liquids.
	It is viso-elastic property that studies deformation of dough including elasticity and flow.
Rhodopsin	A protein involved in vision; it is made in the eye and incorporates a protein called opsin and a form of vitamin A; especially important in night vision.
Rickets	A disease characterized by softening of the bones because of poor calcium deposition. This deficiency disease arises from lack of vitamin D activity in the body.
	The bone abnormalities due to deficiency of Ca, P and vit. D.
Rigor mortis	Stiffening of muscles due to loss of ATP and fall in pH during muscle postmortem.

Terminology	*Definition/Explanation/Principles*
Ripening	It is defined as the sequences of changes in color, flavor, and texture which lead to state at which the fruit is acceptable to eat.
Roasting of groundnut	Dry heat treatment to groundnut for desirable flavor, light roasting.
Ropyness of bread	It is bread fault where sticky, gummy material is developed in center of loaf 1 to 3 days after baking due to contamination of microorganism *B. mesentericus.*
Royal jelly	It is used by bees to include development of larvae into the queen phenotype and is currently sold as a general health tonic.
Saccharin	An alternate sweetener that yields no energy to the body; it is 500 times sweeter than sucrose.
Salt	Generally refers to a mixture of sodium and chloride in a 40:60 ratio.
Salty taste	Only salts are salty; however, not all salts are salty. Some are sweet, bitter, or tasteless. Monovalent cations, especially sodium, can pass directly through ion channels in the tongue, leading to an action potential leading to the salty percept. Sodium chloride in foods may be analyzed using a specific ion electrode. To measure other salts, ion chromatography or atomic absorption emission spectroscopy are generally used.
Sapid	Having the power of affecting the taste receptor.
Sauce	Sauces are mainly two kinds; thin and thick sauces. The thin sauce contains vinegar, extract of spices and herbs. Thick sauce is more viscous and does not flow easily. It has at least 3 per cent acetic acid for improving the keeping quality. Tomato pulp without skin and seeds are used as basic raw material for tomato sauce. Apples, peaches, plums, apricots, mangoes, cauliflower, carrots are also used for preparation of sauce. Onion, garlic, spices, and herbs are used as flavouring agents while vinegar; salt and sugar are added for taste. The mixtures of above all material cooked to get consistency of jam *i.e.* upto 68 to 69 °Brix.
Savory	Appetizing; having an agreeable flavor.
Scalding	It is process of treating birds with hot water and agitating for short period of time. It facilitates removal of feathers by expanding or relaxing muscles that surrounds feathers.
Scale up	Scale up is the study of the problem associated with transferring data obtained in laboratory and pilot equipment to industrial production.

Terminology	*Definition/Explanation/Principles*
Schaal oven test	The sample is stored at about 65 °C and periodically tested until oxidative rancidity is detected.
Scientist P Fizer	In the year 1923 obtained 1st successful plant of citric acid fermentation with the help of sugar and M. O. *Aspergillus niger*.
Score	Which are used to represent their achievement are average scores of the two or more examinations
SCP	Single cell protein *i.e.* protein derived from microbial cell for use as food or a feed supplement.
Screening	The programme which consist of highly selective procedure used for detection and isolation of industrial important microorganisms.
Scurvy	The deficiency disease that results after a few weeks of consuming a diet free of vitamin C; pinpoint hemorrhages on the skin are an early sign.
Scutellum	It is cementing layer which separates endosperm from germ.
Secondary response	Antibody made on second (subsequent) exposure to antigen mostly of class Ig G.
Self-fertile	It refers to ability of plant to mature viable seeds without aid of pollens from some other plant flower.
Semi perishable foods	Food products which contain water in the range of 10-30 per cent are called semi perishable food.
Semolina	Coarsely ground white endosperm chemically same to white flour.
Senescence	It is define as period during which anabolic (synthetic) biochemical processes give way to catabolic (degradative) processes leading to aging and finally the dead of tissue.
	It is defined as period during which anabolic biochemical processes gives way to catabolic processes leading to aging and finally the death of tissue.
Sensory analysis	The science of measuring and evaluating the properties of food products by one or more human senses.
Septicemia	Infection of the bloodstream by microorganisms.
Sequestrants	These are chelating agents or sequestering compounds. They react with trace elements such as iron and copper present in foods and remove them from solution. The trace elements are active catalysts of oxidation and discoloration of food products. Sequestrants such as ethylene diamine tetra acetic acid (EDTA), poly phosphates and citric react with trace elements and inactivate them.

Terminology	*Definition/Explanation/Principles*
Sequestrants	Chemical substances used to chelate or complex free metals in food system.
Serbet (Juice)	It is clear sugar syrup, which has been artificially flavoured.
Settling	It is a method of oil refining. It involves storing heated fats quiescently in tanks with conical bottoms. Water and materials associated with water settle into the cone from where they are drawn off.
Settling down period	Period at which solid surface conditions comes into equilibrium with drying air.
Sequesterant	Substance, which combines with polyvalent metal ions to form a soluble metal complex to improve the quality and stability of products.
Sherbats	This is cooling drink of sweetened diluted fruit juice.
Sherry	It is term given to number of related type of dessert wine originally developed in the area around Serez in the South of Spain.
Shortenings	These are compounded from mixtures of fats prepared by hydrogenation and are called lard compounds or lard substitutes.
Simultaneous reaction	Nutrients converted to products in a variable proportion without accumulation of intermediate
Single whiskey	Single whiskey is the whiskey, either grain or malt, produced by one particular distillery. Blended Scotch whiskey is not a single whiskey.
Size reduction	The breakdown of solid materials through the application of mechanical forces.
Smoke point	The temperature at which the column of solidified oil or fats starts rising in the capillary tube.
Smoking	Decreases the available moisture on surface of the meat, preventing microbial growth and spoilage. It enhances color and falvour of meat.
Soaking of legume	Absorption of water / medium through hull or seed coat.
Sodium-potassium pump	A special transport protein in the plasma membrane of animal cells that transport sodium out of and potassium into the cell against their concentration gradients.
Softening point	The temperature at which the column of solidified oil or fat starts rising in the capillary tube.

Terminology	*Definition/Explanation/Principles*
Soil	It can be defined as superficial earth crust which functions as store lower time providing necessary physical support to plant.
Solubility	The tendency of molecule to dissolve in a liquid.
Solute	A substance that is dissolved in a solution.
Solvent	The dissolving agent of a solution. Water is the most versatile solvent known.
Sour mash	Sour mash is made with a lactic culture and not less than 20 per cent stillage added back to the fermentor and fermented for at least 72 h.
Sour taste	Sourness indicates acidity, through not all acids are sour. The detection of acids facilitates maintaining the body fluid compositional balance. The pH is characteristic of the carbon dioxide levels in blood and cerebrospinal fluid.
Specific gravity	The density of a substance is defined as mass per unit volume. The density of a substance is characteristic property and has a definite value at a given temperature and pressure. The density of one substance in relation to the density of another material (*e.g.* water) is known as specific gravity.
Spirits	Spirits are distilled spirits including all singular whiskey, gin, brandy, rum, cordials, and others made by a distillation process for nonindustrial use.
Spore	Resistant, resting stage of bacteria.
Spray drying	Spray drying is usually applied to fluids high in moisture content. The fluid is divided into droplets by its passage through a spray nozzle at high pressure or its passage through a centrifugal disk at high speed.
Sprouting of legumes	Soaking of legumes in water and then kept for vegetative growth.
Squash	It is essentially strained juice containing moderate quantities of fruit pulp to which cane sugar is added for sweetness.
Stabilizer or Thickener	Substance used to produce viscous solutions or dispersions, to impart body, improve consistency, or stabilize emulsions. In-cludes suspending and bodying agents, setting agents, and bulking agents.
Starter culture	Culture of desirable microbes used to initiate fermentation.
Stature	Height of the body in a standing position.
Stepwise reaction in kinetic pattern	Nutrients completely converted to intermediate before conversion to product.

Terminology	Definition/Explanation/Principles
Sterilization	Sterilization means the destruction of all viable microorganisms by heat treatment.
Steroids	A group of hormones and relate compounds that are derivatives of cholesterol.
Stevens power law, $S = I^n$	The increase in perceived intensity, S, is equal to the concentration, I, to the nth power.
Stillage	Stillage is dealcoholized fermented mash.
Strecker degradation	It involves the interaction of $\propto$-dicarbonyl compounds and $\propto$-amino acids. Volatile products, such as aldehydes, pyrazines and sugar fragmentation products from the strecker reaction may contribute to aroma and flavour.
Stress	A pattern of disruptive psychological and physiological functioning that occurs when an environmental event is appraised as a threat to important goals and one's ability to cope.
Stunning	Of making animal unconscious before slaughtering
Subjective evaluation	Human sense organs for taste, smell, sight, touch and hearing are the ways for subjective evaluation of food.
Superoxide dismutase	An enzyme that can neutralize a superoxide free radical.
Surface Active Agent	Substance used to modify surface properties of liquid food components for a variety of effects, other than emulsifiers. Includes solubilizing agents, dispersants, detergents, wetting agents, rehydrating enhancers, foaming agents, defoaming agents, etc.
Surface Finishing Agents	Substance used to increase palatability, preserve gloss and inhibit discoloration of foods. Includes glazes, polishes, waxes and protective coatings.
Sweating	Moisture condensation on surface of endosperm during wheat conditioning.
Sweet taste	Two types of receptors systems correspond with sweet taste; one responds to certain carbohydrates and the other to high potency sweeteners. The structural requirements for a compound to active sweet receptors have not been fully defined. High pressure liquid chromatography is a key tool used in analyzing sweet components.
Syneresis	The phenomenon of spontaneous exudation of fluid from gel.

Terminology	*Definition/Explanation/Principles*
Synergist	Substance used to act or react with another food ingredient to produce a total effect different from or greater than the sum of the effects produced by the individual ingredients.
Syrup	It is clear sugar syrup, which has been artificially flavoured. It contains 25-30 per cent juice, 59 per cent sugar and 1 per cent citric acid. Syrup used as a sherbet by diluting 1 cup of syrup with 3 cups of water.
Tails	Tails is a residual alcoholic distillate.
Tallow	It is a lipid obtained from beef by the process of rendering.
Tannins	The phenolics with MW 500 to 3000 that form complexes with proteins.
Taste	Chemoreceptive events in the mouth lead to taste perception. Taste is typically described by five modalities coupled with chemesthesis: salt, sour, sweet, and bitter and umami.
TBA (thiobarbituric acid) test	Thiobarbituric acid is a more sensitive reagent than phloroglucinol and gives a yellow coloration with various saturated and unsaturated aldehydes and a red color with a more restricted number.
TD (true digestibility)	N intake – (fecal N – metabolic N) × 100 ÷ N intake.
Tempering	Process of equalizing the moisture of grain by temporary holding the paddy/any grains between drying passes.
Tempering of chocolate	It is the most important process in chocolate manufacturing which decides the quality. It consists of cooling the chocolate with continuous mixing to produce cocoa butter seed crystals and uniform distribution it is a technique of uniform distribution.
Ten State Nutrition Survey (TSNS)	A survey, which was funded by the Department of Health, Education and Welfare (United States) to study the problem of malnutrition in the United States.
Tetany	A syndrome marked by sharp contraction of muscles with failure to relax afterward; usually caused by abnormal calcium metabolism.
Texture	Closely related to the rheology of food materials the texture of food. It is a structural quality of a material.
Texturizers	The chemical substances used to improve texture of food products.

Terminology	Definition/Explanation/Principles
Thermal reactions	Thermal reactions occurring with carbon-carbon bond cleavage yield as primary products volatile acids, aldehydes, ketones, diketones, furans, alcohols, aromatics, carbon monoxide and carbon dioxide.
Thermodynamics	Thermodynamics is the science of energy transfer in relation to the physical properties of substances.
Thickeners	The chemical substances used to improve thickness, viscosity and water holding capacity of liquid or semi-liquid foods.
Thiobarbituric acid (TBA) test	This is one of the most widely used tests for evaluating the extent of lipid oxidation. Oxidation products of unsaturated systems produce a color reaction with TBA. It is believed that the chromagen results from condensation of two molecules of TBA with one molecule of malonaldehyde.
Thiocyanogen value	This value is the result of determining the percentage of oleic, linoleic and linolenic acids, provided that other unsaturated acids are not present and the iodine number and the amounts of saturated acids are known.
Titer test	When molten fatty acids are cooled and begin to solidify, the latent heat of fusion is liberated and a sudden rise in temperature can be observed.
Tocopherols	The chemical name for some forms of vitamin E.
Tomato juice	It is a juice, which contains T.S.S. not less than 5.6 per cent at 20 °C (4-6 per cent salt is added to counteract the astringent taste of the juice and 1 per cent sugar is added to improve the taste).
Tomato ketchup	It is a product made by concentrating tomato juice or pulp (without seed and pieces of skin). Spices, salt, vinegar, onion, garlic are added to the extract that the ketchup contains not less than 12 per cent tomato solids and 28 per cent total solids.
Tomato paste	It is a concentrated tomato juice (without skin and seeds) containing not less than 25 per cent tomato solids. It is further concentrated to 33 per cent tomato solids.
Tomato puree	It is a concentrated tomato pulp without skin or seeds containing not less than 8.37 per cent tomato solids. It is further concentrated to not less than 12 per cent tomato solids. It is called heavy purees.
Total and volatile carbonyl compounds	Methods for the determination of total carbonyl compounds are usually based on measurement of the hydrazones arising from reaction of aldehydes and ketones (oxidation products) with 2,4-dinitrophenylhydrazine.

Terminology	*Definition/Explanation/Principles*
Totipotency	Capacity of cell to regenerate phenotype of the complete and differentiated organism from which it is derived.
Toxin	Any harmful chemical substance released by microbes.
Tracer	Substance added as a food constituent (as required by regulation) so that levels of this constituent can be detected after subsequent processing and / or combination with other food materials.
Transduction	Bacteria phases function as intermediate in transfer of bacterial genetic information from one bacteria to another.
Transpiration	It is important physiological activity of fruit and vegetables in which water loss from lenticels or pores to surrounding atmosphere takes place relates to physiological loss in weight.
Tricep Skin fold Thickness	Measurement, with a calibrated caliper, of thickness of a fold of skin at the upper arm. It is a measurement of subcutaneous fat. Since subcutaneous fat make up about 50 per cent of the adipose tissue stores, tricep skin fold thickness can be an important measurement to estimate body fat reserves of an individual.
Turbidity point	The temperature at which a solution of oil or fat starts showing turbidity when it is cooled slowly.
Turbulent flow	The flow in which the fluid instead of flowing in orderly manner moves erratically in the form of cross and eddies is called turbulent flow.
Ultimate pH	The lowest pH of muscles after slaughter of animal.
Ultra filtration	The process of separating components of a solution largely on the basis of molecular size, utilizing a membrane as a molecular sieve.
Ultra filtration membranes	Membrane is preventing the passage of larger solute molecules in solution by means of filtration through micro pores in a membrane structure.
Umami taste	Umami is the taste of a few amino acids (*e.g.*, glutamate, aspartate, and related compounds) and was classically not included as a taste modality. Sometimes described as savory, brothy, or meaty, it is the dominant taste of such foods as chicken broth, meat, and ageing cheese. Umami perception results from activation of two receptor systems, with one that overlaps with the receptor systems for artificial sweeteners. There are many HPLC systems that integrate sample preparation along with data analysis specific to amino acids analysis.

Terminology	*Definition/Explanation/Principles*
Unbalanced transportation problem	The transportation problem in which the requirement and capacity are not balanced.
Under nutrition	Inadequate intake of one or more nutrients including calories. Chronic under nutrition (chronic malnutrition) refers to a long-term inadequate food intake and is reflected by low height-for-age levels. Acute under nutrition refers to a short-term severe inadequate food intake and is reflected by low weight for height levels.
	State of insufficient supply of essential nutrients.
Under weight	Weight below the average for a given sex, age and height in relation to the NCHS reference tables.
Unit operations	Aerobic fermentation involves unit operation the mixing of three heterogeneous phase microorganisms, medium and air mass transfer of oxygen from the air to the organisms and heat transfer from the fermentation medium.
Unit processes	Fermentation processes can be classified by the reaction mechanisms involved in converting the raw materials into products; these include reductions, simple and complex oxidations, substrate conversions, transformations, hydrolysis, polymerization, complex biosynthesis and formation of cells.
Unsaponifiable matter	This term includes all those constituents of fats, which are not saponified by alcoholic caustic potash. They include sterols, carotenoids, long-chain alcohols and hydrocarbons.
Unsaturated fatty acids	Hydrocarbon chain of fatty acid contains double bonds between carbon atoms.
Uric acid	An insoluble precipitate of nitrogenous waste excreted by land snails, insects, birds and some reptiles.
UV Spectro-photometry	Measurement of absorbance at 2 nm (conjugated dines) and 268 nm (conjugated trienes) is sometimes used to monitor oxidation.
Variability	In a frequency distribution, the separation, dispersion, or spread of the scores on the x-axis.
Variable cost	The cost which varies with the amount produced, raw material, labour.
Vegetable oil	Vegetable oils mostly contain large amounts of oleic and linoleic acids and less than 20 per cent saturated fatty acids. The most important group is cottonseed, corn, peanut, sunflower, safflower, olive, palm, and sesame oils.

Terminology	*Definition/Explanation/Principles*
Verification	Routinely check the system for accuracy to verify that it is functioning properly and consistently.
Very low calorie diet (VLCD)	Known also as protein sparing, modified fast (PSMF), this diet allows a person 400 to 700 kcalories per day, often in liquid form. Of this, 30 to 120 grams are carbohydrate; the rest is high biological value protein.
Very low density lipoprotein (VLDL)	The lipoprotein that initially leaves the liver. It carries both the cholesterol and lipid newly synthesized by the liver.
Virulence	It is the degree or capacity or intensity of pathogencity of an organism *i.e.* to cause the disease in host.
Viscosity	The resistance to flow of a fluid. In the cgs (centimeter-gram-second) system, the absolute unit viscosity is the poise.
Vitamin	An organic molecule required in the diet in very small amounts; vitamins serve primarily as coenzymes or parts of coenzymes.
	Organic compounds occurring in small quantities in the different natural foods and necessary for the growth and maintenance of good health in human being.
Viticulture	It can be defined as the science and art of growing grapes.
Volatibility	The tendency of a molecule to pass in to the vapour state.
Volatile oil	That portion of a botanical that co-distills with water during steam distillation and is generally flavourful.
Wash	Wash is the liquid obtained by fermenting wort with yeast. It contains the beverage spirits and congeners developed during fermentation.
Washing or Surface Removal Agent	Substance used washes or assist in the removal of unwanted surface layers from plant or animal tissues.
Water activity (a_w)	It is a ratio between the partial pressure of water in a sample (P) to the vapor pressure of pure water at the same temperature (Po): $a_w = P/Po$.
	F/Fo = The fugacity of the solvent/The fugacity of the pure solvent.
	Percent equilibrium relative humidity surrounding the product.
Water holding capacity	It is a term frequently employed to describe the ability of a matrix of molecules usually macromolecules to entrap large amounts of water in a manner such that exudation is prevented.

Terminology	*Definition/Explanation/Principles*
Water potential	The physical property predicting the direction, in which water will flow, governed by solute concentration and applied pressure.
Water structure breakers ions	K^+, Rb^+, Cs^+, NH_4^+, Cl^-, Br^-, I^-, NO_3^-, BrO_3^-, IO_3^-, and ClO_4^-
Water structure formers ions	Li^+, Na^+, H_3O^+, Ca^{2+}, Ba^{2+}, Mg^{2+}, Al^{3+}, F^-, and OH^-
Wave length	In measuring light and sound waves, the distance from the crest of one wave to the crest of the next.
Waxes	Esters of long-chain monohydroxy alcohols and higher-chain fatty acids.
Waxy rice	Rice with low amylose content.
Weber's law	Principle stating that the amount by which a stimulus must be increased or decreased to be perceived as different is always a constant proportion of the initial stimulus intensity.
Wet milling of legume	Related to pre-milling treatment, soaking in water.
Wheat shorts	Mixture of bran and germ produced from milled wheat.
Whey	It is by product of cheese and casein and contains approx. 20 per cent of the original milk proteins. These proteins include α-lactpoalbumin, β-lactoglobulin, lactoferrin, lacto-peroxidase, immunoglobulins, glycomacropeptide and a variety of growth factors.
Whitening of rice	The process of removal of bran from brown rice called as whitening of rice.
Wine Gallon	Wine gallon is the measure of actual volume; a U.S. gallon (3.785 L) contains 3785 cm^3 (231.0 cubic in.); a British (Imperial) gallon contains 4546 cm^3 (277.4 cubic in.).
Winterizing	A process where oil is held at 5 °C until crystallization is well advanced, and then filtered in chilled room to remove solids. The winterized oil is used as salad oil.
Work	A force is a means of transmitting an effect from one body to another.
Wort	It is the clear liquid obtained through the filtration of insoluble materials during mashing process of brewing.
Xantham	It is polysaccharides produced by Xanthomonas compestris.

Terminology	*Definition/Explanation/Principles*
Xerophthalmia	A cause of blindness that results from infection of the eye secondary to vitamin A deficiency. The specific cause is a lack of mucus production by the eye, which then leaves it more vulnerable to surface dirt and bacterial infections.
Xylitol	An alcohol derivative of the five-carbon monosaccharide, xylose.
Yeast	A unicellular fungus that lives in liquid or moist habitats, primarily reproducing asexually by simple cell division or by budding of a parent cell.
	Single cell micro organism which multiplies by process of budding. They are single cell protein and rich source of enzymes like lipase, amylase, protease and zymase.
Yogurt	It contains live lactic acid bacteria (probiotic).
Yolk index	It is used to evaluate egg quality height of the yolk in relation to width of yolk. Yolk index is high in fresh egg.
Yo-Yo dieting	The practice of losing weight and then regaining it, only to lose it and regain it again.
Z value	Number of degree required for a specific thermal death time curve to pass through one log cycle.
Z-Score	A score's distance from the mean of the group expressed in units of the standard deviation; used to identify unusual values.
Zwitter ion	Amino acid having net charge zero *i.e.* equal no positive and negative charges.
Zymogen	An inactive form of an enzyme.

APPENDIX–II
Abbreviations Used in Food Science and Technology

Short Form	*Long Form/Full Form*
λ	Latent heat of vaporization, kcal/kg
1-MCP	1-methyl Cyclopropene
AACC	American Association of Cereal Chemists
AAO	Ascorbic Acid Oxidase
ACC	1-aminocyclopropane-1-1carboxylic acid
ACNFP	Advisory Committee on Novel Foods and Processes
ACS	American Chemical Society
ADA	Azodicarbonmide
ADF	Acid Detergent Fiber
ADH	Antidiuretic Hormone
ADI	Acceptable Daily Intake
ADPI	American Dry Products Institute
AECA	Aroma Extract Concentration Analysis
AFLP	Amplified Fragment Length Polymorphism
AGMARK	Agricultural Produce (Grading and Marketing) Act, 1937
AGP	Alpha 1-acid Glycoprotein
AI	Adequate Intake
AICSIP	All India Co-ordinated Sorghum Improvement Project
AIDS	Acquired Immunodeficiency Syndrome
ALP	Alkaline Phosphatase
ALV	Available Lysine Value
AMC	Automatic Machinery and Electronics Inc.
AMCP	Anhydrous Monocalcium Phosphate
AMD	Age-related Macular Degeneration
AML	Amylose Leaching
AMS	Agricultural Marketing Service
ANOVA	Analysis of Variance
AOAC	Association of Official Analytical Chemists

Short Form	*Long Form/Full Form*
AOM	Active Oxygen Method or Atmospheric Oxygen Metabolizing
APEDA	Agricultural and Processed Food Export Development Authority
APHIS	Animal & Plant Health Inspection Service
APMC	Agricultural Produce Market Committees
APP	Acute Phase Proteins
ARI	Acute Respiratory Infection
ASAE	American Aociety of Agricultural Engineers
ASTM	American Society of Testing Materials
ATA	Alimentary Toxic Aleukia
ATP	Adenosine Triphosphate
a_w	Water Activity
BAM	Bacteriological Analytical Manual
BAPN	β-Amino Propionitrile
BATF	Bureau of Alcohol, Tobacco and Firearms
BBT	Bright Beer Tank
BHA	Butylated Hydroxy Anisole
BHI	Body Mass Index
BHT	Butylated Hydroxy Toluene
BIS	Bureau of Indian Standard
BMC	Bone Mineral Content
BMD	Bone Mineral Density
BOAA	β-N-Oxalylamino-L-alanine
BOD	Biological Oxygen Demand
BOP	Broken Orange Pekoe
BOPE	Broken Orange Pekoe Fannings
BRAP	Bilateral Research Activities Program
BSA	Bovine Serum Albumin
BSE	Bovine Spongiform Encephalopathy
BU	Brabender Units
BV	Biological Value
BVA	Brabender Visco Amylogram
BVO	Brominated Vegetable Oils

Short Form	*Long Form/Full Form*
CA	Controlled Atmosphere
CAC	Codex Alimentarius Committee or Citric Acid Cycle
CACP	Commission for Agricultural Costs and Prices
CAF	Calcium Activated Factor
CANP	Calcium Activated Neutral Proteinase
CAP	Controlled Atmosphere Packaging
CAPP	Calcium Acid Pyrophosphate
CAS	Controlled Atmosphere Storage
CBER	Center for Biologics Evaluation and Research
CC	Column Chromatography
CCI	Cotton Corporation of India
CCK	CholecysTokinin
CCMP	Cooked Cured-Meat Pigment
CCP	Critical Control Point
CDC	Centers for Disease Control and Prevention, Atlanta, USA
CDER	Center for Drug Evaluation and Research
CDT	Come-Down Time
CE	Capillary Electrophoresis
CEPCI	Cashewnuts Export Promotion Council of India
CER	Carbon-dioxide Evolution Rate
CFA	Corrugated Fiber Board
CFAM	Cyclic Fatty Acid Monomers
CFC	Chloro Fluoro Carbon
CFTRI	Central Food Technology Research Institute Mysore
CFU	Colony Forming Units
CGC	Canadian Grain Commission
CHARM	Combined Hedonic and Response Measurements
CHD	Coronary Heart Disease
CIP	Clean-in-Place
CLA	Conjugated Linoleic Acid
CM	Carboxymethyl
CM	Chloroform-Methanol

Short Form	*Long Form/Full Form*
CMC	Carboxy Methyl Cellulose
COD	Chemical Oxygen Demand
COSAMB	Council of State Agricultural Marketing Boards
CPg	Specific Heat of Grain, Kacl/kg °C
CPS	Counts Per Second
CPw	Specific Heat of Water, Kacl/kg °C
CRD	Complete Randomized Design
CROs	Contract Research Organizations
CRP	C-Reactive Protein
CRS	Chinese Restaurant Syndrome
CS	Chemical Score
CSA	Cross-Sectional Open Area
CSIR	Council of Scientific and Industrial Research
CTC	Crush Tear Curl or Charge Transfer Certificate
CUT	Come-Up Time
CVA	Cerebrovascular Accident
CVD	Cardio Vascular Disease
CW	Continuous Wave
CWC	Central Warehousing Corporation
d	Level of Absolute Precision
D.B.	Dry Bulb Temperature
d.b.	Dry Weight Basis
DAGs	Diacylglycerols
DAP	Diamino Pimelic Acid
DAP	Diammonium Phosphate
DAS	Diacetone-L-Sorbose
DATE	Diacetyl Tartaric Acid Ester
DATEM	Diacetyl Tartaric Esters of Monoacylglycerols
DBS	Dried Blood Spot
DCPD	Di-calcium Phosphate Dihydrate
DDB	Dialkyl Dihexadecyl Malonate
DDG	Distillers Dry Grains

Short Form	*Long Form/Full Form*
DDM	Dialkyl Dihexadecyl Malonate
DDS	Distillers Dry Solubles
DE	Dextrose Equivalent or Degree of Esterification
DE	Diatomaceous Earth
DEAE	Diethylaminoethyl
DEFF	Design Effect Fortified Food
DEFT	Direct Epifluorescent Filter Technique
DEPC	Diethyl Pyrocarbonate
DFD	Dark, Firm and Dry
DHA	Docosahexaenoic Acid
DHA	Dehydroalanine
DHAA	Dehydroascorbic Acid
DHS	Demographic and Health Survey
DISCUS	Distilled Spirits Council of the United States
DLVO	Derijaguin and Landau and Verwey and Over beek
DM	Degree of Methylation
DMA	Dynamic Mechanical Analysis
DMA	Dimethylamine
DMAPP	Dimethyl allyl Pyrophosphate
DMDC	Dimethyl Dicarbonate
DMI	Directorate of Marketing and Inspection
DMTA	Dynamic Mechanical Thermal Analysis
DNA	Deoxyribonucleic Acid
DO	Dissolved Oxygen
DOPA	Dihydroxy Phenylalanine
DP	Degree of Polymerization or Deep Press or Diastatic Power
DRR	Dough Rate of Reaction
DRV	Daily Reference Values
DS	Degree of Substitution
DSC	Differential Scanning Calorimetry
DSM	Dutch States Mines
DTT	Dithiothreitol

Short Form	*Long Form/Full Form*
DUS	Distingness University and Stability
Dv	Diffusivity, m^2/hr
EA	Emulsion Activity
EAR	Estimated Average Requirement
ECA	Essential Commodity Act.
ECD	Electron Capture Detector
ECOST	European Corporation in the Field of Scientific and Technical Research
EDB	Ethylene Dibromide
EDCT	Ethylene Dichloride Carbon Tetrachloride
EDTA	Ethylene Diamine Tetra Acetic Acid
EFEMA	European Food Emulsifier Manufacture's Association
EGF	Epidermal Growth Factor
Eh	Redox Potential
EHEDG	European Hygienic Equipment Design Group
EIA	Enzyme Immuno Assay
EIC	Export Inspection Council
ELISA	Enzyme Linked-Immuno Sorbent Assay
ELMC	Equilibrium Moisture Content
EM	Electromagnetic
EMA	Equilibrium Modified Atmosphere
EMC	Equilibrium Moisture Content
EME	Electro-magnetic Energy
EMIT	Enzyme Multiplied Immuno Assay Technique
EPA	Environmental Protection Agency
EPA	Eicosapentaenoic Acid
EPA	Esterified Phenolic Acid
EPC	Epicatechin
EPG	Esterified Propoxylated Glycerol
EPI	Expanded Programme on Immunization
EPR	Electron Paramagnetic Resonance
ERH	Equilibrium Relative Humidity

Short Form	*Long Form/Full Form*
ES	Emulsion Stability
ESADDI	Estimated Safe and Adequate Daily Dietary Intake
ESI	Electrospray Ionization
ESLR	Extended Shelf life Refrigerated
ESP	Epithio Specifier Protein
EST	Expressed Sequence Tag
ETO	Ethylene Oxide
EVA	Ethylene Vinyl Acetate
EVOH	Ethylene Vinyl Alcohol Copolymer
FACS	Fluorescence Activated Cell Sorting
FAD	Flavin Adenine Dinucleotide
FAD	Food and Drug Administration
FAME	Fatty Acid Methyl Ester
FAO	Food and Agriculture Organization
FAS	Fetal Alcohol Syndrome
FASEB	Federation of American Societies for Experimental Biology
FCC	Food Chemical Codex
FCI	Food Corporation of India
FDA	Food and Drug Administration
FEMA	Flavour Fragrance Materials Association
FEP	Free erythrocyte protoporphyrins
FFA	Free Fatty Acid
FFDCA	Federal Food Drug and Cosmetic Act
FIA	Fluorescence Immuno Assay
FID	Flame Ionization Detector
FIFRA	Federal Insecticide, Fungicide and Rodenticide Act.
FM	Freezing-Melting
FMA	Fragrance Materials Association
FNB	Food and Nutrition Board Council
FOS	Fructooligosaccharides
FOS	Fructosyl Oligosaccharides
FOSHU	Foods for Specific Health Use

Short Form	*Long Form/Full Form*
FPA	Free Phenolic Acid
FPC	Fish Protein Concentrate
FPO	Food Product Order
FPP	Famesyl Pyrophosphate
FRM	Fat-Reduced Meat
FSIS	Food Safety and Inspection Service
FSO	Food Safety Objective
FTIR	Fourier transformation infrared spectroscopy
GAP	Good Agricultural Practices
GC	Gas Chromatography
GC-MS	Gas Chromatography-Mass Spectroscopy
GCOH	Gas Chromatography Olfactometry of Headspace Samples
G-CSF	Granulocyte Colony-Stimulating Factor
GDL	Glucono-δ-Lactone
GGPP	Geranyl Geranyl Pyrophosphate
GHP	Good Hygiene Practice
GL	Glycolipids
GLC	Gas Liquid Chromatography
GLP	Good Laboratory Practice
GMO	Genetically Modified Organisms
GMP	Good Manufacturing Practice
GMPR	Good manufacturing practice regulations
GOT	Glutamate-Oxalacetate Transaminase
GPP	Geranyl Pyrophosphate
GPT	Glutamate-Pyruvate Transaminase
GPV	Gross Protein Value
GRAS	Generally Recognized as Safe
GRE	Glucocorticoid Response Element
GSC	Granular Stationary Phase
GSH	Glutathione Peroxidase
GTF	Glucose Tolerance Factor

Short Form	*Long Form/Full Form*
Gy	Gray
H	Humidity, kg/kg
HAA	Heterocylic Aromatic Amines
HACCP	Hazard Analysis and Critical Control Points
HADY	High Active Dry Yeast
HART	Hybrid Arrested Translation
HAZOP	Hazard and Operability Study
Hb	Hemoglobin
HCN	Hydrogen Cyanide
HDA	Homochiral Derivatizing Agents
HDI	Human Development Index
HDL	High Density Lipoprotein
HDP	Heat Pump Dehumidifier
HEMF	4-hydroxy-2-(or 5-) ethyl-5-(or 2-) methyl-3 (2H) Furnone
HFCS	High Fructose Corn Syrup
HH	Household
HHS	Health and Human Services
HIC	Hydrophilic Interaction Column
HIV	Human Imuno-Deficiency Virus
HKI	Helen Keller International
HLB	Hydrophilic Lipophilic Balance
HM	High-Methoxyl
HMF	Hydroxy Methyl Furaldehyde
HMF	Hydroxymethyl Furfural
HMFP	High Moisture Fruit Products
HMG CoA	Hydroxy Beta-methyl Glutaryl CoA
HMP	Hexose Monophosphate Pathway
HMW	High molecular Weight
H_o	Humidity of Atmospheric air, kg/hr m^2
HOCl	Hypochlorous Acid
HPC	Hydroxypropyl Cellulose
HPCE	High Performance Capillary Electrophoresis

Short Form	*Long Form/Full Form*
HPLC	High Pressure/Performance Liquid Chromatography
HPLC	High-Pressure Liquid Chromatography
HPMC	Hydroxypropyl Methyl Cellulose
HRGC	High-Resolution Gas Chromatography
HSH	Hydrogenated Starch Hydrolysates
HTC	Hard-to-Cook
HTS	High Throughput Screening
HTST	High Temperature Short Time
IADY	Instant Active Dry Yeast
IARI	Indian Agricultural Research Institute
IBC	Iodine Binding Capacity
IBPA	Insoluble Bound Phenolic Acid
IC_{50}	Inhibitory Concentration
ICAR	Indian Council of Agricultural Research
ICC	International Association of Cereal Science and Technology
ICCIDD	International Council for the Control of Iodine Deficiency Disorders
ICUMSA	International Commission for Uniform Methods of Sugar Analysis
ID	Iron Deficiency
IDA	Iron Deficiency Anemia
IDD	Iodine Deficiency Disorders
IDF	International Dairy Federation
IDF	Insoluble Dietary Fiber
IDL	Intermediate Density Lipoprotein
IEF	Isoelectric Focusing
IF	Intrinsic Factor
IFS	International Foundation for science
IFTM	Instron Food Testing Machine
IGF	Insulin like Growth Factor
ILPS	International Lecithin and Phospholipids Society
ILSI	International Life Science Institute
IMF	Intermediate Moisture Foods
IMMPaCt	International Micronutrient Malnutrition Prevention and Control Program, CDC

Short Form	*Long Form/Full Form*
IMP	Industrial Membrane Process
IMS	Immuno Magnetic Separation
INACG	International Nutritional Anemia Consultative Group
INCAP	Institute of Nutrition of Central America and Panama
IOM	Institute of Medicine
IPNSs	Integrated Plant Nutrition Systems
IPRs	Intellectual Property Rights
IQF	Individually Quick Frozen
IR	Infra Red
IRDA	Insurance Regularity & Development Authority
ISFE	International Society of Food Engineering
ISI	Indian Standards Institution
ISM	Industrial Scientific and Medical Applications
ISO	International Organization for Standards
ISTA	International Seed Testing Association
IU	International unit
IV	Iodine Value
IVACG	International Vitamin A Consultative Group
IVPD	In Vitro Protein Digestibility
JCI	Jute corporation of India
K	Constant
K	Drying constant, 1/hr
KAP	Knowledge, Attitudes, and Practices
kD_a	Kilo Dalton
KI	Potassium Iodide
KV	Kilo Volt
LAB	Lactic Acid Bacteria
LBG	Locust Bean Gum
LBW	Low Birth Weight
LCD	Liquid Crystal Display
LCE	Low-Cost Extrusion cooking
LCFA	Long-Chain Fatty Acids
LCP	Liquid Cyclone Process

Short Form	*Long Form/Full Form*
LCR	Ligase Chain Reaction
LDL	Low Density Lipoproteins
LEC	Low-cost Extrusion Cookers
LF	Lactoferrin
LH	Low Hydroxy group
LIFDCs	Low-Income Food Deficit Countries
LISA	Low Input Sustainable Agriculture
LM	Low Methoxyl
LMP	Low Methoxy Pectin
LMW	Low Molecular Weight
LNG	Liquefied Natural Gas
LOAEL	Lowest Observed Advance Effect Level
LOEL	Lowest Observed Effect Level
LOX	Lipoxygenase
LP	Lactoperoxidase
LP	Lacto Peroxidase or Lipid Peroxidase
LPC	Leaf Protein Concentrates
LSE	Lant Stanol Esters
LTI	Lysine Tri-isocyanate
LUVs	Large Unilamellar Vesicles
LYC	Lycopene Cyclase
M	Molarity
MA	Modified Atmosphere
MAGs	Monoacylglycerols
MAHPD	Modified Atmosphere Heat Pump Dehumidifier
MALDITOF	Matrix-Assisted Laser-desorption Ionization Time-of-Flight
MAP	Modified Atmosphere Packaging
MAPit	Micronutrient Action Plan Instructional Tool
MAS	Monoacetone-L-Sorbose
Mbo	Oxymyoglobin
MC	Methyl Cellulose
MCC	Micro-Crystalline Cellulose

Short Form	*Long Form/Full Form*
MCH-PCR	Magnetic Capture Hybridization Polymer Chain Reactions
MCP	1-methyl cyclopropene
MCP	Mono-Calcium Phosphate
MCTs	Medium-Chain triacylglycerols
MDA	Malondialdehyde
MDSC	Modulated Differential Scanning Calorimetry
MetMb	Metmyoglobin
MF	Modified Filtration or Microfiltration
MFGM	Milk Fat Globule Membrane
MHP	Modified Humidity Packaging
MI	Micronutrient Initiative
MICS	Multiple Indicator Cluster Survey
MIG	Mercury-in-Glass
MLVs	Multilamellar Vesicles
MM	Micronutrient Malnutrition
MPC	Milk Protein Concentrates
MPEDA	Marine Products Export Development Authority
MR	Moisture Ratio
MRI	Magnetic Resonance Imaging
MRL	Maximum Pesticide Residue Limits
MRP	Maximum Retail Price
MS	Methyl Sulphonate
MS	Moles of Substitution or Mass Spectrometry
MSG	Mono Sodium Glutamate
MSR	Multistage Recycle Designs
MTD	Maximum Tolerated Dose
MUFA	Monounsaturated Fatty Acid
MVS	Machine Vision System
MWCO	Molecular Weight Cutoff
MWM	Molecular Weight Marker
N	Normality
NA	Not Analyzed

Short Form	*Long Form/Full Form*
NACMCF	National Advisory Committee on the Microbiology Criteria for Foods
NACMF	National Agricultural Cooperative Marketing Federation
NAD	Nicotine Amide Dinucleotide
NAFED	National Agricultural Finance and Economical Development
NAIP	National Agricultural Innovative Project
NAS	National Academy of Sciences
NASBA	Nucleic Acid Sequence-Based Amplification
NATP	National Agricultural Technology Project
NBI	Nitrogen Balance Index
NCA	N-carboxy-alpha-Amino Acid Anhydride
NCCF	National Consumers Cooperative Federation
NCDC	National Cooperative Development Corporation
NCHS	National Center for Health Statistics
NCTGF	National Cooperative Tobacco Growers Federation
ND	Not Detected
NDDB	National Dairy Development Board
NDF	Neutral Detergent Fiber
NDGA	Nordihydro Guaiaretic Acid
NDMA	N-nitrosodimethylamine
NDPE	Net Dietary Protein Energy Ratio
NDPV	Net Dietary Protein Value
NDR	Neutral Detergent Residue
NDUS	Novelty Distingness, Uniformity and Stability
NF	Nanofiltration
NFC	Not from Concentrate
NFDM	Non Fat Dry Milk
NFE	Nitrogen Free Extract
NHB	National Horticulture Board
NIF	Nasal Impact Frequency
NIH	National Institute of Health
NIN	National Institute of Nutrition
NIS	Nitrogen Solubility Index

Short Form	*Long Form/Full Form*
NISCAIR	National Institute of Science Communication and Information Resources
NLEA	Nutrition Labeling and Education Act.
NMN	Nicotinamide Mononucleotide
NMR	Nuclear Magnetic Resonance
NNMB	National Nutrition Monitoring Bureau
NOAEL	No Observed Advance Effect Level
NOEL	No Observed Effect Level
NPAGE	Non-Denaturing Polyacrylamide Gel-Electrophoresis
NPN	Non-Protein Nitrogen
NPR	Net Protein Ratio
NPU	Net Protein Utilization
NPV	Net Protein Value
NR	Not Recorded or Never-ripe
NRCS	National Research Centre for Sorghum
NSI	Nitrogen Solubility Index
NSP	Non-Starch Polysaccharides
NSP	Non Storage Protein
NVMCE	Nonvolatile Methylene Chloride Extract
OIE	Office International Epizootics
OMFs	Oscillating Magnetic Fields
ONC	Ocean Nutrition Canada
OP	Orange Pekoe
ORP	Oxidation-Reduction Potential
OSHA	Occupational Safety and Health Administration
OUR	Oxygen Uptake Rate
P	Phosphorus
PA	Phytic Acid or Polyamides
PABA	Para-Amino Benzoic Acid
PACMS	Primary Agricultural Cooperative Marketing Societies
PAHs	Polycyclic Aromatic Hydrocarbons
PAL	Phenylalanine Ammonialyase

Short Form	*Long Form/Full Form*
PAO	Palm Acid Oil
PATH	Program for Technology in Health
PBB	Polybrominated Biphenyls
PC	Polycarbonate
PCA	Perchloric Acid
PCB	Polychlorinated Biphenyls
PCCMP	Powdered Cooked Cured-Meat Pigment
PCM	Protein Calorie Malnutrition
PCMB	Para-Chloro-Mercuric Benzoate
PDA	Photodiode Array Detection or Potato Dextrose Agar
PDAs	Personal Digital Assistants
PDCAAS	Protein Digestibility-Corrected Amino Acid Score
PDCB	Partially Defatted Chopped Beef
PDI	Protein Dispersibility Index
PDS	Phytoene Desaturase
PE	Polyethylene
PEF	Pulsed Electric Field
PEM	Protein Energy Malnutrition
PER	Protein Efficiency Ratio
PET	Polyethylene Terephthalate
PF	Pekoe Fanning's
PFA	Prevention of Food Adulteration Act.
PFAD	Palm Fatty Acid Distillate
PFDA	Pure Food and Drug Act
PG	Propyl Gallate
PGE	Polyglycerol Esters
PHT	Post Harvest Technology
PI	Protein Isolate
PIC	Paired Ion Chromatography
PIT	Phase Inversion Temperature
PKC	Palm Kernel Cake
PKOL	Palm Kernel Olein

Short Form	*Long Form/Full Form*
PL	Phospholipids
PME	Pectin Methyl esterase
PMO	Pasteurized Milk Ordinance
POD	Peroxidase
ppm	Parts Per Million
PPO	Polyphenol Oxidase
PPO	Propylene Oxide
PPPP	Phytoene Pyrophosphate
PPS	Probability Proportionate to Size
PR	Protein Rating
PRE	Protein Retention Efficiency
PS	Protein Score
PS	Polystyrene
PSE	Pale, Soft and Exudative
PSU	Primary Sampling Unit
PSY	Phytoenesynthase
PT	Press-twist
PTKs	Protein Tyrosine Kinases
PU	Pasteurization Unit
PUFA	Poly Unsaturated Fatty Acids
PV	Peroxide Value
PVC	Polyvinyl Chloride
PVDC	Polyvinylidene Chloride
PVOH	Polyvinyl Alcohol
PVP	Polyvinyl Pyrrolidone
PW	Propanol-Water
QA	Quality Assurance
QAE	Quaternary Aminoethyl
QC	Quality Control
QPM	Quality Protein Maize
QUATS	Quaternary Ammonium compounds
R	Solvent-to-Seed Flour Ratio

Short Form	Long Form/Full Form
R	Regression Coefficient
RA	Risk Analysis
RAPD	Randomly Amplified Polymorphic DNA
RAR	Retinoic Acid Receptor
RBP	Retinol Binding Protein
RCA	Rolling Circle Amplification
RD	Registered Dietitian
RDA	Recommended Dietary Allowance
RDI	Recommended Dietary Intake
RDI	Reference Daily Intake
Rf	Response Factor
RFLP	Restricted Fragment Length Polymorphism
RFLP-PCR	Restricted Fragment Length Polymorphism-Polymerase Chain Reaction
RH	Relative Humidity
RI	Refractive Index
RIA	Radio Immuno Assay
RIAU	Research Institutions and Agricultural Universities
RNA	Ribonucleic Acid
RNI	Recommended Nutrient Intakes
RO	Reverse Osmosis
ROS	Reactive Oxygen Species
RQ	Respiratory Quotient
RSV	Respiratory Syncytial Virus
RTD	Resistance Temperature Device
RTD	Resistance Thermometer Device
RTE	Ready-to-Eat
SALP	Sodium Aluminum Phosphate
SAM	S-adenosyl-L-methionine
SAMB	State Agricultural Marketing Boards
SAPP	Sodium Acid Pyrophosphate
SAS	Statistical Analytical System
SAS	Sodium Aluminum Sulfate

Short Form	*Long Form/Full Form*
SCC	Somatic Cell Count
SCCMO	Special Commodity Cooperative Marketing Organizations
SCF	Supercritical Fluid
SCFA	Short-Chain Fatty Acids
SCMF	State Cooperative Marketing Federation
SCP	Single Cell Proteins
SD	Standard Deviation
SDA	Strand Displacement Amplification
SDAM	State Directorate of Agricultural Marketing
SDF	Soluble Dietary Fiber
SDS-PAGE	Sodium Dodecylsulphate Polyacrylamide Gel-Electrophoresis
SEM	Scanning Electron Micrograph/Microscopy
SEPC	Silk Export Promotion Council
SF	Swelling Factor
SF	Serum Ferritin
SFDA	Small Farm Development Agency
SFEs	Sucrose Fatty Acid Esters
SFPE	Sucrose Fatty Acid Polyester
SGA	Small for Gestational Age
SGE	Starch Gel Electrophoresis
SH	Sulfhydryl
SHMP	Sodium Hexametaphosphate
SIP	Sterilization-in-Place
SMB	Simulated Moving Bed
SMER	Specific Moisture Extraction Rate
SMEs	Sucrose Monomers
SMFs	Static Magnetic Fields
SMUF	Simulated Milk Ultrafilterate
SNIF	Surface of Nasal Impact Frequency
SO	Soybean Oil
SOD	Super Oxide Dismutase
SOPP	Sodium-O-phenylphenate

Short Form	*Long Form/Full Form*
SOPs	Standard Operating Procedures
SP	Sulphoproyl
SP	Storage Protein
SPC	Standard Plate Count
SPS	Sanitary and Phytosanitary
SSL	Sodium Stearoyl Lactylate
SSLL	Sodium Stearoyl Lactoyl Lactate
SSP	Shelf-Stable Products
STC	State Trading Corporation
STP	Standard Pressure and Temperature
SUVs	Small Unilamellar Vesicles
SWCS	State Warehousing Corporations
SWM	Standards of Weight and Measures
TAC	Trialkoxycitrate
TAGs	Triacylglycerols
TATCA	Trialkoxy Tricarballylate
TBA	Thio Barbituric Acid
TBHQ	Tetra Butylated Hydro Quinone
TBT	Technical Barriers to Trade
TCA	Trichloro Acetic Acid
TCA	Trichloroanisol/Trichloroacetic Acid/Tricarboxylic Acid
TCD	Thermal Conductivity Detector
TD	True Digestibility
TDF	Total Dietary Fiber
TDI	Tolerable Daily Intake
TEF	Toxic Equivalency Factor
TFA	Trifluoro Acetic Acid
TFA	Transfatty Acids
TFC	Thin Film Composite
TFR	Tempeh Fish and Rice Mixture
TfR	Transerrin Receptor
THBP	2,4,5-Trihydroxy ButyroPhenone

Short Form	*Long Form/Full Form*
TLC	Thin Layer Chromatography
TMA	Transcription-Mediated Amplification
TMA	Thermo-Mechanical Analysis
TMA	Trimethylamine
TMAO	Trimethylamine Oxide
TMCT	Thermal Mechanical Compression Test
TNF	Tumor Necrosis Factor
TOFMS	Time of Flight Mass Spectrometry
TOS	Total Organic Solids
TQM	Total Quality Management
TRIFED	Tribal Cooperative Marketing Federation
T-RNA	Tetrahymena Relative Nutritive Value
TS	Total Solids
TSAI	Transition State Analog Inhibitors
TSH	Thyroid-Stimulating Hormone
TSH	Thyroid Stimulating Hormone (or thyrotropin)
TSNS	Ten State Nutrition Survey
TSP	Texturized Soya Proteins
TTB	Tax and Trade Bureau
TTI	Time-Temperature Integrator/Time-Temperature Indicator
TTT	Time-Temperature-Tolerance
TVP	Textured Vegetable Proteins
TWT	Traveling Wave Tube
UF	Ultra Filtration
UHT	Ultra High Temperature
UI	Urinary Iodine
ULO	Ultra-Low Oxygen
ULV	Unilamellar Vesicles
UNICEF	United Nations International Children Fund
UNICEF	United Nations Children's Fund
UNU	United Nations Union
UP	Utilizable Proteins

Short Form	*Long Form/Full Form*
USDA	United State Department of Agriculture
USFDA	United State Food and Drug Administration
USI	Universal Salt Iodization
USS	United States Standards
UTLIEF	Ultra Thin Layer Isoelectric Focusing
UV	Ultra Violet
V/V	Volume-to-Volume
VAD	Vitamin A Deficiency
VFA	Volatile Fatty Acids
VFD	Variable Frequency Drive
VFMPT	Vacuum Freezing Multiple-Phase Transformation
VLCD	Very Low Calorie Diet
VLDL	Very Low Density Lipoproteins
VP	Vacuum Packaging
VSP	Vacuum Skin Packaging
VVM	Vessel Volume Ratio
W.B.	Wet Bulb Temperature
W/V	Weight-to-Volume
W/W	Weight-to-Weight
WARC	World Administrative Radio Conference
WBP	Water Binding Potential
WFI	Water-for-Injection
WHC	Water Holding Capacity
WHO	World Health Organization
WLF	Williams-Landel-Ferry
WOF	Warm-Over Flavour
WRA	Women of Reproductive Age
WSC	World Summit for Children
WSSN	World Standards Services Network
WTO	World Trade Organization
XME	Xenobiotic Metabolising Enzymes
XOD	Xanthine Oxidase
XRE	Xnobiotic Response Element
Z	Z-score
ZDS	ξ-Carotene Desaturase

APPENDIX–III
Conversion Factors (Weights and Measures)

Length	1 m	10^{10} A	39.37 in	3.2808 ft
Area	1 m^2	10^4 cm^2	1550.0031 in^2	10.7639 ft^2
Volume	1 m^3	1000 L	10^6 cm^3	264.17 gal (US)
Mass	1 kg	2.2046 lbm	35.274 oz	1.1023 x 10^{-3} ton (short)
Density	1 g/cm^3	1000 kg/m^3	62.428 lbm/ft^3	8.3454 lbm/gal (US)
Velocity	1 m/s	3.6 km/h	2.2369 min/h	196.85 ft/min
Force	1 kg-m/s^2 (N)	0.2248 lbf	10^5 g.cm/s^2 (dyn)	7.2330 lbm.ft/s^2 (pdf)
Pressure	1 atm	101.325 kg/ms^2 (pa)	14.696 lbf/in^2 (psi)	760 mm Hg
Energy	1 kg.m^2/s^2 (J)	9.4782 x 10^{-4} Btu	0.2388 cal	0.7376 ft.lbf
Power	1 J/S (W)	1 kg/m^2/s^3	3.4121 Btu/h	1.341 x 10^{-3} hp
Thermal conductivity	1 W/m.k	0.5778 Btu/h.A.°F	2.388 x 10^{-3} cal/s.cm.°C	6.9335 Btu.in/h.ft^2.°F
Specific heat	1 KJ/kg.K	0.2388 cal/g.°C	0.2388 Btf/lbm.°F	185.86 ft.lbf/lbm.°F
Viscosity (dynamic)	1 pa.S	1 kg/m.s	1000 cp	0.6720 lbm/ft.S
Temperature	°C=5/9 (F-32)	°F = 9/5 (°C) + 32	K = °C + 273.15	°R = °F + 45967

APPENDIX–IV
Equipments and Machinery Used in Fruit and Vegetable Processing Industry

Washing Equipments

1. Over-head sprayer of water
2. Continuous woven wear belt
3. Rotating rod-cylinder with helical screw
4. Cement or galvanized iron tank

Sorting and Grading Equipments

1. Continuous broad belt made of woven metal
2. Roller grader
3. Screen grader
4. Rope or cable grader

Coring and Peeling Equipments

1. Core remover
2. Cutting knife
3. Peeling knife
4. Pitting knife
5. Corer or seed remover
6. Potato peeler

Juice/Pulp Extraction Equipments

1. Halving and burring machine
2. Continuous screw expeller press
3. Plunger-type press
4. Roller-type press
5. Grater
6. Crusher
7. Basket press
8. Rack and cloth press
9. Hydraulic press

Straining/Screening Equipments

1. Pulper with brush paddles
2. Filter press

Deaeration and Pasteurization

1. Deaerator
2. Flash-pasteurizer

Bottling Equipments

1. Bottle washing machine
2. Bottle filling machine
3. Crown corking machine

Canning Equipment

1. Cans
   (*a*) Acid-resistant cans
   (*b*) Sulphur resistant cans
2. Vacuum can tester
3. Can reformer
4. Can opener
5. Refractometer
6. Retorts

Equipments for Making Jams, Jellies and Marmalade

1. Steam jacketed kettles
2. Jelmeter/viscosimeter
3. Jelly thermometer
4. Cookers

Driers

1. Tray drier
2. Puff drier
3. Foam mat drier
4. Drum drier
5. Spray drier
6. Vacuum drier
7. Freeze drier

Other Equipments and Accessories

1. Stainless steel sieves
2. Electronic balances
3. Avery weighing scale
4. Salometer
5. Preparation tables
6. Aprons, rubber gloves
7. Laboratory glassware

APPENDIX–V
Addresses for Suppliers of Equipments/Machinery

Food and Beverage Processing/Preservation Equipments/ Machinery

M/s. Adam Fabriwerk Pvt. Ltd., 203, Rajguru Apts., New Nagardas Road, Andheri (E) Mumbai-400069, Maharashtra

M/s. A. E. C. Instruments, 6, Akashdeep Welfare Society, Near Salvi Chawl, Shankarwadi, Jogeshwari (E), Mumbai – 400 060, Maharashtra.

M/s. Akshay Industries, 101, Rajshree Indl. Estate II, Chitalsar Manpada, Ghodbunder Road, Thane-400607, Maharashtra

M/s. Alok Technical & Marketing Service Pvt. Ltd., P.O. Box-8, Chandra Bldg., Kalkaji Temple, New Delhi-ll0019

M/s. Alven Foodpro System (P) Ltd., 24, Goldfield Plaza, 45, Sassoon Road, Pune-411 00 1, Maharashtra

M/s. Apple Bakery Machinery Pvt. Ltd. 20/1, Hansraj Damodarwadi, 1st Flr. Off. Kennedy Bridge Opera House, Mumbai – 400004, Maharashtra.

M/s. Asian Chemical Works (Bombay) Pvt. Ltd., Asian House, 29, Ramakrishna Mandir Road, Andheri-Kurla Road, Andheri (E), Mumbai-400059, Maharashtra

M/s. B. Sen Barry and Co., 65/11,New Rohtak Road, Karol Bagh, New Delhi-110005

M/s. Bajaj Maschinen Pvt. Ltd., 0-14, Lajpat Nagar-II, New Delhi-l10024

M/s. Baker Enterprises, 23, Bhera Enclave, Near Peera Garhi, New Delhi-110087

M/s. Cantech Machines, 13, Vora Bhavan, Maheshwari Udyan, King's Circle, Matunga (c. Rly), Mumbai-400019, Maharashtra

M/s. OSI Industries, C-10, Devatha Plaza, Residency Road, Bangalore-560025, Karnataka

M/s. Eastend Engineering Co., 173/1, Gopal Lal Thakur Road, Kolkata-700016, West Bengal

M/s. Environmental Products (India) Pvt. Ltd. 160/3, Rajani House, Opp. Don Bosco School, L. T. Road, Borivali (W), Mumbai – 400 091, Maharashtra.

M/s. Fab Flavours & Fragrances Pvt. Ltd. S. P. Mukerjee Marg, New Delhi-110006.

M/s. Filtron Engineers Ltd., 117-A, Vithalwadi Road, Pune-411030, Maharashtra

M/s. FMC Food Tech, 7, Shivaji Housing Society, Pune-411053, Maharashtra

M/s. Frigoscandia Winner Food Process System Ltd., Shreesh Chambers, 3rd Floor, 25/1, Yeshwant Niwas Road, Indore-452003, M.P.

M/s. Ganesh Benzoplast Ltd. Ganesh House, A1/A2, Gurudutt CHS Ltd. Near Jankalyan Bank, J. B. Nagar, Andheri (E), Mumbai-400059, Maharashtra.

M/s. Gardners Corporation, 6, Doctors Lane, Near Gole Market, Post Box 299, New Delhi- 110001

M/s. Guru Nanak Engg. & Foundary Works (Regd.), 166, Indl. Focal Point, Mehta Road, Amritsar-143001, Punjab

M/s. Hindusthan National Glass & Industries Ltd. 2, Red Cross Place, Kolkata-700 001 (W.B.).

M/s. HMT Ltd. (Food Processing Mlc Div.), H-2, MIDC, Chikalthana Indl. Area, P.B. No. 720, Aurangabad-431210, Msharashtra

M/s. Ion Exchange (India) Ltd. Ion House, Dr. E. Moses Road, Mahalaxmi, Mumbai-400011, Maharashtra.

M/s. Jay Chem Marketing, 101, Labh Sarita, Opp. Manek Nagar, M.G. Road, Kandivali (W), Mumbai-400067, Maharashtra.

M/s. K.S. Seetharamiah & Sons Pvt. Ltd., 29/1, Jaraganahalli, 10th Km. Karakapura Road, Bangalore-560078, Karnataka

M/s. Malladi Specialities Ltd. 9, GST Rd, St. Thomas Mount, Chennai-600016.

M/s. Mather and Platt (lndia) Ltd., 805-806, Ansal Bhawan, 16, Kasturba Gandhi Marg, New Delhi-110001

M/s. M.M.M. Buxabhoy and Co., 141-A, Sarang Stree, 1st Floor Near Crawford Market, Mumbai-400003, Maharashtra.

M/s. Micron Industries Pvt. Ltd. R-710, TTC Indutrial Area, MIDC Rabale, Navi Mumbai, Maharashtra.

M/s. Mojj Engg. Systems (P) Ltd., 81-B/15, M.LD.C., Opp. Morris Electronics, Bhosari, Pune-411026, Maharashtra

M/s. Mukul Brothers Engg. Works, P.B. No. 325, Kishan Flour Mill Compound, Tirthankar Mahavir Mart (Rly. Road), Meerut-250002, Ll.P

M/s. Pharmalab Engg. India Ltd., Star Metal Compound, L.B.S. Marg, Vikhroli (W), Mumbai-400083, Maharashtra

M/s. Quasar Engineers, Plot No. 53, Sector 'A' Indl. Area, Sanwer Road, Indore-452003, M.P.

M/s. Satellite Plastic Industries, 2A, Court Chambers, 35 New Marine Lines, Mumbai-400020, Maharashtra.

M/s. Shiva Engineers, Patel Avenue, Plot No. 165, Flat No. 1, Right Bhusari Colony, Paud Road, Pune-411038, Maharashtra.

M/s. Sri Rajalakshmi Commercial Kitchen Equipments Pvt. Ltd. 57, (1st Floor) Silver Jubilee Park Road, Bangalore-560002, Karnataka.

M/s. Sri Rajalakshmi Industrial Agency, 57(30/1), Silver Jubilee Park Road, Rajalakshmi Corner House, Post Box No. 6690, Bangalore- 560002, Karnataka

M/s. Stern Ingredients India Pvt. Ltd. 211 Nimbus Center, Oberoi Complex Off. Link Road, Andheri West, Mumbai – 400 053, Maharashtra.

M/s. Raylons Metal Works, Ramakrishna Mandir Road, Kondivita Village, opp. Marol Bazar, J.B. Nagar, Andheri (E), Mumbai-400059, Maharashtra

M/s. Riddhi Pharma Machinery Ltd. Om-Shivam Buiding 2, Tarun Bharat Sahar Road, Andheri (E), Mumbai-400009, Maharashtra.

M/s. Rita Bottling Machines Ltd., 1B & 1C, Suvarna Darshan, 47, 2nd Main Road, Gandhi Nagar, Adyarn, Chennai-600020, Tamil Nadu

M/s. Uma Brothers, C-110, Bhaveshwar Plaza, 189, L.B.S.Marg, Ghatkopar (W), Mumbai-400086, Maharashtra.

M/s. Ultra International Ltd. 304, AVG Bhavan, M-3 Connaught Circus, New Delhi - 110001.

M/s. V.S. Enterprises, Gala No. 7, Laxmi Industrial Estate Penkar Pada Road, Near Dahisar Check Naka, Mira, Thane-401104, Maharashtra.

M/s. Veenu Hitech, Plant Manufacturing Pvt. Ltd., F-6 St. Soldier Tower, G-Block, Vikas Puri, New Delhi-110 018

Canning Equipments/Machinery

M/s. Asian Consolidated Industries Ltd., "Asian House", 0-193, Okhla Indl. Area, Phase-I, New Delhi-110020

M/s. Ganga Singh Engg. Works P, Ltd., No.1, Vishal Indl. Estate, Village Road, Bhandup (W) Mumbai- 400078, Maharashtra

M/s. Quality Equipment Co., 89, Mahavira Street, Haiderpur Indl. Area Delhi-110052

M/s. Recon Machine Tools Pvt. Ltd., 37, Sarvodaya Indl. Estate, Mahakali Caves Road, Andheri (E), Mumbai-400093, Maharashtra

M/s. Sangram Engg. Ltd., B-5, Super Con, Opp. LT.L, Aundh, Pune-411007, Maharashtra.

M/s. Thakar Equipment Co., 66, Okhla Indl. Estate, New Delhi-l10020

Drying/Dehydration Equipments/Machinery

M/s. Admir Enterprises, Plot No. IIE, 4, Shivaji Nagar Govandi, Mumbai-400043.

M/s. Aifso Industrial Equipment Co., B/13, Veena Beena Apts, P. Thakrey Marg, Sewri (W), Mumbai 400015, Maharashtra

M/s. Aratic India Engg. Pvt. Ltd., 20, Rajpur Road, Delhi-110054

M/s. Bombay Industrial Engineers. 430, Hind Rajasthan Chambers, D.S. Phalke Road, Dadar (C Rly.), Mumbai-400014. Maharashtra

M/s. Bry-Air India Pvt. Ltd., 419-420, Udyog Vihar, Phase-Ill, Gurgaon, Haryana

M/s. Burman Plant & Machinery Co. Pvt. Ltd., 36, Sarkar Lane, Kolkata-700007, West Bengal

M/s. CM. Equipments & Instruments (India) Pvt. Ltd., B-194, 5th Main Road. P.B. No. 5847, Peenya 2nd Stage, Bangalore-560058. Karnataka

M/s. Eec Cee & Co., 1, Anant Indl. Estate, Opp, Cemet Fruit & Chemicals, Rakhial Ahmedabad-380023, Gujarat

M/s. Orbit International Technologies (P) Ltd., 404, Taramandal Complex, Secretariat Road, Hyderabad -500004, A.P.

M/s. Ratan Equipments 69, Lake View Road, Kamakati St., West Mambalam, Chennai-600003, Tamil Nadu

M/s. Raylon Engg. Works, 31 A, Ghanshyam Indl. Estate., Near Veera Desai Road, Andheri (W), Mumbai-400058, Maharashtra

M/s. Wintech Taparia Ltd., 25/1, Yeshwant Niwas Road, Shreesh Chambers, III Floor, Indore-452003, M.P.

Product Specific Equipments/Machinery

Jam Making

M/s. Foram Foods Pvt. Ltd., 397, Swami Vivekanand road, Vile Parle (W), Mumbai-400056, Maharashtra

M/s. Inventure India BV, 24, Gold Field Plaza, 45, Sasso on road, Pune-411001, Maharashtra

Pickle and Chutney Making

M/s. Geeta Food Engineering, Plot C-7/1 TIC Area, Pawana M.ID.C., Thanc-Belapur Road, Behind Savita Chemical Ltd., Navi, Mumbai-400705, Maharashtra

M/s. Techno Equipments, Saraswati Sadan, Girgaum Court, Girgaum, Mumbai-400004, Maharashtra

Tomato Processing

M/s. Goma Engineering Pvt. Ltd., Majiwada, Behind Universal Pertol Pump, Thane-40060 1, Maharashtra

M/s. K.S. Seetharamiah & Sons. Pvt. Ltd., 29/1, Jaraganahalli, 10th Km. Kanakapura Road, Bangalore-560078, Karnataka

M/s. Penwalt India Ltd., 507, Kakad Chambers, 32, Dr. Annie Besant Road, Worli, Mumbai-400018, Maharashtra

Mushroom Processing

M/s. Jwala Engg. Co., 12, Surve Indl. Estate, Sonawala Cross Road No.1, Goregaon (E), Mumbai-400063, Maharashtra

M/s. Wintech Taparia Ltd., 25/1, Yeshwant Niwas Road, IIIrd Floor, Shreesh Chambers, Indore-452003, M.P.

Wrapping/Filling/Packaging Equipments/Machinery

M/s. Aarkay Wrapping Machines Pvt. Ltd., **1,** Hormurz, 131, August Kranti Marg, Mumbai-400036, Maharashtra

M/s. Abhay & Abhay Pvt. Ltd., B-84-1, Okhla Indl. Area, Phase-Il, New Delhi-ll0020

M/s. APT Packaging, 184/5007, Pantnagar Ghatkopar (E), Mumbai-400075, Maharashtra

M/s. Bombay Engineering Industry, R. No.6 (Estns.), Sevanthibhai Bhavan, Chimatpada, Marol Naka, Andheri (E), Mumbai-400059, Maharashtra

M/s. Compack Systems, E-211, F.F. Complex, Okhla III, New Delhi-ll0020

M/s. Debes Industries, 11, Govt. Place East, Kolkata-700069, West Bengal

M/s. E.C. Packaging Pvt. Ltd., 14/7 Mile Stone, Mathura Road Faridabad- 121003, Haryana

M/s. Europack Machines (I) Pvt. Ltd., Akash Business Centre, CST Road, Kurla (W), Mumbai-400070, Maharashtra

M/s. Multipack Systems Pvt. Ltd., 2nd Floor, Patrict Complex, Ellora Park, Vadodara-390007, Gujarat

M/s. Packaging Machines of India, 8-3-229/5/3/7, Jai Bharani Nagar Yusufguda Check Post, Hyderabad, A.P.

M/s. Reliance Packaging, 155-A, Ekta Enclave, Peeragarhi, Main Rohtak Road, New Delhi–110041

Refrigeration/Cold Storage Equipments/Machinery

M/s. Airtech Engineers, B-19, Okhala Indl. Area, Phase-II, New Delhi-110020

M/s. Anand Refrigeration Co. Pvt. Ltd., D-Lajpat Nagar-I, New Delhi-110024

M/s. Blue Star Ltd., Kasturi Bldg., Mahan T. Advani Chowk, Jamshedji Tata Road, Mumbai-400020, Maharashtra

M/s. Carrier Refrigeration (P) Ltd., 700/701 A, Lado Sarai, Aurobindo Marg, Mehrauli, New Delhi-11 0030

M/s. Greenfield Agencies, Shop No.6, Anuradha, Opp. Golf Club, Rest House, Tidke Colony, Nasik-422002, Maharashtra

M/s. Industrial Refrigeration Pvt. Ltd., 901, Maker Chambers-V, Nariman Point, Mumbai-400021, Maharashtra

M/s. Kooling System, 35, Mannar Reddy Street, T. Nagar, Chennai- 600017, Tamil Nadu

M/s. Western Refrigeration Ltd., 4, Ready Money Terrace, Dr. A.B. Road, Worli, Mumbai-400018, Maharashtra

Packaging Materials

M/s. Alu Foil Products Pvt. Ltd., 208, Mukti Chambers, 2nd Floor, 4, Clive Row, Kolkata-700001, West Bengal

M/s. Arora Box & Carton Pvt. Ltd., HR 12, Gali No.10, Anand Parbat Indl. Area, New Delhi-II 0005

M/s. Jain. Flexipack Pvt. Ltd., M.A. C-9/9, Sector-5, Rohini, New Delhi-110085

M/s. Jayco Trading Co., 506, Kakad Market Bldg., 306, Kalbadevi Road, Mumbai-400002, Maharashtra

M/s. Oriental Containers Ltd., P.O. Box No. 6584, 1076, Dr. E. Moses Road, Worli, Mumbai-400018, Maharashtra

M/s. Packwell Industries 3746, Netaji Subhash Marg, New Delhi-110 002

M/s. Panchmahal Polypack (P) Ltd., 36, Mithila Society, Opp. Shreyas School, Ambawadi, Ahmedabad-380015, Gujarat

M/s. Swan Packaging, 53/54, Unique Indl. Estate, Dr. RP. Road, Mulund (W), Mumbai-400080, Maharashtra

M/s. Vindhyachal Process Corporation, 116, Malviya Nagar, Bhopal-462003, M.P.

Books

M/s. Bio-Green Books, 4760-61/23, Ansari Road, Darya Ganj, New Delhi-110002, Phone: +91-011-23245578, 23244987, E-mail: info@biogreenbooks.com; Website: www.biogreenbooks.com

APPENDIX–VI
Project Reports (Standard Format) of Processing Industries

1. Primary Information

(*a*) Name of Unit: ....

(*b*) Units' address: ....

(*c*) Constitution: ....

(*d*) Name of proprietor: ....

(*e*) Birth Date: .... Age: ....

(*f*) Qualification: ....

(*g*) Nature of business: ....

(*h*) Registration No. of business: ....

(*i*) Loan for: ....

(*j*) Loan amount: ....

2. Introduction

In this section the information on following points should be included:

(*a*) Experience of the proprietor in this field

(*b*) Facilities and arrangement for product sale

(*c*) Information regarding machinery requirement and their availability as well as importance

(*d*) Availability of the skilled and unskilled labour

(*e*) Information regarding self-sale counter

(*f*) Importance of this business to the society

3. Scope of Business

Give the location of the processing unit and availability of raw material in that area. Provide the information regarding marketing of the processed product.

4. Manufacturing Process

Give the details of manufacturing process/technology employed.

5. Land

Land for proprietor owns unit or it is in partnership give details. If in partnership then make agreement and submit with this proposal.

6. Building

Give the details about building construction (shed, RCC or load bearing etc.).

7. Water Supply

Water supply from bore well or Municipal Corporation etc.

8. Electric Power Supply

MSEB connection deposit, main switchboard, wiring, fitting, tubes, fans, labour charges etc.

9. Printing

For labeling, advertisement, specification on the product etc.

10. Furniture

Expenditure occurred on tables, chairs, cupboards, stools etc.

11. Quantitative Details

(Rs. in Lacs)

Sl.No.	*Description*	*1St Year*	*2nd Year*	*3rd Year*	*4th Year*	*5th Year*
1.	Installed capacity 300 days & 20 hrs					
2.	Utilization of capacity (per cent)					
3.	Raw material consumption at installed capacity (Rs./Ton)					
4.	Sale at installed capacity (Rate Rs./Ton)					
5.	Raw material consumption at utilized capacity					
6.	Sales at utilized capacity					

12. Calculation of Working Capital

Sk.No.	*Particular*	*Holding Period (days)*	*Amount*
1.	Raw material		
2.	Consumable packing		
3.	Finished goods		
4.	Other Exp.		
5.	Debtors		
Gross current assets:			
	Less credits		
	Net working capital		
	Own margin	35 or 25 per cent	
	Bank Finance	56 or 75 per cent	

13. Cost of Project and Sources of Finance

Sl.No.	*Item*	*Cost Already Incurred*	*Cost to be Incurred*	*Total*
A.	**Cost of project**			
1.	Land			
2.	Site development			
3.	Civil & Technical work			
4.	Plant & Machinery			
5.	Total working capital			
6.	Preliminary & preoperative Exp. (Bank fee, mortgage fee, Prof. Fee etc.)			
	Total			
B.	**Source of finance**			
1.	Equity capital			
2.	Subsidy/Quasi equity			
3.	Others			
4.	Unsecured Loan			
5.	Term Loan			

14. Project Profit and Loss Account

(Rs. in Lacs)

Sl.No.	Particular	1st Year	2nd Year	3rd Year	4th Year	5th Year
1.	Gross sales					
2.	Job work charges					
A.	**Net sales**					
B.	**Cost of production**					
1.	Raw material (as per qty. details)					
2.	Power & Fuel					
3.	Direct labour & Wages					
4.	Consumables stores					
5.	Repair & Maintenance					
6.	Other Mfg. Exp.					
7.	Depreciation					
8.	Add opening stock in process & finished goods					
9.	Deduct closing stock in process & finished goods					
C.	Cost of sales					
D..	Gross profit (A- C)					
E.	Interest					
F.	General & administrative over					
G.	Profit before Tax					
H.	Provision for Tax					
I.	Net Profit					
J.	Depression added back					
K.	Net cash accruals					
L.	**Repayment obligations**					
M.	Debt service ratio (Average DSCR 3.87)					

15. Projected Balance Sheet

(Rs. in Lacs)

Sl.No.	Particular	1st Year	2nd Year	3rd Year	4th Year	5th Year
A.	**Assets**					
1.	Fixed assets					
2.	Investment/L. C. Deposits					
3.	Current assets					
	a. Raw material, packing Exp.					
	b. Finished goods					
	c. Debtors					
	d. Other current assets					
	Total					
B.	**Liabilities**					
1.	**Equity capital**					
2.	Profit & loss A/C					
3.	Unsecured Loans					
4.	Quasi equity					
5.	Subsidy (Fr. MFPI)					
6.	Secured Loans					
7.	Working limits Fr. Bank					
	a. Term Loan					
	b. Creditors					
	c. Other current Liabilities & Provision					
	Total					

16. Projected Cash Flow and Fund Flow Statement

(Rs. in Lacs)

Sl.No.	Particular	1St Year	2nd Year	3rd Year	4th Year	5th Year
A.	**Sources**					
1.	Net cash accruals					
2.	Subsidy					
	Total					
B.	**Application**					
1.	**Repayment of Term Loan**					
2.	Repayment of unsecured & quasi equity					
3.	Increase in investment					
4.	Increase in W. C.					
	Total					

17. Projected Breakeven

(Rs. in Lacs)

1.	Fixed cost (Average)	
2.	Average selling price	
3.	Average variable cost (per ton)	
	a. Materials	
	b. Labours	
	c. Other overheads	
4.	Contribution/Unit	
5.	Breakeven qty	
6.	Breakeven value	
7.	Percentage on installed capacity	

18. Repayment Schedule of Term Loan and Interst on Term Loan

Period	*Principle Amount*	*Interest*	*Total*
Interest in .... per cent, period in quarters			
1st Year			
I Quarter			
II Quarter			
III Quarter			
IV Quarter			
Total			
2nd Year			
I Quarter			
II Quarter			
III Quarter			
IV Quarter			
Total			
3rd Year			
I Quarter			
II Quarter			
III Quarter			
IV Quarter			
Total			
4th Year			
I Quarter			
II Quarter			
III Quarter			
IV Quarter			
Total			
5th Year			
I Quarter			
II Quarter			
III Quarter			
IV Quarter			
Total			
Grand Total			

19. Calculation of Depreciation as per IT Rules and WDV Method

Period	*Building*	*Plant and Machinery*	*Total*
Initial			
1st year			
Total			
2nd year			
Total			
3rd year			
Total			
4th year			
Total			
5th year			
Total			

20. List of Plant and Machinery

Sl.No.	*Name of Plant or Machinery*	*Price (Rs.)*	*Quantity (Nos.)*	*Total Amount (Rs.)*
1.				
2.				
3.				
4.				
5.				
6.				
7.				
8.				
9.				
10.				
	Total			

21. Total Interest Payable Yearwise

Period	*Term*	*Working*	*Other*	*Total*
@ p. a.	11 per cent	15 per cent	18 per cent	
1st year				
2nd year				
3rd year				
4th year				
5th year				
Total				

22. Salary and Wages Details

Sl.No.	*Particular*	*Nos.*	*Salary P.M.*	*Total*
1.	Operator			
2.	Skilled workers			
3.	Unskilled workers			
	Total			

Proprietor/Partner

(Name & Address of the Unit)

Index

P

Q

U

V

W

X

Z

Dryer

Screw type juicer

Fruit Pulper

Kettle

Steam jacketed Kettle movable

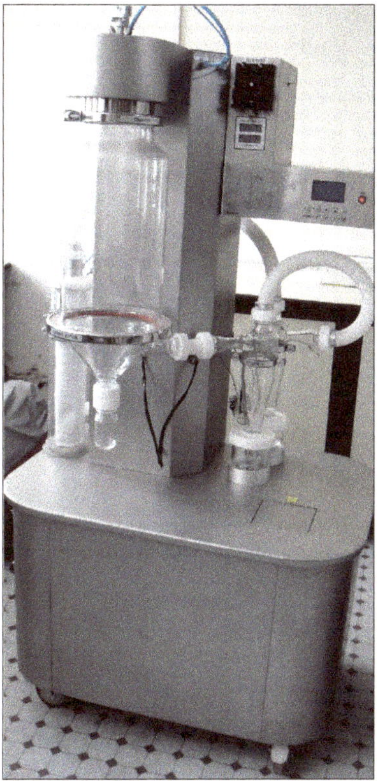

Freeze drier

Retoard

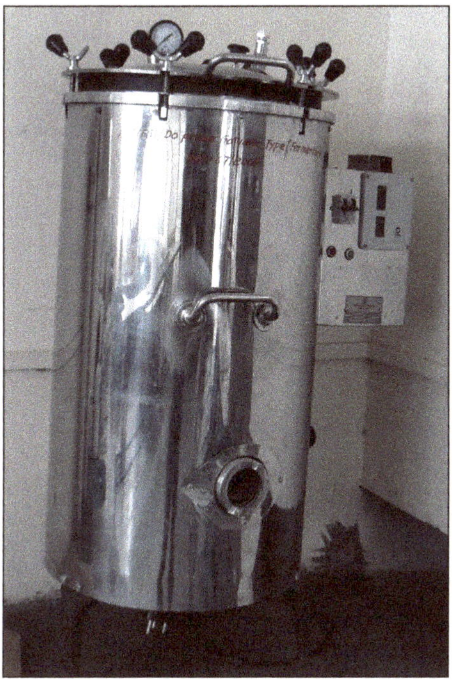

Autoclave

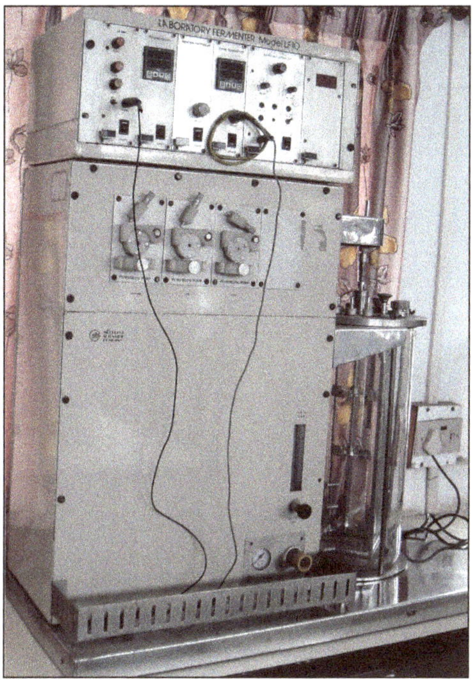

Fermentor

Potato Slicer

Potato Peeler

Fruit mill machine

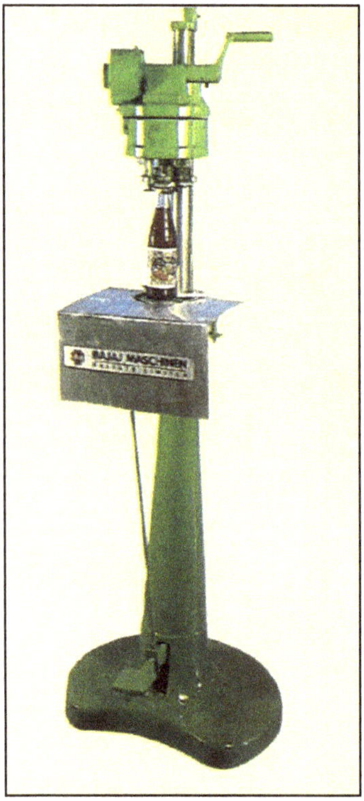

Cap Sealer

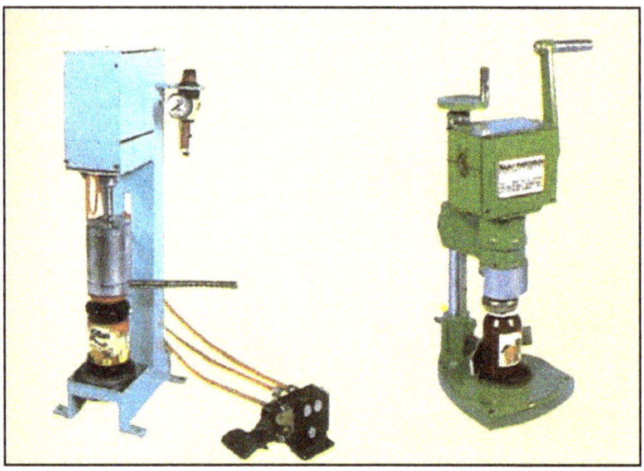

Cap sealer (Pneumatic) Cap Sealer (Mechanical)

Roasing machine

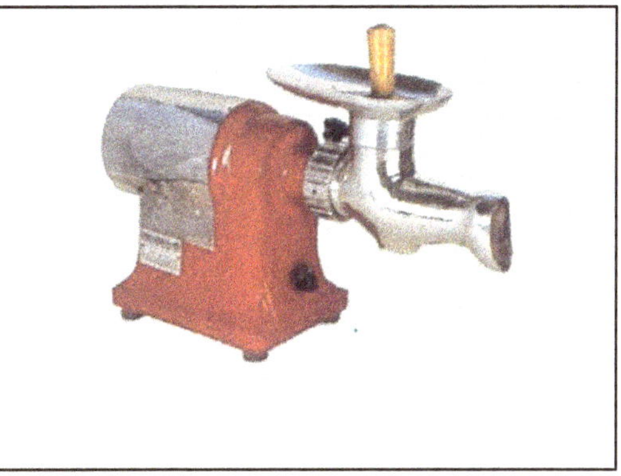

Screw type juice extractor

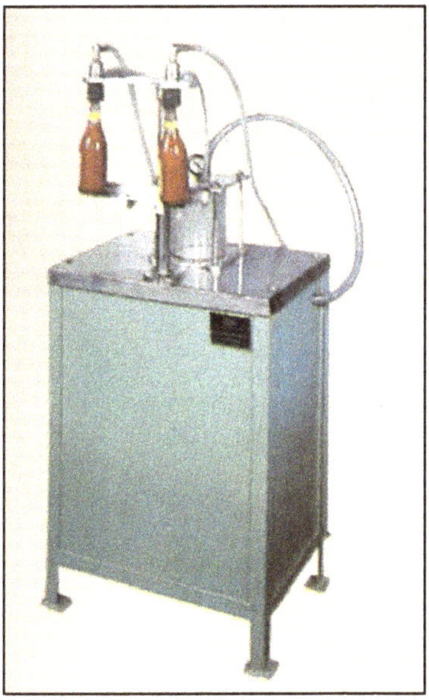

Vacuum filling

Fruit Pulper

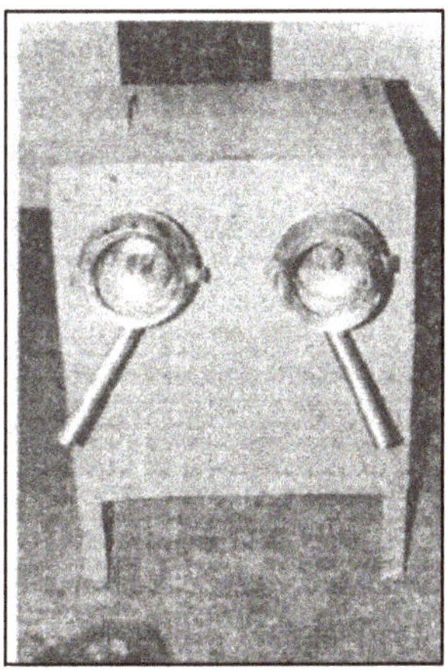

Cone-type juice extractor

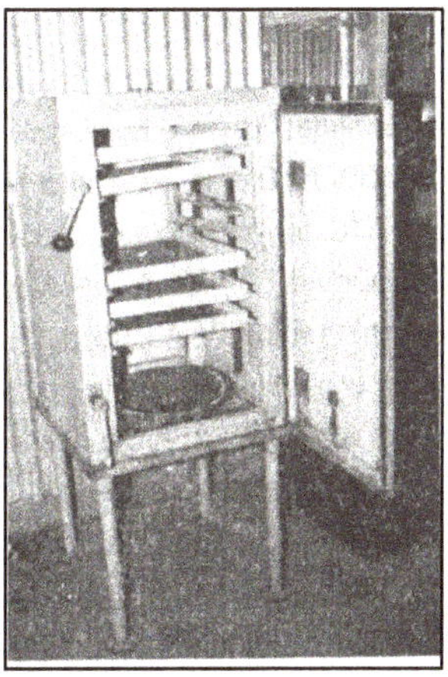

Sulphur fumigation box

Basket press

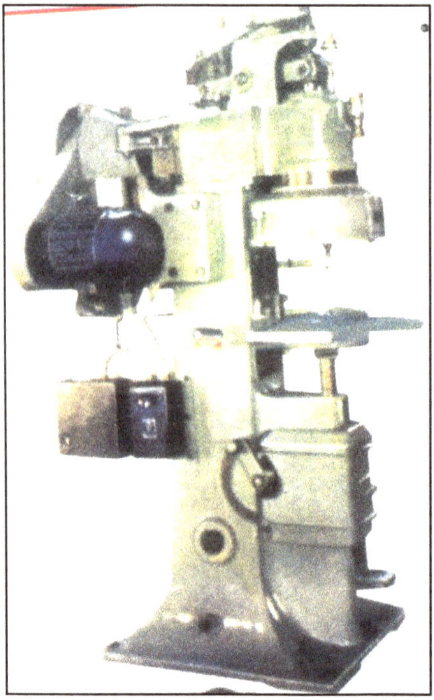

Canning machine for Fruits & Vegetables Industry

Dried figs

Ber Candy

Grape raisins

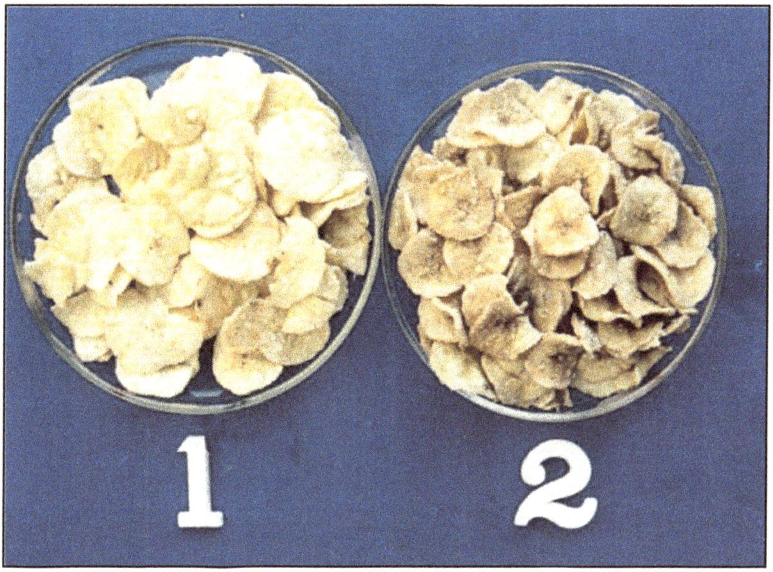

Banana chips

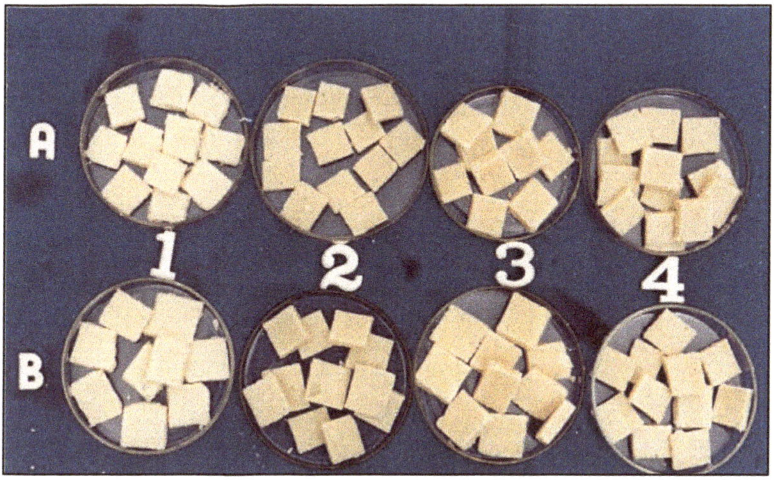

Banana toffee

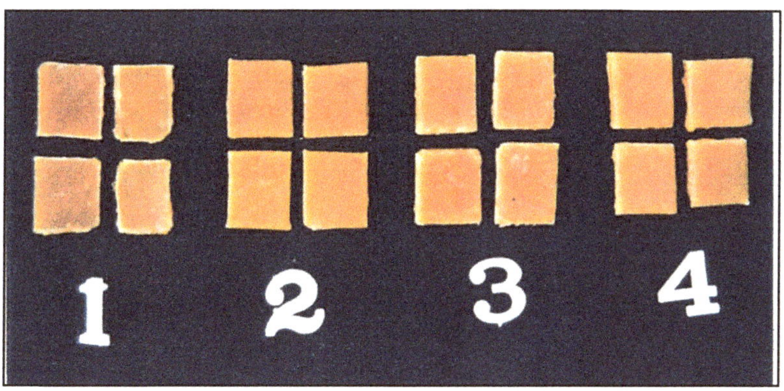

Papaya toffee

Pomegranate wine

Fig toffee

Tamarind powder

Dried Methi

Dried onion

Dried spinach

Pomegranate Anardhana

Deanna Fig

Poona fig

Carrots

Guava

Ginger candy

Potato

Carrots tutti-frutti

Pumpkin

Dry Ginger

Custard apple

Strawberry

Pomegranate

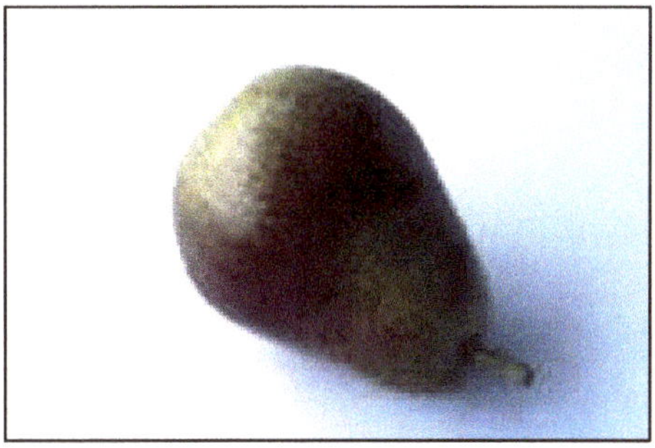

Pear

Tomato

www.ingramcontent.com/pod-product-compliance
Ingram Content Group UK Ltd.
Pitfield, Milton Keynes, MK11 3LW, UK
UKHW022217051026
15624UKWH00001B/212

9 789351 240761